D1538863

Historical
GIS Research
in Canada

Canadian History and Environment Series

Alan MacEachern, Series Editor

ISSN 1925-3702 (Print) ISSN 1925-3710 (Online)

The Canadian History & Environment series of edited collections brings together scholars from across the academy and beyond to explore the relationships between people and nature in Canada's past. Published simultaneously in print and open-access form, the series then communicates that scholarship to the world.

Alan MacEachern, Director
NiCHE: Network in Canadian History & Environment
Nouvelle initiative canadienne en histoire de l'environnement
http://niche-canada.org

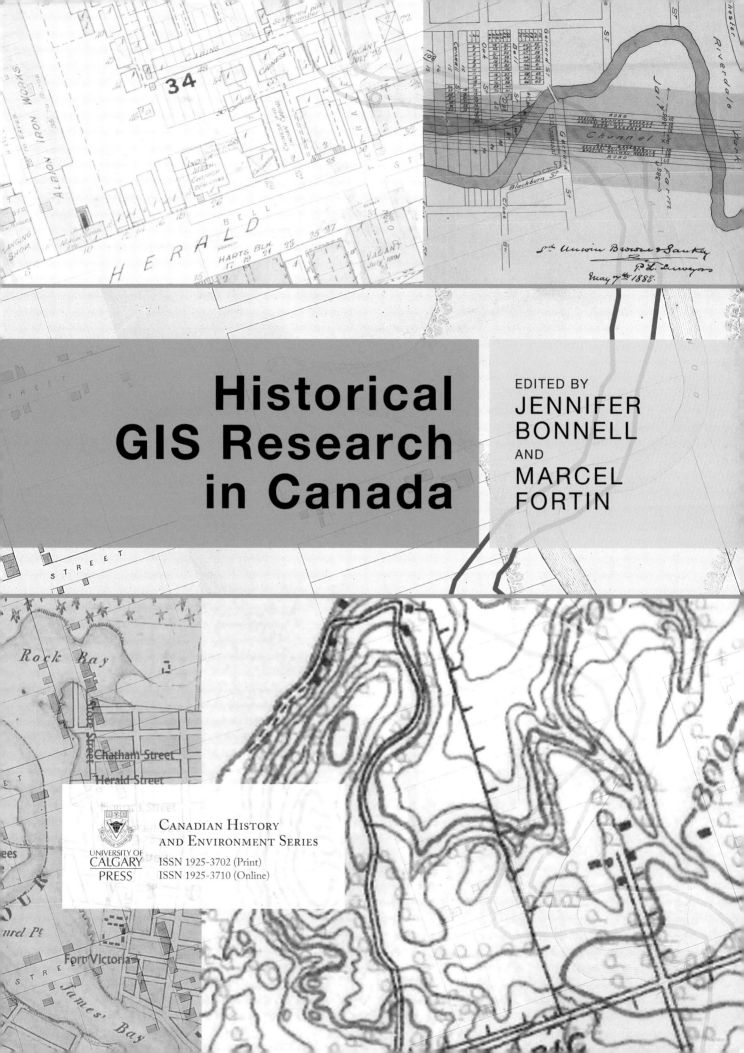

Historical GIS Research in Canada

EDITED BY

JENNIFER BONNELL

AND

MARCEL FORTIN

CANADIAN HISTORY
AND ENVIRONMENT SERIES

UNIVERSITY OF CALGARY PRESS

ISSN 1925-3702 (Print)
ISSN 1925-3710 (Online)

© 2014 Jennifer Bonnell and Marcel Fortin

University of Calgary Press
2500 University Drive NW
Calgary, Alberta
Canada T2N 1N4
www.uofcpress.com

This book is available as an ebook which is licensed under a Creative Commons license. The publisher should be contacted for any commercial use which falls outside the terms of that license.

LIBRARY AND ARCHIVES CANADA CATALOGUING IN PUBLICATION

Historical GIS research in Canada / edited by Jennifer Bonnell and Marcel Fortin.

(Canadian history and environment series, 1925-3702 ; 2)
Includes bibliographical references and index.
Issued in print and electronic formats.
ISBN 978-1-55238-708-5 (pbk.).—ISBN 978-1-55238-756-6 (open access pdf).—
ISBN 978-1-55238-744-3 (eDistributor pdf).—ISBN 978-1-55238-750-4 (html).—
ISBN 978-1-55238-751-1 (mobi)

1. Historical geographic information systems—Canada. 2. Canada—
Historical geography. 3. History—Data processing. I. Fortin, Marcel, 1967-,
editor of compilation II. Bonnell, Jennifer, 1971-, editor of compilation
III. Series: Canadian history and environment series ; 2

G70.217.H57H57 2013   910.285   C2013-907480-5
                                                            C2013-907481-3

The University of Calgary Press acknowledges the support of the Government of Alberta through the Alberta Media Fund for our publications. We acknowledge the financial support of the Government of Canada through the Canada Book Fund for our publishing activities. We acknowledge the financial support of the Canada Council for the Arts for our publishing program.

This book has been published with the help of a grant from the Canadian Federation for the Humanities and Social Sciences, through the Awards to Scholarly Publications Program, using funds provided by the Social Sciences and Humanities Research Council of Canada.

This book has been published with support from NiCHE: Network in Canadian History & Environment, using funds received through the Social Sciences and Humanities Research Council of Canada's Strategic Knowledge Clusters program.

Printed and bound in Canada by Friesens
♻ This book is printed on Sterling Premium Matte paper
Cover images: Clockwise, see Fig 1.4 (p.7), Fig. 3.6 (p.56), Fig. 3.3 (p. 51), Fig. 5.3 (p. 95), and Fig. 1.3 (p.7).

Cover design, page design, and typesetting by Melina Cusano

CONTENTS

ACKNOWLEDGMENTS

The editors wish to thank the Network in Canadian History & Environment (NiCHE) and the University of Toronto Libraries for supporting the development of the Don Valley Historical Mapping Project. Thanks to Helen Mills and the late Terry McAuliffe of Lost Rivers for sharing research on points of interest throughout the valley. For her careful work and attention to detail, thanks to Jordan Hale. We are grateful to Derek Hayes and the Toronto Public Library for access to their collection of map images. This book benefitted from the thoughtful feedback of two anonymous reviewers, and from the guidance and support of series editor Alan MacEachern. We also wish to thank Ruth Sandwell and Bill Turkel for their encouragement and continued support, without which this book would not have been possible. Bonnell acknowledges the support of the Social Sciences and Humanities Research Council during her research on the history of Toronto's Don River Valley. We are indebted to Scott McKinnon and Anne Fortin for their patience and support as this project developed.

The authors of Chapter 1 would like to thank Shannon Lucy, Kate Martin, Tylor Richards, Jenifer Sguigna, Kathleen Traynor, and Kevin Van Lierop for their invaluable assistance in conducting and presenting this research.

Daniel Macfarlane would like to thank his co-authors (particularly Jim Clifford for his tutelage) for their help, the co-editors of this volume for their insights and patience, and the Network in Canadian History and Environment (NiCHE) for support. Colleen Beard wishes to acknowledge the assistance of several people in developing the Welland Canals Google Earth project: Sharon Janzen for her Google Earth Pro technical expertise; Brock University Geography internship students and Map Library staff who assisted with digitizing collections, creating map overlays, and adding content; and Brock University Communications Department staff, who assisted with editing audio recordings. Photographs for this portion of Chapter 2 were obtained from Brock University Special Collections and Archives, Niagara Falls Public Library, and several personal collections.

As Stephen Bocking and Barbara Znamirowski discovered, a paper that discusses an HGIS project can only ever represent a fraction of the work involved in such a project. They wish to acknowledge the assistance of several members of the Maps, Data, and Government Information Centre of the Trent University Library: Tracy Armstrong, Mike Kyffin and David Lang, whose GIS skills and creativity were essential to the project's technical development. Siobhan Buchanan and Kathryn McLeod provided valuable research assistance. The editors of this volume provided helpful comments on an earlier draft. John Wadland also provided extensive and thoughtful comments. Bocking and Znamirowski also wish to thank the following institutions for their assistance: Trent Valley Archives, Trent University Archives, University of Toronto Map and Data Library, Western University Map Library, McMaster University Map Library, Brock University Map Library, Kawartha Land Trust, Land Information Ontario, National Air Photo Library, and Library and Archives Canada.

Joanna Dean and Jon Pasher would like to acknowledge the Social Sciences and Humanities Research Council for funding the geospatial analysis in Chapter 6, which was completed at the Geomatics and Landscape Ecology Laboratory at Carleton University. Air photographs for this chapter were provided courtesy of the National Air Photo Library and the City of Ottawa. Digital mapping was produced by Michelle Leni.

Daniel Rueck is grateful to Louis-Jean Faucher for designing the maps in Chapter 7. He would also like to thank the editors, as well as Stephen Bocking, Barbara Znamirowski and Sherry Olson for their helpful comments and suggestions. The research for this chapter was funded in part by the Social Sciences and Humanities Research Council, the Bibliothèque et Archives nationales du Québec, and the Osgoode Society for Canadian Legal History. Rueck is also grateful to the many Kahnawá:kehró:non who contributed to his research by offering critical insights and many forms of hospitality and generosity.

Francois Dufaux and Sherry Olson would like to thank the Bibliothèque et archive nationale du Québec (BAnQ), Ville de Montréal, and the Social Science and Humanities Research Council of Canada on behalf of the MAP project, 'Montréal l'avenir du passé'.

Matthew G. Hatvany would like to thank Donald Cayer, Chrystian Careau and the staff of the Université Laval cartographic laboratory for their assistance with the photointerpretation, data collection and mapping.

Joshua D. MacFadyen and William (Bill) M. Glen would like to acknowledge the Social Sciences and Humanities Research Council of Canada Fellowships Program for supporting the research in Chapter 10. Access to the inventories and the aerial photographs was given courtesy of the Prince Edward Island Department of Environment, Energy and Forestry.

Sally Hermansen and Henry Yu would like to thank Graeme Wynn for introducing them.

Special thanks to students Jeremy Alexander, Maria Ho and Edith Tam at the University of British Columbia, and Oliver Khakwani and Stephanie Chan at the Stanford Spatial History Lab. Yu would also like thank collaborators Peter Ward, Eleanor Yuen and Phoebe Chow at UBC, the Stanford Spatial History Lab, his Chinese Canadian Stories research team and their academic collaborators at Simon Fraser University and community partners across Canada. Research for this chapter was conducted with generous financial support from the Community Historical Recognition Program and the Social Sciences and Humanities Research Council.

Ruth Sandwell would like to acknowledge the financial support of SSHRC Standard Grant for research completed in her chapter.

Byron Moldofsky would like to acknowledge the support for this chapter provided by the Department of Geography and Program in Planning, University of Toronto, by the Canadian Foundation for Innovation which was the major source of funding for the CCRI, by the Social Sciences and Humanities Research Council and Statistics Canada for their joint initiative for Access to Research Data Centres, and by the University of Alberta and its Library for hosting the CCRI data for 1911 and subsequent years as they become available.

Thanks to Don Lafreniere for help in compiling the references in the Appendix.

Introduction

Jennifer Bonnell and Marcel Fortin

This book originated in a collaborative project. The Don Valley Historical Mapping Project, completed in 2009, involved the collection and synthesis of a wide range of historical maps and other documents to produce a historical GIS of Canada's most urbanized watershed.[1] As producers of this publicly accessible database of geospatial information, we came to learn the practical benefits and creative possibilities that followed from collaboration across disciplinary and professional divides. Seeking information on technical approaches and source materials, and the particular challenges of doing historical GIS in the Canadian context, we looked to other Canadian studies for insights. This book is the result of those investigations.

Historical GIS is a relatively new tool in historical scholarship, and this is especially true in Canada. The purpose of this collection is to showcase work going on across Canada by historians and other researchers focusing on diverse periods and topics in the study of Canada's past. The book is both a collection of case studies, and a reflection on the process and practice of doing historical GIS. Building on the work of Anne Kelly Knowles' 2002 and 2008 collections of innovative studies in historical GIS research,[2] and a number of Canadian studies that have employed historical GIS (see Appendix A), this book showcases the range of possibilities available when historians, geographers, and other researchers use GIS to develop and enrich their analyses. More specifically, this collection brings into focus the particular challenges – and opportunities – inherent to conducting historical GIS research in Canada. Contributors highlight the benefits of working with GIS technology in an academic context where few historians have incorporated historical GIS approaches into their work. They discuss the challenges of applying GIS to historical sources, and the difficulties of obtaining access to historical source materials within the context of Canada's information access policies and restrictive copyright legislation. Readily apparent in the results of their analyses are the rewards of working with GIS technology to visualize past places and societies at differing scales. Much more than a map-making tool, GIS emerges here as a powerful new approach to historical inquiry.

Collaboration is a fundamental characteristic of this work. Traditionally viewed as solitary practitioners, historians are, more and more, finding opportunities to work together and to reach across disciplinary boundaries in order to apply new approaches to their investigations. Most of the chapters in this collection emerge from collaborative projects between historians, geographers, map and GIS librarians, and, in a few cases, scientists and professionals from other disciplines. Many have relied on supports from technical staff, graduate students, and archivists. The projects they describe range in size and scope from large, well-funded, multi-year initiatives to small, low-budget projects completed by a few researchers in a matter of weeks. Throughout the collection, contributors discuss the challenges associated with building multi-disciplinary projects, and the possibilities they bring for historical scholarship.

While this book is indicative of new trends in historical research, it also demonstrates the new territory librarians are charting in their role as information professionals. No longer stereotyped as gatekeepers of paper collections, librarians now advocate for the digitization, use, and dissemination of books, maps and other collections. Increasingly, they are getting involved in research projects where they can lend technical and subject expertise. As

research collaborators with a professional mandate to facilitate access to information, librarians are changing the nature and the outcomes of academic research.

Geographic Information Systems (GIS) have a relatively long history in Canada, dating back to the 1960s. Some of the earliest computer-generated maps and geographic analysis tools were developed through what was called the Canadian Geographic Information System (CGIS). The computerized manipulation and management of geographic information began in both the United States and Canada at around the same time. In the United States, the Harvard Laboratory for Computer Graphics began developing automated methods for creating maps using computer technology in the mid-1960s. In Canada, the CGIS was created with the purpose of managing the Canada Land Inventory (CLI) maps and data. The CLI covered over 2.5 million square kilometres of land and water, categorizing land use according to its capability for agriculture, forestry, wildlife, recreation, and wildlife. Over 1,000 mapsheets at the 1:250,000 scale were created over the lifetime of the project. By the start of the 1970s, the CGIS was a fully functional geospatial data management tool for much of the Canadian territory, making Canada a forerunner in the development of GIS technology.[3]

Modern GIS are not created as mapping tools alone. Geographic Information Systems, as the name implies, are actual computer systems. That is, they are made up of several components, including computer hardware, software, geospatial data and other information. The systems are used to integrate, analyze, and display spatially referenced information. Outputs and analyses can vary from maps and atlases, to graphs, official reports, and traditional scholarly communication. In scholarly research, GIS can be used to tackle questions, to solve problems, and to reveal geospatial relationships and patterns.

Historical GIS takes the power of geographic analysis and applies it to the realm of historical research. Historians have long used maps to investigate, research, and teach history, but GIS greatly enhances the potential of this work by enabling the manipulation, analysis, and output of location information within the historical landscape. As Ian Gregory and Paul Ell explain, "In GIS the map is no longer an end product; it is now a research tool.... [S]patial patterns within the data can be repeatedly re-explored throughout the research process."[4] With its capacity to integrate and layer divergent source materials and tie them to specific locations in space, GIS effectively enables the creation of new source materials. As American historian Richard White observes, GIS software "allows the orientation and co-ordination of dissimilar things – an aerial photograph and a map, for example – in terms of a single location. It allows us to merge things created at dramatically different times to create what are, in effect, new modern images which potentially reveal things about the past that the original artifacts did not."[5] By enabling researchers to undertake a geospatial interpretation of historical questions, HGIS becomes a powerful method for historical investigation. As White argues, HGIS (and spatial history more broadly) is "not about producing illustrations or maps to communicate things that you have discovered by other means. *It is a means of doing research*; it generates questions that might otherwise go unasked, it reveals historical relations that might otherwise go unnoticed, and

it undermines, or substantiates, stories upon which we build our own versions of the past."[6]

To date, some of the most important work in the field has involved the production of massive repositories of geospatial data at the national level. In the United States, the National Historical Geographic Information System project compiled and made accessible all available aggregate census information for the United States between 1790 and 2000.[7] In the UK, the work of Ian Gregory and Humphrey Southall on the *Great Britain Historical GIS* and the *Vision of Britain Through Time* projects has generated an impressive repository of freely accessible historical source materials georeferenced to current coordinates, enabling researchers to visualize and interpret past landscapes.[8] These projects have provided the geospatial data to seed many smaller initiatives. David Rumsey's massive private collection of historical maps and images, many of which have been georeferenced for easy importation into GIS applications, is another notable resource for historical GIS projects.[9]

In March 2004, the Newberry Library in Chicago hosted "History and Geography: Assessing the Role of Geographical Information in Historical Scholarship." The conference was a watershed moment for HGIS, bringing together respected scholars from around the world to talk about their work using GIS for historical research. Among the presenters were contributors to Anne Kelly Knowles' 2002 ESRI Press collection *Past Time, Past Place: GIS for History*. The conference was an inspiration to many historians, historical geographers, and librarians. It led to a second ESRI publication, *Placing History: How Maps, Spatial Data, and GIS are Changing Historical Scholarship* (2008). These two books, together with Ian Gregory and Paul

Ell's 2008 *Historical GIS: Technologies, Methodologies and Scholarship*, form the core texts for historical GIS scholarship.[10]

In recent years, there has been a discernible "spatial turn" in the practice of history. Historians of diverse fields and periods have given greater emphasis to the geographical context of their investigations, drawing upon a range of spatial methods and technologies to interpret source materials in new ways. This work takes its most creative expression in initiatives such as Richard White's Spatial History Project at Stanford University, where a collaborative community of scholars are engaged in "creative visual analysis" to further historical research, and by regional initiatives such as Eric Sanderson's Manahatta project, which uses georeferenced historical data and other techniques to allow viewers to visualize the natural history of Manhattan.[11] Within this context, the term "spatial history" has come to describe a range of approaches, including historical GIS, spatial statistics, and data visualization. As a result of these developments, spatial history is sometimes used interchangeably with HGIS. For the purposes of this collection, we will use HGIS to refer to the specific method of applying GIS technology to historical research questions; references to spatial history will suggest a broader palette of approaches that include but are not limited to HGIS.

In Canada, the HGIS landscape is very different than in the UK and the United States. Since the 1990s, only a small number of historical geographers and historians have used GIS technology to create geospatial representations of past landscapes and societies in Canada. Notable projects include *Montréal, l'avenir du passé*, a comprehensive HGIS database of Montreal led by McGill geographer Sherry Olson and Memorial University historian Robert Sweeny.[12] Jason Gilliland's work at the University of Western Ontario, with Olson on the spatial history of Montreal, and more recently with Don Lafreniere on the spatial history of London, Ontario, has also made important contribution to our understanding of Canada's urban history.[13] The relative scarcity of significant historical GIS initiatives in Canada has been due in part to the absence of a national-level repository of geospatial data on par with the national historical GIS initiatives in the United States and the UK. While the federal government's GEOGRATIS and GEOBASE data portals provide an ever-expanding collection of geospatial information for researchers, they lack historical content.

Beginning in 2003, the Canadian Century Research Infrastructure (CCRI) project set out to fill some of these gaps by digitizing and compiling Canadian census data from 1911 to 1951 into a set of interrelated databases. A multi-disciplinary, multi-institutional project led by principal investigators Chad Gaffield and twelve other leading academics, with funding from the Canada Foundation for Innovation (CFI), the CCRI will eventually incorporate census data from 1871 to 2001 with companion geospatial data for census subdivisions. (For more on the CCRI and its significance, see Chapter 1.) Despite the great promise of these initiatives, the lack of a "one stop shop" for geospatial data, coupled with restrictive and confusing Canadian copyright legislation (as we will discuss later in this introduction), continues to present obstacles to the growth of historical GIS research in Canada.

Compared with even ten years ago, however, an assessment of the status of historical GIS research in Canada reveals a growing number of projects investigating questions in Canadian social, cultural, and environmental history – many of which are represented in this collection. A number of factors have contributed to this upswelling of interest in historical GIS. Olson, Sweeny, and Gilliland's work has captured the interest of a new generation of graduate students and new scholars. The influence of digital humanities scholars such as William J. Turkel of the University of Western Ontario (UWO) is also notable. Through his award-winning blog *Digital History Hacks* (2005–2008) and his co-authored ebook *The Programming Historian*, Turkel has inspired many historians to take up computer programming in their pursuit of historical questions.[14] Not to be elided here either is the influence on new scholars of the Network in Canadian History and Environment (NiCHE), directed by UWO historian Alan MacEachern with funds from the Social Sciences and Humanities Research Council (SSHRC). Since 2007, NiCHE has been sponsoring digital historical projects (including more than one project in this collection) that have helped change the way historians conceive of and carry out their work. NiCHE has encouraged the use of computing widely in historical research and in 2008 collaborated with the University of Toronto Libraries to sponsor a comprehensive two-day Historical GIS workshop for doctoral students from across the country.

While there is evidently an increased appetite for HGIS research in Canada, the practical challenges involved in launching new projects often prevents them from getting off the ground. The expense of these projects is often the first hurdle. Time, expertise, and research assistant fees are quantifiable demands that require financial or in-kind contributions,

ideally some combination of the two. Commercial software such as ESRI's ArcGIS itself requires a considerable outlay of expense, softened in many university environments by educational licensing agreements. Open source applications, such as Quantum GIS (QGIS), can assist in reducing project expenses. Newly accessible applications such as Google Earth have also contributed to the "democratization" of HGIS research, providing stripped-down GIS technology to a wider range of users, at no cost. This collection showcases two examples of researchers using Google Earth to explore historical questions in innovative ways.

Access to data is another problem that stymies the development of HGIS projects in Canada. Despite Canada's status as forerunners in the development of GIS technology, stringent restrictions on access to data have caused GIS, and consequently HGIS projects, to suffer. The delay is probably due to what UBC geographer Brian Klinkenberg would describe as Canada's "spatial-data culture," one that is radically different from the United States, where geospatial data are seen as a public good and are in most circumstances made publicly available upon creation. The difference between the two countries is in large part due to the double burden of Canadian government policies and Crown Copyright legislation. The impact of these has influenced the type and extent of the use of GIS in both the business sector and in academia, as neither could afford the acquisition of data at critical stages in the development of GIS technology. Klinkenberg argued in a 2003 article that the political culture of the 1980s had a significant effect on the developing spatial-data culture in Canada: that "the government owned spatial data through Crown copyright and could control their use

– even after allowing a third party access to it – had a dramatic and lasting impact on our spatial-data policies at the critical time at which policies were being developed."[15]

The situation has changed dramatically over the last few years. The government of Canada was one of the first and certainly the most important government in the country to make large amounts of geospatial data available to Canadians for free. On the recommendations of a report commissioned by Natural Resources Canada's Geoconnections and penned by KPMG, the Canadian Geospatial Data Policy study put forth that all government data in Canada should either be reduced dramatically in price or made available to the public without charge. Natural Resources Canada responded by opening access to their massive National Topographic Database, at first just to the academic community in 2004, and then to the entire world in 2007 through their web portals Geogratis.ca and Geobase.ca. Subsequently, a number of other government bodies at the provincial and municipal level have made their data available, taking their cue from a global open data movement.

While these moves toward more liberal access to data are all to be applauded, opportunities lost over the years have created a situation where not only has there been an almost debilitating delay in the use, development, teaching, and research with geospatial data and technology in Canada, but the restrictive attitude towards data, as described by Klinkenberg, has spilled over into non-governmental segments of society. Data access limitations and the protective nature of data creators have also pervaded the academic community. Data are viewed as a commodity in Canada rather than a public good, and governments, businesses, and

academics alike have safeguarded their data as a matter of practice. Rare is it that geospatial data projects are conceived, let alone completed, in academia, with the intention of making resulting data available to the public, other than through traditional scholarly communication outputs. This proprietary attitude to data persists in Canadian scholarly work despite the fact that Tri-Council policy surrounding funding for academic research in Canada has specified for many years that data created with their financial support be made publicly accessible via a research library.

Libraries and archives are not immune, either, from practices that restrict the use of valuable sources of information. Cost recovery and even profit are the norm in most institutions, despite persistent questions about the ethics and financial viability of charging for publicly held information. Users of historical sources must, more often than not, pay handsomely for copies of digital reproductions. And when researchers have received digital images through purchase, frustrations often mount further as use restrictions, particularly for maps, are often built into the licences and contracts that researchers must sign. In many cases, for example, institutions forbid the use of their images on the web or as high-resolution reproduction without further cost.

As Klinkenberg has pointed out, Canada's copyright laws have not helped the historical researcher. Several issues have combined to confuse and stifle the use of reproductions of historical sources. Not only do the users of material get confused between American and Canadian copyright laws, but librarians and archivists also get confused by complicated and outdated legislation that fails to address the specifics about working with maps or GIS data. Fire insurance plans, some of the most important cartographic sources for HGIS projects in urban areas, exemplify this problem, caught as they are in the confusion between copyright regulations and side-deals between big public institutions and corporations. Produced for most urban areas across the country, fire insurance plans are richly detailed documents that provide excellent, scaled renderings of street grids, building outlines and construction materials, as well as other structural land uses. Access to and restrictions on the use of these sources have been a concern and an impediment for both librarians and researchers for years. Fear of copyright infringement has led many libraries and archives to restrict and in some cases eliminate the photocopying and scanning of all fire insurance plans for and by researchers. There is a great deal of confusion and misunderstanding on this issue, which accentuates the difficulties of accessing many fire insurance plans and atlases in collections across Canada.

The complicating factor in this story is that SCM Risk Management Services currently claims copyright on all fire insurance maps published by Charles E. Goad and all successor companies (such as the Underwriters' Survey Bureau and the Canadian Underwriters Association), regardless of date of publication. Under Canadian Copyright law, as most map librarians in Canada interpret it, any map that is fifty years old and a day is considered to be outside of copyright protection. That is, only if no cartographer is named for the map in question. The map library community does not consider Charles E. Goad and successor companies to be the cartographers of fire insurance plans. SCM and previous successor companies do not interpret the law in the same manner,

and thus a stalemate existed for many years over what could and could not be done with these invaluable sources of historical information. In 1993, the Ontario Archives, Library and Archives Canada, and CGI, SCM's predecessors, reached a compromise agreement wherein the two government institutions would restrict the reproduction of fire insurance maps less than ninety years old (rather than fifty years). As a result of this agreement, many other institutions, including most university libraries, implemented the same restrictions. In the decades that have followed, many institutions have openly questioned the agreement and its particular compromise. In 2010, a municipal government in Quebec, following an investigation by their lawyers, concluded that they were within the bounds of the law in allowing the duplication of any fire insurance plans older than fifty years.

Although challenges persist in access to reproductions of fire insurance plans across Canada, promising developments surrounding other source materials have made the creation of historical GIS datasets easier than ever before. For most HGIS projects, paper reproductions of historical maps constitute the main data source. The rich paper map collections held in library and archival map collections across the country can be seen, in this context, as historical GIS projects waiting to happen. Before affordable scanners were available, equipment to convert paper maps to digital data was not only very expensive and rare, it was also difficult to operate. Digitizing tables were the domain of specialists only. Many library and archive facilities are now equipped with large-format scanners, which allow for most sheet maps to be scanned in mere seconds, making geospatial data creation much more possible and expedient than before. More and more, Canadian historians are recognizing these opportunities to take their analyses in new directions. They are using GIS to integrate diverse source materials, to reveal spatial patterns that were otherwise invisible, and to challenge existing interpretations of the past.

The thirteen chapters in this collection are the result of people working together to combine approaches from diverse disciplinary backgrounds in order to better understand the historical geography of Canada's past. Twenty-seven contributors share their insights here, including fourteen historians, seven geographers, five librarians, and a forester. Some are pioneers in historical GIS approaches; many worked with GIS for the first time. Established professionals experiment with new approaches to their research questions, and graduate students and new scholars familiarize themselves with the possibilities of GIS technology early in their careers. Librarians collaborate with academics to repurpose and showcase their collections and reach beyond the academy to make information accessible to community groups.

This book is intended for people who are new to GIS or just beginning to familiarize themselves with the technology as a method of historical inquiry. Accessible prose and richly illustrated descriptions also make this collection a useful resource for undergraduate teaching. Individual case studies provide compelling teaching documents for upper-level undergraduate courses in Canadian social, cultural, and environmental history, historical geography, and for upper-level methods courses in geography, history, information studies, and other disciplines. Reflections on the process of multidisciplinary collaboration will also be of interest to scholars and practitioners planning large interdisciplinary projects involving historical GIS.

In keeping with its subject, this book is organized geographically, moving across the country from west to east and culminating with three chapters that take a pan-Canadian approach. Beginning on the west coast, Chapter 1 employs historical GIS to study the implications of racial discrimination in late nineteenth-century Victoria, British Columbia. Here a partnership between historians John Lutz, Patrick Dunae, and Megan Harvey, and geographers Jason Gilliland and Don Lafreniere produces a study that challenges existing understandings of racial discourse by probing relationships between race and space in one of the key nodes of the British Empire at the peak of its power.

Chapter 2 takes us to Ontario, and the work of two projects that make use of accessible and easily mastered Google Earth technology to document aspects of the history of transportation infrastructure on the Great Lakes. Brock University map librarian Colleen Beard produces a rich resource for regional heritage planning and history enthusiasts with her Google Earth representation of the historic Welland Canals. At the other end of Lake Ontario, historians Daniel Macfarlane and Jim Clifford use Google Earth to map the evolution of the St. Lawrence Seaway and Power Project. Both projects demonstrate the versatility of Google Earth as a freely available tool for analyzing and displaying the geospatial relationships of past events and places.

The urban landscape of Toronto is the subject of Chapters 3 and 4. In Chapter 3, historian Jennifer Bonnell and map and GIS librarian Marcel Fortin explore the results of a small HGIS initiative as an example of what can be accomplished with relatively limited funds and the partnership of an academic library in seeking out and compiling geospatial resources. The Don Valley Historical Mapping Project assembled and digitized a wide range of geographical documents for the Toronto area to produce a publicly accessible database of the valley's industrial and environmental history. In Chapter 4, historian Andrew Hinson teams up with librarians Jennifer Marvin and Cameron Metcalf to investigate socio-economic relationships emerging from pew seating records at Toronto's Knox Presbyterian Church in 1882. By correlating address data from pew records with census data and city assessment rolls, they create a street-level picture of the social status of congregants outside the church walls and make some surprising discoveries about social positioning within.

In Chapter 5, Stephen Bocking and Barbara Znamirowski reflect upon their work developing a regional atlas of environmental history for south-central Ontario. GIS technology, they find, enables the telling of key "stories" in Canadian environmental history, such as the expansion of agricultural and urban settlement, the rise and decline of resource industries, and the emergence of conservation. Joanna Dean and Jon Pasher use historical aerial photographs to measure canopy cover in selected Ottawa neighbourhoods in Chapter 6, correlating their results with social indices to show how street trees are an environmental benefit that is socially produced and unevenly distributed.

Chapter 7 takes us to a new time and place, to nineteenth-century Kahnawá:ke, a Mohawk community near Montreal, and the effects of a Euro-Canadian surveying initiative upon the culturally distinct land practices of the

Kahnawá:ke residents. Using GIS technology to analyze the resulting survey documents, historian Daniel Rueck reveals a palimpsest of two radically divergent land practice regimes. His work draws to our attention important cultural considerations in the use of GIS to explore the environmental histories of indigenous people, and the cultural assumptions it has the capacity to reinforce. In Chapter 8, architectural historian François Dufaux and geographer Sherry Olson uncover some of the choices made and constraints faced by individual actors in rebuilding the St. Mary neighbourhood of Montreal (now Notre Dame East) after a devastating fire in July 1852. Drawing upon the geospatial databases of the Montreal H-GIS *Montreal, l'avenir du passé* (MAP), and incorporating an impressive range of source materials, including builders' specifications and loan contracts from the period of the fire, and a later set of architectural drawings and expropriation documents, Dufaux and Olson sketch a succession of imagined futures for this Montreal neighbourhood.

Moving further east, historical geographer Matthew Hatvany uses historical GIS to challenge scientific and lay observations of a generalized erosion of coastlines in the St. Lawrence Estuary in Chapter 9. His detailed time-series analysis of salt-marsh growth and erosion in Kamouraska County, Quebec, presents a different picture of environmental change that raises important epistemological questions about scientific ways of understanding dynamic environments, and the value of historical approaches to environmental questions. In Chapter 10, historian Joshua MacFadyen and forester William Glen examine historical forest inventories, compiled using GIS and aerial photography, to call into question previously held estimates of agricultural activity, its eco-logical impact, and the rates of forest regrowth on Prince Edward Island farmland.

Chapters 11, 12, and 13 adopt a pan-Canadian approach. In Chapter 11, geographer Sally Hermansen and historian Henry Yu use data collected as a result of Chinese Head Tax legislation to map the origins and destinations of Chinese migrants to Canada between 1910 and 1923. By using historical GIS to reveal the patterns that emerged between Chinese counties of origin and Canadian destinations, Hermansen and Yu construct an "imagined geography" of aspirational mobility. Historian Ruth Sandwell charts the expansion of the electrical grid across Canada from the 1920s to the 1950s and correlates it with data on population density and changing domestic fuel use in Chapter 12. Part of a larger study on the social history of energy in Canada, this chapter explores the profound importance of space and place in understanding the extraction, production, processing, transportation, and consumption of fuels in Canada, even though the hallmark of "modern" fuels is their apparent "placelessness" and near-invisibility (and literal invisibility in the case of electricity) to consumers. In Chapter 13, cartographer Byron Moldofsky introduces the Canadian Century Research Infrastructure (CCRI) project and its significance for historical research in Canada.

NOTES

1   See http://maps.library.utoronto.ca/dvhmp/.

2   Anne Kelly Knowles, *Past Time, Past Place: GIS for History*, illustrated ed. (Redlands, CA: ESRI Press, 2002); Anne Kelly Knowles and Amy Hillier, eds., *Placing History: How Maps, Spatial Data, and GIS Are Changing Historical Scholarship* (Redlands, CA: ESRI Press, 2008).

3   All the data compiled as part of the CLI are still available for use from the Government of Canada's website Geogratis.ca. For further information on the CLI, see http://res.agr.ca/cansis/nsdb/cli/intro.html. For a good introduction to the history of the CGIS, see Roger Tomlinson, "The Canada Geographic Information System," in Timothy W. Foresman, ed., *The History of Geographic Information Systems: Perspectives from the Pioneers* (Upper Saddle River, NJ: Prentice Hall, 1998): 21–32. Foresman's collection provides a useful overview of the history of Geographic Information Systems and key individuals and organizations involved in its development. On concurrent developments at Harvard, see also Nicholas R. Chrisman, *Charting the Unknown:*

How Computer Mapping at Harvard Became GIS (Redlands, CA: ESRI Press, 2006).

4   Ian N. Gregory and Paul S. Ell, *Historical GIS: Technologies, Methodologies, and Scholarship* (Cambridge: Cambridge University Press, 2008), 10.

5   Richard White, "What Is Spatial History?" (Stanford Spatial History Lab: Working Paper, February 1, 2010), http://www.stanford.edu/group/spatialhistory/media/images/publication/what%20is%20spatial%20history%20pub%20020110.pdf.

6   Ibid.

7   See https://www.nhgis.org/.

8   See, respectively, http://www.port.ac.uk/research/gbhgis/ and http://www.visionofbritain.org.uk/.

9   See http://www.davidrumsey.com/.

10 Gregory and Ell, *Historical GIS*. See also Ian Gregory, *A Place in History: A Guide to Using GIS in Historical Research* (Oxford: Oxbow, 2003).

11 The Manahatta project has since been expanded as the Welikia Project, which encompasses the

natural history of all of New York City, including the Bronx, Queens, Brooklyn and Staten Island, and surrounding waters. See http://welikia.org/.

12   The MAP project is accessible at http://www.mun.ca/mapm/. A great deal of scholarship has resulted from this project, among them Sherry Olson and Patricia Thornton's recent book, *Peopling the North American City: Montreal 1840–1900* (Montreal: McGill-Queen's University Press, 2011). A full listing of scholarship resulting from the MAP project can be found in Appendix A. For an overview, see Jason Gilliland and Sherry Olson, "Montreal, l'avenir du passe," *GEOinfo* January–February (2003): 5–7.

13   See Mathew Novak and Jason Gilliland, "Trading Places: A Historical Geography of Retailing in London, Canada." *Social Science History* 35, no. 4 (2011): 543–70. For a complete list of historical GIS research conducted on Canadian topics, and by Canadians, consult Appendix A.

14   Respectively, http://digitalhistoryhacks.blogspot.ca/ and http://niche-canada.org/programming-historian.

15   B. Klinkenberg, "The True Cost of Spatial Data in Canada," *The Canadian Geographer/Le Géographe Canadien* 47, no. 1 (2003): 41.

Turning Space Inside Out: Spatial History and Race in Victorian Victoria

John S. Lutz, Patrick A. Dunae, Jason Gilliland, Don Lafreniere, and Megan Harvey

In describing Rachel Whiteread's sculpture 'House,' cultural geographer Doreen Massey inadvertently captured what we propose to do with this research project: "it turns space inside out … it exposes the private sphere to public view and thereby to questioning and understanding."[1]

Whiteread made concrete casts of the inside of an archetypal London row house which was being prepared for demolition; when the buildings came down, the outside disappeared and what had once been empty space had become solid. Her sculpture (Fig. 1.1) was an interpretive act of an artist fixing a space in time.

Like Whiteread, we too want to "turn space inside out" and interpret the relationship of space to time as we build a spatial history of race between 1861 and 1911 in one of the key nodes of the British Empire at the peak of its power. Building a spatial history using the tools of the scholar "not only creates the possibility of history becoming more collaborative," observed Richard White, "it virtually necessitates it."[2] This project is an exemplar of this necessity. Since 2007, an interdisciplinary team of cultural historians, urban geographers, and GIS technicians, both faculty and students, have been creating and utilizing historical GIS to "expose the private sphere to public view" to get

Fig. 1.1. Rachel Whiteread's 1993 House. (Photo by Sue Omerod; courtesy of Gagosian Gallery, London. (c) 2011. Rachel Whiteread.)

a better understanding of how race and racial space operated in Victorian-era Victoria, British Columbia.[3]

What makes this unusual among historical GIS projects is that it also experiments with bringing this methodology into conversation with a type of scholarship that has lived in isolation and sometimes open hostility for decades: interpretive discourse analysis. Bringing these skills and methods together, we believe we can show that a spatialized "geohumanities"[4] promises something substantively different from our previous ways of knowing space and time while it offers new insights about the

dominant interpretation of racism in British Columbia history.

This chapter illustrates some of that promise by presenting new conclusions about how race was created and experienced and documenting some of the "aha! moments" when never-before-seen patterns jumped off the computer screen and reframed our historical understanding. For example, in the early phase of this long-term project we have dispelled the myth of Chinatown as a Forbidden City and have documented the "vanishing" of Indigenous people from the city of Victoria between 1861 and 1911.

As a contribution to interdisciplinary collaboration, we also reflect on four interrelated challenges: 1) spatializing historical censuses and other data to exact street addresses over a forty-year period when for part of the era no street addresses existed; 2) working with multiple partial data sets including census, directories, tax assessments, photographs, and fire insurance plans; 3) adding "discourse analyses" to GIS mapping; and, finally, 4) transcending disciplinary boundaries.

CONCEPTS

For the cultural historian of race, the addition of spatial analysis offers a unique opportunity to empirically test conclusions inferred from the study of text. Such testing is rarely performed but is particularly useful when it comes to studying sensitive topics such as race, where a key problem is that people do not always say what they mean and what they say is not always what they do. What we currently believe about the history of racism in Canada is based on what social elites said in the press, the courts, and government. Relatively little is known about the lived experience of race, either among the articulate minority or the largely silent majority. In this research project, we use census data and GIS to spatialize the lived experience of race and test it against the public discourse.

Thanks in large measure to the growing appreciation of the insights of Henri Lefebvre alongside Michel Foucault's metaphor of "archaeologies," the humanities have taken what some have called the "spatial turn" in the last decade or so.[5] Lefebvre's key insight was that space is a social product, created and made to seem real by everyday social practices. "Space," he wrote, "is permeated with social relations; it is not only supported by social relations, but it is also producing and produced by social relations."[6] Foucault, with his metaphor of archaeologies of knowledge, has provided a language to explore knowledge spatially, while his related work has influenced an earlier "cultural turn" in history, which focused specifically on the ways in which knowledge and power are seen as constituted through discourse. Race, in particular, is seen by scholars in this tradition as a system created by European writers who created a classification that placed themselves at the top and authorized themselves to seize the land, resources, and, in some cases, the bodies of others lower on their scale.[7]

There is already a rich post-colonial literature that focuses on the discursive practices of racism and colonialism.[8] Another growing literature looks empirically at how race was spatially deployed in different urban contexts using census and geographic information systems and a variety of complex measures.[9] Our project hopes to inaugurate a literature connecting the two and testing the assumptions of the one against the other.

In British Columbia, the research on race and racism assumes that race "trumped class" as the 'primal line' that segmented populations and that the primary focus of immigrants was to create a 'White Man's province.'[10] But historians have taken the pronouncements on race in the media and government documents as describing racial relations on the ground. They have not had the tools to seriously investigate how extensively 'popular attitudes' about race were shared. We do not yet know how people lived their racial pronouncements nor how or if elite racial ideas were shared by the wider population. This disjuncture between what we claim to know and how people lived is understandable in the context of the sources that have, until now, been available to study race. Only the elite have left writings for posterity, and even when they recorded voices of ordinary citizens, in trial transcripts for example, these are filtered and reformulated. In their attempt to expose and excavate the racism of the period, scholars have read the papers looking for evidence of racist language, and they found what they were looking for.

We have taken two steps that diverge from this model. First, we ask what race meant in Victorian Canada by exploring how it was lived and experienced spatially in the city of Victoria. Focusing on space allows us to get at one important aspect about how race was lived. Equipped with Canadian census data spatially referenced within a historical Geographic Information System (GIS), we have begun to survey the main streets and back alleys and open up the garrets and lavish parlours of Victorian Victoria in order to determine the extent to which personal lives and private spaces were

Fig. 1.2. Detail of "View of Victoria," June 13, 1860, with the Songhees village in the foreground. (Painting by H. O. Tiedemann, lithograph by T. Picken; Library of Congress Geography and Map Division, Washington, G3514.V5A35 1860 .T5.)

organized by race in Victoria between 1861 and 1911. With GIS, we can determine physical space occupied and the relative proximity of different racial groups in a precise way. Second, this project takes advantage of the availability of recently digitized newspapers to search, not only for negative commentary on racialized groups, but for all commentary in order to give a fuller picture of the nature of racial discourse. By combining these two modes of analysis, we can answer new questions about the nature of those life-spaces as well as if, and how, changes in colonial society and discourse restructured those worlds.

Victoria (Fig. 1.2) is an ideal location to explore these ideas. Urbanist Jane Jacobs observed: "It was in outpost cities that the spatial order of imperial imaginings was rapidly ... realised," and, as recent work by Penelope Edmonds and Renisa Mawani has shown, such settler-cities have distinctive patterns and morphologies.[11] Moreover, racism is not just an historic phenomenon here or elsewhere. We live every day with the legacy of the period we are interested in.

Victoria, although unique in many respects, was like other imperial centres in that it was "a site of contradiction between a persistent localism and the context ... of a global empire."[12] Despite public attempts to be "more English than the English," with its gentlemen's clubs, teas at the Empress Hotel, and a city newspaper proudly named "*The British Colonist*," Victoria was in fact a place of inter-racial mingling, cohabitation, miscegenation, and hybridity – a meeting place of indigenous people and diverse immigrant groups.[13] From our work to date we know that Caucasians not only lived and worked alongside Chinese, Black, Hawaiian, and Indigenous peoples, they slept with, married, and raised children and chickens with them. This raises the question: did Victorian British Columbians 'live race' the way elites talked about race?

New Insights I: Vanishing Indians

One of the key ways in which the spatialization of data has changed our thinking about the settler colonies lies in the ability to map where each individual person lived and to observe who they lived with. This provides a unique lens into how the city of Victoria worked as a site of settler colonialism. The seed of the city was the Hudson's Bay Company Fort Victoria, established near several Indigenous communities in 1843. Many of the nearby Indigenous people relocated to live beside the fort within a few years of its completion, and they vastly outnumbered the settler population well into the next decade. In addition to the local people, Indigenous people from all over the west coast, from Puget Sound to Alaska, began in the late 1840s to make seasonal visits to trade and work in Victoria. Several thousand of these visiting native people would camp seasonally in shanty towns on the edge of Victoria, a village which, until 1858, could count its settler population in the hundreds.[14]

When we compared the maps of the 1850s and 60s with our georeferenced census in the GIS showing where Aboriginal people lived in the period 1881–1911, we were shocked by what we saw. The Indigenous people living in what is now Victoria, who had numbered in the thousands in the 1840s to early 1860s, had vanished by 1901.

In 1861, there were three Indigenous "communities" in what is now Victoria: the Songhees Indian Reserve located across the harbour from the booming town, the shanty towns of migrant workers around Victoria, which ballooned in the summer and dwindled in the winter, and the Indigenous residents of the town itself. A visitor to Victoria recorded

that year that "Indians … are seen everywhere throughout the town – in the morning carrying cut wood for sale; the women, baskets of oysters, clams etc.… In his account of Victoria in the 1860s, Edgar Fawcett emphatically stated: "Indians performed all the manual labor."[15] In April 1862, the *British Colonist* reported that "the Indians have free access to the town day and night. They line our streets, fill the pit in our theatre, are found at nearly every open door during the day and evening in the town; and are even employed as servants in our dwellings, and in the culinary departments of our restaurants and hotels."[16]

Victoria in the 1840s to 1860s was very much an Indian town, but the nominal data in a municipal census taken in 1871, the year British Columbia joined Canada, suggests the situation had changed (see Table 1.1).[17]

The census was a limited one, just capturing the head of household by name, the number of adults by race in the household, the number of children by gender, and occasionally a street name to identify residence location. We can see that adult urban Victoria was 10 per cent Aboriginal, 5 per cent coloured, 7 per cent Chinese, and 78 per cent white. There were 45 per cent more Aboriginal women than Aboriginal men, largely reflecting the Indigenous wives or common-law wives of white settlers and a number of sex trade workers. Of the 1,083 households enumerated, sixty-six contained only Aboriginal people, and another thirty-six contained white males and Aboriginal females. There were thirty-six native women living in households with no men. Of these Indigenous urbanites, a third came from Fort Rupert on Northern Vancouver Island, a third from Haida Gwaii (Queen Charlotte Islands) and the remainder from eleven other First Nations

Table 1.1 Victoria Municipal Census 1871

(Does not include Indians living on the Songhees Reserve)

	White Males	White Females	Chinese Males	Chinese Females	Coloured Males	Coloured Females	Native Males	Native Females	Children All races	Total
No.	1716	1188	219	29	119	77	144	210	1231	4933
Percent of Adults	46	32	6	1	3	2	4	6		

communities, including the Songhees community across the harbour. Where streets are listed with census entries, Johnson Street (Fig. 1.3) appears especially significant for Aboriginal residents: all native people living without whites, and a third of those co-habiting with whites, occupied this street.[18] Across the harbour, the Indian superintendent reported in 1876 that there were 182 Indians living on the Songhees Reserve, including fifty-five adult males and sixty-two adult females.[19] The shanty towns that had housed visiting migrant Indigenous people in the 1850s had been razed as a settler response to the 1862 smallpox epidemic and were never substantially rebuilt. Afterwards, visiting Indigenous people expanded the town's "Indian Quarter."

Ten years later, when Johannes Jacobsen first visited Victoria in 1881, he was surprised that "the streets of this town swarmed with Indians of all kinds."[20] When Franz Boas visited in 1884, he recalled: "The Indians are at present in the habit of living part of the year in Victoria, Vancouver, or New Westminster, working in various trades: in saw-mills and canneries, on wharves, as sailors, etc.... They have their own quarter in every city." He goes on: "Walking around the suburbs of Victoria, we come to that part of the town exclusively inhabited by Indians." By the time Boas arrived, the native quarter had shifted north from Johnson Street, away from town centre to Herald Street (Fig. 1.4) and was significantly reduced in size.[21] The federal census of 1881 shows 172 people whose origin is listed as "Indian" in the three wards that make up urban Victoria, of which 122, or 71 per cent lived in the "Johnson Street Ward," which included Herald Street, forty-one in the James Bay Ward, and two in the Yates Street Ward. The census showed sixteen mixed-race families with an Indigenous wife and non-Indigenous husband, twelve of those in the Johnston Street Ward. Another 180 Indians resided on the Songhees reserve.

While the 1891 federal census did not ask questions about race, origin, and ethnicity, we are lucky in Victoria to have an 1891 municipal census that does. The Municipal Check Census was initiated because Victorians believed they had been under-enumerated by the federal census. The municipal census counted over 23,000 people compared to the federal count of 16,841. The Municipal Census recorded 377 Indigenous people in Victoria – only 1.6 per cent of the population. The Indian Quarter was still centred on Herald Street, but the

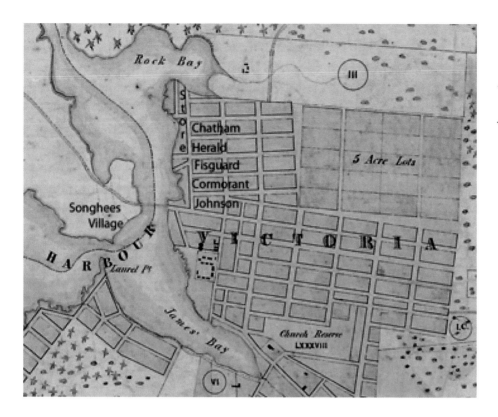

Fig. 1.3. Victoria 1860 showing Johnson and Herald Streets. (Adapted from "Victoria District," 1860, B.C. Land Title and Survey Authority, 28T2.)

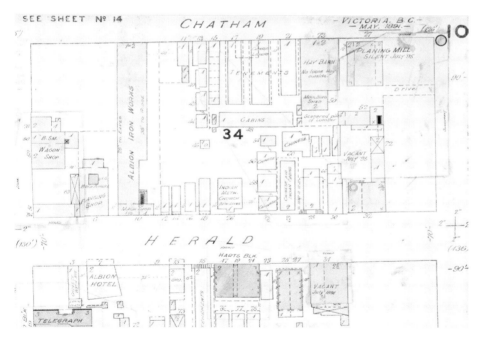

Fig. 1.4. Herald and Chatham Streets, 1891, from fire insurance plan "Victoria, B.C." (Charles Goad and Company [Montreal, 1891]. Courtesy of Royal B.C. Museum, BC Archives.)

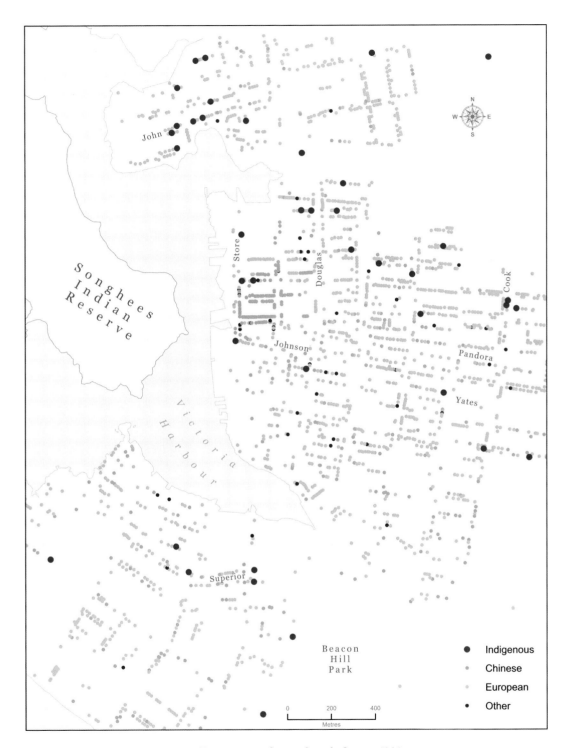

Fig. 1.5. Location of residences by race, Victoria, 1901. Source: Canada Census, 1901.
Also available on viHistory.ca.

Fig. 1.6. Store Street encampment 1904 (Image PN 1433; courtesy of Royal B.C. Museum, B.C. Archives.)

municipal census also notes an Indian camp on Store Street (Fig.1.6).[22]

When the next census was taken in 1901, questions about race and "colour" were asked. The number of those who identified their race as North American Indian in Victoria had dropped to only fifty-three people (and, of these, twenty-seven were living on sealing ships in the harbour). Actual residents included only nineteen "North American Indian" women and six men living in the city of Victoria outside of the Songhees reserve. Only seven were living in mixed-race families. Added to the North American Indians, another 153 were classified as of the "red colour," apparently meaning they had at least one Indigenous ancestor. In

total, "red people" were a mere 1 per cent of the population of the city (Fig. 1.5). Johnson Street, which had been Victoria's Indian Street in the 1860s, was by 1901 a white commercial district. Fires that broke out in the Indian and Chinese quarter in August 1904 and July 1907 finished the clearances.[23] In 1901, the Indian Agent recorded 101 Indians on the Songhees reserve. Taking the reserve population and all the North American Indians in the city, the Indigenous population of Victoria had declined 80 per cent in the half-century since the 1850s.

By 1911, even the Indian Reserve was gone – relocated outside the urban boundaries. Since the 1860s, efforts had been made to relocate the Indian Reserve on Victoria Harbour, and

Fig. 1.7. Number of Aboriginal Individuals and Mixed Race Families in Victoria, 1871–1911. Source: viHistory.ca from Victoria municipal censuses 1871 and 1891 and federal censuses, 1881, 1901, and 1911. (Department of Indian Affairs, Annual Reports, 1879, 1891, 1901, 1911.) 1871 does not include Indigenous children; mixed race families not available for 1891.

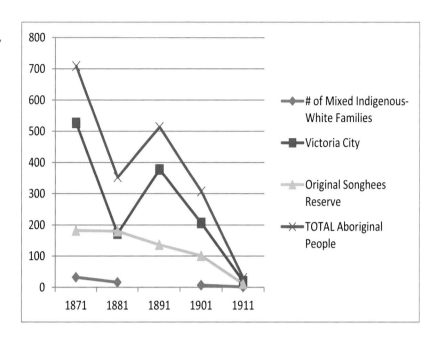

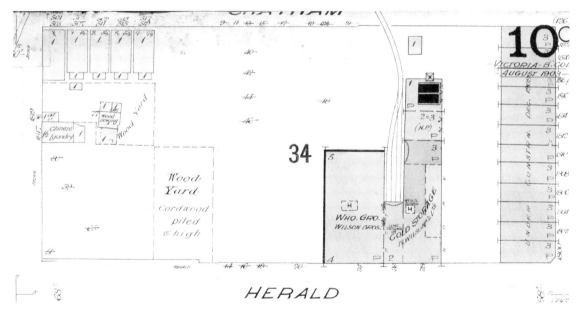

Fig. 1.8. Herald Street, 1907. (*From* Insurance Plan of Victoria, BC, 1903 *[revised 1907]; Charles Goad and Company [Montreal and Toronto, 1907]. Courtesy of Royal B.C. Museum, B.C. Archives.)*

finally in 1910 an agreement was reached with the Songhees and the federal and provincial governments.[24] By the time of the 1911 census, only three Indian families still lived on the site of the former reserve (Fig 1.7), and only one family of four whose racial ancestry was indicated as North American Indian lived in the city itself (Fig. 1.8). With three Indigenous sealers recorded on ships and one Native woman married to an Englishmen, out of the total population of Victoria of 28,500 those with a racial ancestry as North American Indian only counted as nineteen in 1911 or less than one tenth of one per cent of the population.[25] The Indians had vanished.

New Insights II: The un-Forbidden[26]

Victoria's Chinatown is perhaps the most striking symbol of racial space in the history of the city. Canada's oldest Chinatown was established in the colonial era and is traditionally seen as a racially segregated ghetto, an enclave that provided Chinese residents with a refuge from the hostility they routinely endured from the surrounding white community. It was regarded as a "Forbidden City," an inscrutable place closed to outsiders.[27] But through the use of our historical GIS we upset some long-standing assumptions about Chinese space in the city and the interactions between Chinese and non-Chinese residents. An historical GIS of Victoria's Chinatown, focusing on the year 1891[28] and derived from spatially referenced records, challenges and contradicts assumptions that have been derived mainly from narrative records. We are not disputing the prevalence of racial prejudice during the period, but our evidence suggests that the virulent rhetoric of

racism evident in traditional sources – such as Royal Commission reports on Chinese immigration and newspaper accounts of anti-Chinese labour agitators – may not have been actualized on the ground. From our GIS perspective, the alleged gulf between East and West in Victoria was not nearly as pronounced as historians have averred. Our research reveals a Chinatown that was a transactional community, predominantly but not exclusively Chinese.

In many ways, we are exploring new terrain, but in other respects we are following lines of inquiry that geographers and historians delineated in the 1970s, when they began making connections, in David Cannadine's memorable phrase, between "shapes on the ground" and "shapes in society."[29] Of course, we are also informed by a newer generation of historians and social scientists who have drawn upon social and cultural theory and theoretical concepts of spatiality in historical cities.[30] In practice, we are following the lead of Sherry Olson, whose pioneering work on nineteenth-century Montreal is a model of spatial history and building on the expertise that our team mates developed while constructing an historical GIS of London, Ontario.[31]

Our historical GIS of Victoria revealed trends and phenomena that we had not expected from the existing literature. By mapping the location of the Chinese and others, we were struck by the fact that Chinatown – far from a "Forbidden City" – was a porous multi-ethnic community. The time-spaces that are captured in the historical GIS of Victorian Victoria's Chinatown show a neighbourhood that was not all Chinese (Fig. 1.9). With a quarter of the population not Chinese, we have to start to question the idea of a forbidden city.

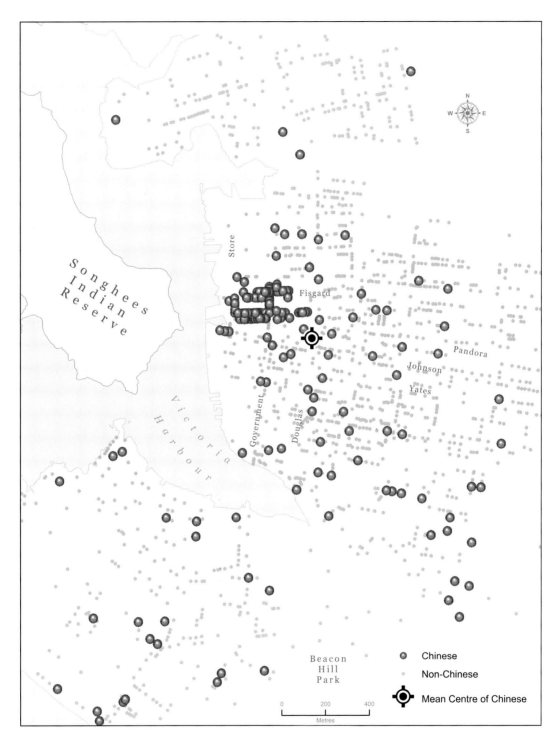

Fig. 1.9. Victoria population, 1891, showing residences of Chinese and non-Chinese. Source: Canada Census, 1891 with addresses provided by the City of Victoria 1891 Check Census in the BC Archives (Add Mss 1908) and Williams' Illustrated Official British Columbia Directory for the Cities of Victoria, Vancouver, Nanaimo and New Westminster, 1892 *compiled by the firm of R. T. Williams of Victoria. The directory was compiled in the fall of 1891. Both the census and directory are available on line at www.vihistory.ca.*

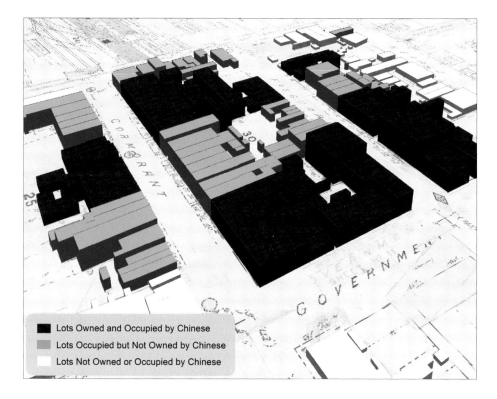

Fig. 1.10. Map of Chinatown ownership, 1891. Source: City of Victoria Archives, Tax Assessment Rolls. The assessment rolls are available online at www. viHistory.ca.

When we linked the tax records to the map, it revealed that 60 per cent of Chinatown lots and buildings were owned by white Victorians, who offered long-term leases to Chinese restaurant operators, launderers, boarding house keepers, merchants, and manufacturers (Fig. 1.10). Property owners in Chinatown included prominent figures, such as Joseph Carey, a land surveyor and former mayor, and Amor de Cosmos, a journalist and politician who opposed Chinese immigration twenty years earlier. Spatializing the census data allows us to see that several of the streets in the neighbourhood were heterogeneous, with white workers living beside Chinese shops, Indigenous women living with white men and between Chinese families. When we also map where the Chinese lived outside Chinatown, we see a city in which Chinese lived in every neighbourhood, including the wealthiest, where they worked in the homes and gardens. Our findings suggest that Chinatown witnessed daily ebbs and flows of both Chinese and Euro-Canadian settlers as they exchanged goods, services, and cultural events with one another.

New Insights III: Adding talk to space

The surprises of the vanishing Indians and the un-forbidding Chinatown are significant in their own right, but they also open up new questions of how to explain what we can now see. And, like the result of any inquiry, they ought to be tested against other forms of evidence that might challenge or confirm them. One such check on the idea of the "un-forbidden city" is to see if the public discussion about race verifies the patterns of spatial relationships. Michel Foucault asked historians to map "discursive practices in so far as they give rise to a corpus of knowledge" about race, among other things. Drawing on Foucault in part, but more substantively on the work of Teun Van Djik and Margaret Whetherelland Jonathan Potter, we wanted to see if "mapping the language of racism" would support the conclusions of our maps of racial space.[32]

When it comes to understanding public discourse about race, we have some excellent studies. Thinking primarily about the discourse around the Chinese in British Columbia, the work of Patricia Roy is foremost among them, but others like Renisa Mawani, Kay Anderson, David Lai, and Peter Ward have greatly added to our knowledge.[33] These scholars have done the hard work of reading newspapers, government documents, and, to a lesser degree, private manuscripts and have found them full of racist rhetoric. British Columbia, they found, was a racist society: white British Columbians looked down on Chinese; racism seemed to get worse over the last half of the nineteenth century, and, starting in the late 1870s, it manifested itself in discriminatory legislation against the Chinese.

The evidence from our GIS mapping suggested that the racial landscape in Victoria was more nuanced and racial boundaries more porous than standard historical accounts have indicated. To test these conclusions and look for an explanation, we turned to the discourse. The advent of fully digitized newspapers offered another methodology that was not previously possible. We reasoned that, if we systematically searched for all mentions of Chinese and examined the context, instead of just the negative references, we would get a fuller view

of racial discourse than earlier textual studies have given.

As a check on our conclusions drawn from the GIS, we sampled the major colonial/provincial paper, the *British Colonist*, in the decades 1861 to 1911, coinciding in the later period with census years. We read every word in every paper from a two-week period in the spring and a two-week period in the fall, a sample of 8 per cent in the chosen years. We identified every mention of Chinese or any synonym and did a close reading to identify the context of the reference: was it an editorial or an advertisement? Was the subject labour, crime, or disease? In addition, we coded the references as negative, neutral, or positive, a reflection on whether they were actively engaged in shifting the discourse to a more negative or positive portrayal of the Chinese. If neither, we assumed that the author did not choose at that time to push or reinforce elements of the discourse, leaving it instead to interact with the reader's existing ideas of "Chineseness."[34]

At the most basic level, we wanted to see what terminology was used to describe the Chinese and to see if these were negatively loaded, neutral, or positive, and if they changed over time. We can demonstrate this numerically but the impact is clearer using the conversion algorithms of Wordle, a text-mapping application that displays the size of a word as a representation of its relative frequency of use.[35] We found, to our surprise, that the national term "China" or "Chinese" was used much more frequently than negatively charged synonyms like "Chinaman," "coolie," or others (Fig. 1.11). By comparison, negative terms almost never occurred.

At the next level, we wanted to determine the parts of the paper (news, editorials,

Chinaman

Coolie China

Oriental

Chinese

Chinamen

Fig. 1.11. Wordle figure showing terms for Chinese in the British Colonist; sample weeks, 1861–1910, with size proportional to frequency of occurrence.

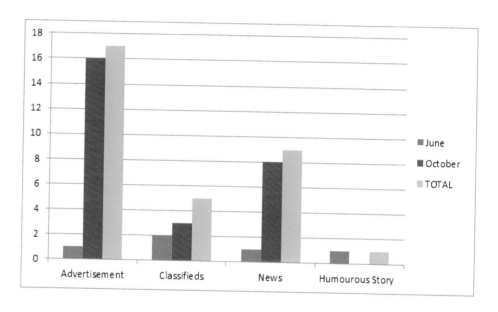

Fig. 1.12. Frequency of references to Chinese (all synonyms) in the British Colonist; sample weeks, 1891.

advertisements, or feature stories) in which references to the Chinese occurred. Again, we were surprised. The most frequent occurrence was in advertising: Chinese merchants advertising their wares or non-Chinese merchants advertising Chinese goods, like rice or crockery (Fig. 1.12).

Third, we wanted to see what proportion of the characterizations of the Chinese were negative, and if any were positive or neutral. Again the results surprised us. While negative characterizations were more common than positive, overall, loaded language was rarely

used. The discourse around the Chinese was overwhelmingly neutral: it did not try to move public opinion or reinforce either negative or positive views (Fig. 1.13).

Even at this surface level, the mapping of racial talk in the newspapers offers a confirmation of some of the conclusions suggested by our historical GIS research and a partial explanation for the openness of Chinatown. Where there was charged language, the preponderance of the negative supports the conclusions of earlier scholars that this was a society structured by race in which the white

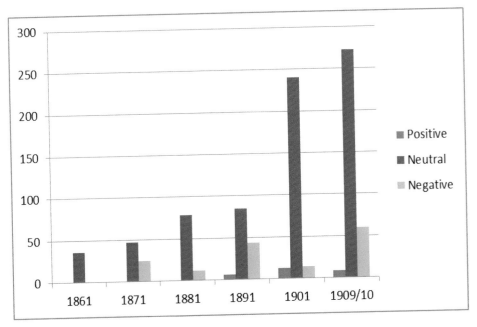

Fig. 1.13. References to Chinese in the British Colonist; *sample weeks, 1861–1910.*

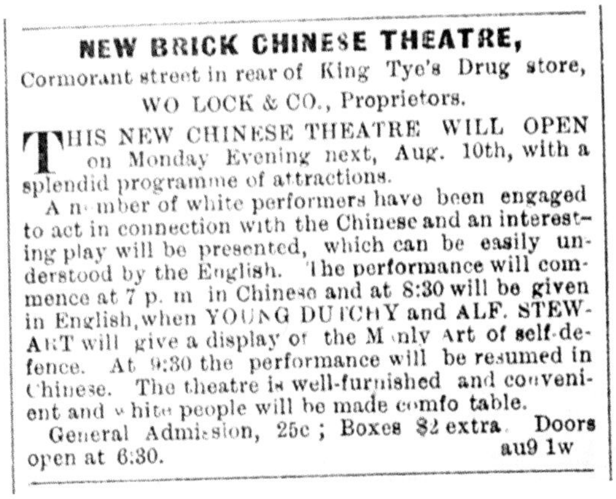

NEW BRICK CHINESE THEATRE,

Cormorant street in rear of King Tye's Drug store,
WO LOCK & CO., Proprietors.

THIS NEW CHINESE THEATRE WILL OPEN on Monday Evening next, Aug. 10th, with a splendid programme of attractions.

A number of white performers have been engaged to act in connection with the Chinese and an interesting play will be presented, which can be easily understood by the English. The performance will commence at 7 p. m in Chinese and at 8:30 will be given in English, when YOUNG DUTCHY and ALF. STEW-ART will give a display of the Manly Art of self-defence. At 9:30 the performance will be resumed in Chinese. The theatre is well-furnished and convenient and white people will be made comfo table.

General Admission, 25c ; Boxes $2 extra. Doors open at 6:30. au9 1w

Fig. 1.14. Chinese theatre advertisement from the British Colonist, *August 9, 1885, p. 2.*

immigrants worked to maintain a hierarchy that put them at the top. However, the relative lack of any charged language and the apparent success of both Chinese and white merchants in selling Chinese goods to the colonial population suggests a vibrant transactional space where Chinese and other Victorians frequently met and exchanged goods and services in a regime more structured by the impersonal relations of capitalism than the embodied and personal relationships of race. The presence of both positive and negative references suggests competing ideas of Chineseness among the immigrant elite of the colony. Here we can only hint at the complexity of the analyses, but the key point is that the discourse describes a racial and commercial regime totally consistent with the un-forbidden city, in which Chinese and white lived, worked, and shopped alongside each other, just as they bought and sold from each other.[36] On occasion, they shared the same amusements (Fig.1.14).

CHALLENGES: BUILDING AN HISTORICAL GIS

There are numerous obstacles to trans-disciplinary scholarship, including learning and effectively using new methods, finding data that can be reliably analyzed through different disciplinary tools, and joining disparate data sets. Methodologically, we faced numerous challenges in building an historical GIS with data that was often not easily spatialized. We drew on a large pool of data that we began assembling several years ago for an online digital archive of Vancouver Island called *viHistory*. The *viHistory* website was developed as a joint initiative by the University of Victoria and Vancouver Island University and was launched in 2001.[37]

Thanks to the work of the Canadian Families Project and the Canada Century Infrastructure Project and the ongoing development of the *viHistory* Project, we have the complete federal census for Victoria digitized for 1881, 1891, 1901, and 1911, as well as municipal censuses for 1871 and 1891. To these we have linked directory entries for the corresponding years and tax assessments in 1891 and 1901. We know a lot about everyone the census caught (and even about those whom it didn't[38]), including their race, family composition, who their neighbours were, what their jobs and religion were, and even in 1901 and 1911, income and weeks of employment. Such a detailed, individual-level, and long-run data set in digital form allows a wide range of inquiries into the lives of Victorians.

The Dominion census was designed so that one person was identified as the head of the household with other residents being assigned relative positions, such as such as wife of head, son of head, lodger, and so forth. Census households or families were numbered consecutively so that census family number 2 was next to census family 1, and physically located in space between 1 and 3. By connecting the head of a census household to a polygon, we were able to map everyone in that household. The census information in 1901 and 1911 included addresses, so linking people to their home spaces required a digitized map of cadastral lots for the period. We started with a modern GIS base map, generously provided by a local government authority, the Capital Regional District [CRD] of Victoria, and geo-referenced archival maps to it.[39] By digitally scanning and geo-referencing these cartographic records to a modern cadastre of the city, we were able to establish property boundary lines with precision.[40] Contemporary fire insurance plans, drawn to a scale of one inch to fifty feet, provided information about addresses, streets, buildings, land use, urban infrastructure, and utilities. Conveniently for us, nineteenth-century fire maps of Victoria also often indicate occupancy and, reflecting some of the discriminatory attitudes of the era, they pointedly indicated buildings that were occupied by Chinese and Aboriginal people.[41] Street addresses in Victoria were completely renumbered in 1907 so a conversion table had to be created to link old to new addresses, and street name changes had to be researched so the data could be shown on a modern map.

Spatializing information for 1881 and 1891 was more challenging because in 1881 street numbers had not been assigned and in both 1881 and 1891 enumerators did not record the civic addresses of the households they documented. Fortuitously, for 1891 we were able to determine many civic addresses of people in the federal census by consulting the municipal check census. Although the check census collected much less information about each household, it did include addresses. We supplemented the check census with addresses from *Williams' Illustrated Official British Columbia Directory*, compiled in autumn 1891 and published for 1892.[42]

We added property ownership records with the help of undergraduate student research assistants, who transcribed property tax records from large, leather-bound volumes in the City of Victoria Archives. These not only indicated the names of property owners and the assessed value of city lots, they also provided information about the location, dimensions, and legal descriptions of city properties. We linked that information to polygons on our historical, geo-referenced cadastre of Victoria.

To link the people enumerated in 1881 to their "time-space," we first looked for matches between the census and the city directory compiled that year. The directory listed a street but no street number for heads of families and businesses.[43] We then looked for matches between the 1881 census and the 1891 census, for which we had already derived addresses. Where we found matches of people living on the same street as they had a decade before, we attached them to the same address. For the households between known addresses, we used the knowledge that the enumerators proceeded systematically to place them in order on the block. Fire insurance plans from after 1885 gave us the addresses for those lots. Between the maps, the census, the directories, and the assessment roles, we mapped over 80 per cent of Victorians to their lots with a high degree of confidence, giving us an entirely new view of racial space in Victoria.

CHALLENGES OF TRANSDISCIPLINARY RESEARCH

While spatializing the data presented challenges that we have largely surmounted, spatializing the historians on the team and historicizing the geographers is an ongoing process. We have found ourselves negotiating the challenges of transgressive scholarship, most particularly between humanists attentive to particular places and fetishizing context and geographers anxious to compare spaces with a variety of methods and analytical tools, including statistical measures of diversity that are abstracted from context. Peta Mitchell has called our attention to the challenges of finding a common language in that space "where the disciplinary strata of geography and history abut and blur,"[44] and it is true that we have each had to learn a new language to better communicate and new cultural practices.

Each member of our team has brought a skill and knowledge set to the project that has allowed for a scale and scope of conceptualization that exceeds the capacity of any given scholar. Still, the specialization meant that the geographers were much more skilled in using the tools of the GIS and the historians were quite hazy about what could and could not be accomplished. Likewise, the statistical measures of dispersion, density, and dissimilarity and the ability to use and critique them were much more in the repertoire of the geographers. The historians were located in Victoria and consequently were much more attuned to the particularities of place; they were also more immersed in different humanities literatures around race, space, and colonialism. A different and perhaps more nuanced use of the tools might emerge if all the skills were in the hands of a single scholar or if we all got better at using the tools and sharing the disciplinary culture of the other.

One of the cultural differences is the conventional practice of counting scholarship and sharing credit. In the social sciences model of geography, the main measure of output is refereed journal articles, and an article in an edited collection is accorded little weight. In the humanities, chapters in edited collections carry the same credit as journal publications, and books are the ultimate gold standard. In the social sciences model, everyone with a significant involvement in building a project gets authorial credit for any of the outcomes of that project, their relative contribution indicated by how close or far they are from being "first author" or presenter. In the humanistic rituals, only those actually involved in writing a publication or conference presentation gets authorial credit, with the rest of the team getting an acknowledgment within the publication/presentation itself. We have adopted history's disciplinary practice of contributing to edited collections and geography's convention of sharing credit in the hope that there is lots of credit to share!

We have also taken a page from another discipline, ethnography, to overcome the cross-cultural barriers imposed by disciplinary practices. We have done some participant observation of the exotic other: the historians both enrolling in week-long intensive GIS courses; the geographers coming to Victoria to get local historical tours and to investigate the traditions of hospitality among historians. We read each other's work from the project and beyond, and each of us has moved inexorably from a spot central to our disciplines to a location closer to the geohumanities.

CONCLUSION

Where Peta Mitchell sees "the inescapable stratification of disciplinary ground," we have found the little- or un-tilled spaces between the disciplines to be the most fertile places to put our attention, and the results seem to be rewarding the efforts. In his survey of the implications of GIS for history, David Bodenhamer offers a vision of "a unique post-modern scholarship, an alternate construction of the past that embraces multiplicity, simultaneity, complexity, and subjectivity."[45] This may be a heavier load than the marriage of history and geography can carry but we do see grounds for optimism. We are still finalizing the GIS and the discourse data and are poised to be able to answer a variety of questions we could not before, but already the results have surprised us in important ways.

We hope that we have lived up to Douglas Richardson's recent invitation to show "how integrative digital technologies such as GIS might substantively or qualitatively impact scholarly research."[46] Our new research confirms the work of the pioneering scholars in this field, that Victoria and British Columbia were colonial spaces very much structured on a scaffolding of racist ideas constructed spatially and discursively in the late nineteenth century. Our contribution, we think, is to open up this black box of racism a little and ask how racism was expressed at different times and in one very crude but fundamental way, experienced in where people lived and who they lived with.

The pattern of racial space that emerges from the GIS time-lapse "snapshots" taken with each decennial census shows distinctive racial enclaves, shifting, growing, and shrinking. We have found striking differences in the history of the Indigenous people and the Chinese immigrants. The regime of racialized space occupied by Indigenous people reversed over the Victorian era, from one where a few whites had a foothold in an Aboriginal space to a situation in which a few Indigenous people had only the slimmest toehold in white space. The hybrid, private space of mixed marriages and inter-racial children that expanded in the 1860s was also practically gone by 1911, a reminder of Lefebvre's injunction that social space is something we construct and which others construct about us. Space, as Doreen Massey says, "is always mobile, always changing, always open to revision, and potentially fragile. We are always creating, in other words, not just a space, a geography of our lives, but a *time*-space for our lives."[47] Functionally, the Chinese immigrants replaced Indigenous people in many labour markets just as Chinatown, spatially, displaced the Indian Quarter. Unlike the Aboriginal population which mixed households with European settlers extensively in the mid-century, there are very few instances of households with Chinese and non-Chinese living together – with the exception of the many Chinese servants living in white households. At the level of intimate space, racial space was highly segregated when it comes to the Chinese. At the level of public space, however, we see something quite different. We have revealed a Chinatown that is at the opposite end of the scale from what the Forbidden City would suggest.

The results of the discourse study confounded. We expected to find lots of negative racialized references and, while we found some, when we looked for all the references to Chinese, and not just the negative ones, most of the discourse was not racialized one way or the other. Chinese merchants advertised their wares; white merchants advertised their exotic imports from China. News stories that mentioned Chinese tended to avoid racializing discourse. Editorial commentary, letters, and features, in the later years, tended to be where we saw racialized language more, but this was a minority of commentary, and, when the discourse was racialized, 15–20 per cent of it was positive. A full consideration of the discourse suggests that when Chinese entered the public sphere it was not as a demonized other but most often as a fellow resident, a neighbour, merchant, or labourer.

Testing the discourse analyses and GIS against each other shows a consistency between them. Both suggest a racially divided city but one whose divisions were more of a continuum, with more severe boundaries on the register of intermarriage and sexual intercourse, and much greater openness on the level of commercial intercourse, with a whole range of public and residential interactions in between. Joining the two approaches over a fifty-year time span opens a door into space-time, or, to use Mikhail Bakhtin's term, into a "chronotope," where "spatial and temporal indicators are fused into one carefully thought-out, concrete whole." In such an analysis, "time, as it were, thickens, takes on flesh, becomes artistically visible; likewise space becomes charged and responsive to the movement of time, plot and history." We can, like Rachel Whiteread's House, "turn space inside out."[48]

NOTES

1 Doreen Massey, "Space-Time and the Politics of Location," *Architectural Digest* 68, nos. 3&4 (1998): 34. Whiteread's concrete cast of the inside of an entire Victorian terraced house was completed in autumn 1993, exhibited at the location of the original house – 193 Grove Road – in East London (all the houses in the street had earlier been knocked down by the council) and demolished on 11 January 1994. It won both the Turner Prize for best young British artist in 1993 and the K Foundation art award for worst British artist.

2 Richard White, "Foreword," in Anne Kelly Knowles, *Placing History: How Maps, Spatial Data and GIS are Changing Historical Scholarship* (ESRI, 2008), xi.

3 The project was funded for three years by a SSHRC Standard Research Grant. The team includes historians John Lutz and Megan Harvey at the University of Victoria, Patrick Dunae, emeritus of Vancouver Island University, and geographers Jason Gilliland and Don Lafreniere at the University of Western Ontario (UWO). Kevin Von Lierop at UWO acted as our GIS technician for the early part of the project and we have benefited from the research help and insights of the following students: Kate Martin, Tylor Richards, Julie Ruch, Karla Partel, Kim Madsen, Vanessa Dunae, Shannon Lucy, Carley Russell, Kathleen Trayner, Ryan Primrose, and Alan Kilpatrick.

4 We take the term from Michael Dear et al., *Geohumanities: Art, History, Text at the Edge of Place* (New York: Routledge, 2011).

5 Sarah Luria, "Geotexts," in *Geohumanities*, ed. Dear et al., 67; Michel Foucault, "Space, Knowledge and Power," *The Foucault Reader*, ed. Paul Rabinow (New York: Pantheon, 1984); Charles Withers, "Place and the 'Spatial Turn' in Geography and History," *Journal of the History of Ideas* 70, no. 4 (2009): 637–58.

6 Henri Lefebvre, "Space, Social Product and Use Value," in J.W. Freiberg, ed., *Critical Sociology: European International Perspectives* (New York: Irvington/Wiley, 1979): 286; see also *The Production of Space*, trans. Donald Nicholson-Smith (Oxford: Blackwell, 1991), 73.

7 Stuart Hall and Paul du Gay, eds., *Questions of Identity* (London: Sage, 1996); Agnes and Brian Smedley, *Race in North America: Origin and Evolution of a Worldview* (Boulder, CO: Westview, 2012); Margaret Wetherell and Jonathan Potter, *Mapping the Language of Racism: Discourse and the Legitimation of Exploitation* (New York: Columbia University Press, 1992).

8 The literature was launched by Edward Said, *Orientalism* (New York: Vintage, 1978). For an excellent survey, see Patrick Wolfe, "Land, Labour, and Difference: Elementary Structures of Race," *American Historical Review* 106, no. 3 (2001): 866–905.

9 For example: Amy Hillier, "Redlining in Philadelphia," in *Past Time, Past Place: GIS for History*, ed. Anne Knowles (Redlands, CA: ESRI Press, 2002), 79–92; Amy Hillier, "Spatial Analysis of Historical Redlining: A Methodological Exploration," *Journal of Housing Research* 14, no. 1 (2003): 137–67, and Jason Gilliland, Don Lafreniere, Sherry Olson, Pat Dunae, and John Lutz, "Residential Segregation and the Built Environment in Three Canadian Cities, 1881–1961," paper presented to the European Social Science History Conference, Ghent, Belgium, April 2010.

10 Cole Harris was among the first scholars to bring the insights of Lefebvre and Foucault to the shores of British Columbia in *Making Native Space, Colonialism, Resistance, and Reserves in British Columbia* (Vancouver: UBC Press, 2002) and "How Did Colonialism Dispossess? Comments from an Edge of Empire," *Annals of the Association of American Geographers* 94, no. 1 (2004): 165–182. Major works on the history of race in British Columbia include: W. Peter Ward, *White Canada Forever: Popular Attitudes and Public Policy towards Orientals in British Columbia (*Montreal and Kingston: McGill-Queen's University Press, 2002); Patricia E. Roy, *A White Man's Province: British Columbia Politicians and Chinese and Japanese Immigrants, 1858–1914* (Vancouver: UBC Press, 1990). Patricia F. Roy, *The Oriental Question: Consolidating a White Man's Province, 1914–41* (Vancouver: UBC Press, 2003); Patricia E. Roy, *The Triumph of Citizenship: The Japanese and Chinese in Canada, 1941–67* (Vancouver: UBC Press, 2007); Timothy

J. Stanley, *Contesting White Supremacy: School Segregation, Anti-Racism, and the Making of Chinese Canadians* (Vancouver: UBC Press, 2011).

11 Jane Jacobs, *Edge of Empire: Postcolonialism and the City* (New York: Routledge, 1996), 3; Renisa Mawani, *Colonial Proximities: Crossracial Encounters and Juridical Truths in British Columbia, 1871–1921* (Vancouver: UBC Press, 2009); Penelope Edmonds, *Urbanizing Frontiers: Indigenous Peoples and Settlers in 19th Century Pacific Rim Cities* (Vancouver: UBC Press, 2010).

12 Felix Driver and David Gilbert, eds., *Imperial Cities: Landscape, Display and Identity* (Manchester: Manchester University Press, 1999).

13 Edmonds, *Urban Frontiers*, 227–29; Mawani, *Colonial Proximities*; Adele Perry, *On the Edge of Empire. Gender, Race, and the Making of British Columbia, 1849–1871* (Toronto: University of Toronto Press, 2001); Jean Barman, "Taming Aboriginal Sexuality: Gender, Power, and Race in British Columbia, 1850–1900," *BC Studies* 115/116 (1997/98): 237–66.

14 John Sutton Lutz, *Makuk: A New History of Aboriginal–White Relations* (Vancouver: UBC Press, 2007), 73–85.

15 Sophia Cracroft, *Lady Franklin Visits the Pacific Northwest, February to April 1861 and April to July 1870*, ed. Dorothy Blakey Smith (Victoria: Provincial Archives of British Columbia Memoir XI, 1974), 79; Edgar Fawcett, *Some Reminiscences of Old Victoria* (Toronto: William Briggs, 1912), 84.

16 *British Colonist*, April 28, 1862, 2.

17 The municipal census data of 1871 was summarized and printed in the 1872 Dominion Sessional Papers. See Canada, *Sessional Papers*, 1872, Vol. 6, No. 10: "British Columbia. Report by the Hon. H. L. Langevin, C. B., Minister of Public Works." The *Sessional Papers* did not include this level of detail transcribed from the manuscript census by Hugh Armstrong and digitized for the *viHistory* website.

18 British Columbia Archives (BCA), GR 428, Vancouver Island. Police and Prisons Department, Esquimalt, 1862–1868, City of Victoria 1871 Municipal Census, online database on Patrick Dunae, ed., Vihistory.ca <http://www.vihistory.ca>. The data on community origin is not complete so the

numbers represent the fractions of those reporting a community origin.

19 I. W. Powell, in Canada, Sessional Papers, Department of Indian Affairs, Annual Report, 1877, 32–34.

20 J. A. Jacobsen, *Alaskan Voyage, 1881–83: An Expedition to the Northwest Coast of America*, translated from the German text of Adrian Woldt by Erna Gunther (Chicago: University of Chicago Press, 1977); BCA, A/E/Or3/C15, Alexander Campbell, "Report on the Indians of British Columbia to the Superintendent General of Indian Affairs," October 19, 1883.

21 Franz Boas, *The Ethnography of Franz Boas: Letters and Diaries of Franz Boas Written on the Northwest Coast from 1886 to 1931*, Ronald P. Rohner, ed. (Chicago: University of Chicago Press, 1969), 28.

22 See Patrick A. Dunae, "Making the 1891 Census in British Columbia," *Histoire sociale–Social History* 31, no. 62 (1998): 234–36.

23 *British Colonist*, August 10, 1904, 1, 5, 6; July 24, 1907, 1.

24 Grant Keddie, *Songhees Pictorial: A History of the Songhees People as seen by Outsiders, 1790–1912* (Victoria, BC: Royal British Columbia Museum, 2003) and Renisa Mawani, "Legal Geographies of Aboriginal Segregation in British Columbia: The Making and Unmaking of the Songhees Reserve, 1850–1911," in *Isolation, Places and Practices of Exclusion*, eds. Carolyn Strange and Alison Basher (London: Routledge, 2003), 173–90.

25 Online at Patrick Dunae, ed., *ViHistory* <http://www.vihistory.ca>

26 This section draws on Patrick A. Dunae, John S. Lutz, Donald J. Lafreniere, and Jason A. Gilliland, "Making the Inscrutable Scrutable: Race and Space in Victoria's Chinatown, 1891," *BC Studies* 169 (Spring 2011): 51-80.

27 This interpretation is emphasized by geographer David Chuenyan Lai in *Chinatowns: Towns within Cities in Canada* (Vancouver: UBC Press, 1988) and *Forbidden City within Victoria: Myth, Symbol and Streetscape of Canada's Earliest Chinatown* (Victoria: Orca Books, 1991). Historians have also represented Victoria's Chinatowns as a defensive ghetto, sealed off from the surrounding host community. See Roy, *A White Man's Province,*

and Ward, *White Canada Forever*. The Chinese quarter in Vancouver has also been regarded as a rigidly segregated community. See Kay Anderson, *Vancouver's Chinatown: Racial Discourse in Canada, 1875–1980* (Montreal and Kingston: McGill-Queen's University Press, 1991).

28   Although our study focuses on 1891, we observe that the patterns hold true for 1881, 1901, and 1911 as well.

29   David Cannadine, "Residential Differentiation in Nineteenth Century Towns: From Shapes on the Ground to Shapes in Society," in *The Structure of Nineteenth Century Cities*, J. H. Johnson and C.G. Pooley, eds. (London: Croon Helm, 1982), 235–52.

30   Colin G. Pooley, "Space, Society and History," *Journal of Urban History* 31 (2005): 753–61; Edmonds, *Urban Frontier*; Philip Ethington, "Placing the Past: 'Groundwork' for a Spatial Theory of History," *Rethinking History* 11, no. 4 (2007): 465–94; Deryck Holdsworth, "Historical Geography: New Ways of Imaging and Seeing the Past," *Progress in Human Geography* 27, no. 4 (2003): 486–93.

31   We have benefited from foundational studies by Sherry Olson, including "Occupations and Residential Spaces in Nineteenth Century Montreal," *Historical Methods* 22 (Summer 1989): 81–99; "Ethnic Partition of the Workforce in 1840s Montreal," *Labour / Le Travail* 53 (Spring 2004): 1–66; Jason Gilliland and Sherry Olson, "Montréal, l'avenir du passé," *GEOinfo* (January–February): 5–7; Sherry Olson and Patricia Thornton, *Peopling the North American City: Montreal 1840–1900 (*Montreal and Kingston: McGill-Queen's University Press, 2011). See also Jason Gilliland and Sherry Olson, "Residential Segregation in the Industrializing City: A Closer Look," *Urban Geography* 33 (January 2010): 29–58; Jason Gilliland and Sherry Olson, "Claims on Housing Space in Nineteenth-Century Montreal," *Urban History Review / Revue d'histoire urbaine* 26, no. 2 (1998): 3–16; and Jason Gilliland, "Modeling Residential Mobility in Montréal, 1860–1900," *Historical Methods* 31 (January 1998): 27–42. For early work on the London HGIS, see Jason Gilliland and Mathew Novak, "On Positioning the Past with the Present: The Use of Fire Insurance Plans and GIS

for Urban Environmental History," *Environmental History* 10, no. 1 (2006): 136–39; Mathew Novak and Jason Gilliland, "'Buried beneath the waves': Using GIS to Examine the Social and Physical Impacts of a Historical Flood," *Digital Studies* 1, no. 2 (2010) <http://www.digitalstudies.org/ojs/index.php/digital_studies/issue/view/20>; Donald Lafreniere and Jason Gilliland, "Beyond the Narrative: Using H-GIS to Reveal Hidden Patterns and Processes of Daily Life in Nineteenth Century Cities," paper presented to the Association of American Geographers, Seattle, April 2011 and Don Lafreniere and Jason Gilliland, "A Socio-Spatial Analysis of the Nineteenth-Century Journey to Work in London, Ontario," paper presented to the Social Science History Association, Chicago, Illinois, November 2010.

32   Wetherell and Potter, *Mapping the Language of Racism*; Teun Adrianus van Dijk, *News as Discourse* (Hillsdale, NJ: L. Erlbaum, 1988).

33   Roy, *A Whiteman's Province*; Mawani, *Colonial Proximities*; Anderson, *Vancouver's Chinatown*; Lai, *Chinatowns*; and Ward, *White Canada Forever*.

34   Our methodology is fully described in a paper by Megan Harvey and John Lutz forthcoming in *Victorian Review*.

35   Jonathan Feinburg, "Wordle," in Julie Steele and Noah Illinsky, eds., *Beautiful Visualization* (Sebastopol, CA: O'Reilly Media, 2010).

36   We explore this in more detail in Dunae et al., "Making the Inscrutable Scrutable."

37   The *viHistory* website is located at http://www.vihistory.ca. It is edited by Patrick Dunae and maintained by the Humanities Computing and Media Centre at the University of Victoria. The attribute data used in our HGIS of Victoria are freely accessible to researchers on this open-source website.

38   We are able to capture the additional population caught by the municipal censuses and directories.

39   Geo-referencing is an intricate and exacting procedure, wherein a series of X/Y, longitude/latitude ground control points from high-precision, modern GIS maps are applied to digital images of historical maps. Using a technical process known as polynomial transformation, and informally

called rubber-sheeting, the historical maps are digitally stretched and skewed to correspond to their real-world spatial coordinates.

40   The exacting work of geo-referencing the cadastral layers, delineating polygons and linking attribute data to parcels was done by GIS technician Kevin Van Lierop and co-author Donald J. Lafreniere in the Human Environments Analysis Laboratory in the Department of Geography at the University of Western Ontario. We used ArcGIS, the popular GIS software from ESRI, in this project.

41   On the function and value of these spatially related records, see Diane L. Oswald, *Fire Insurance Maps: Their History and Application* (College Station, TX: Lacewing Press, 1997). See also Gilliland and Novak, "Positioning the Past with the Present."

42   R. T. Williams, ed., *Williams' illustrated official British Columbia directory, 1892; under the patronage of the Dominion and provincial governments, as well as the various municipalities throughout the province, containing general information and directories of all the cities and settlements in British Columbia, with a classified business directory* (Victoria, BC: Colonist Printers, 1892).

43   *The British Columbia Directory for the Year 1882–83*, published in Victoria by the firm of R. T. Williams in 1882 but compiled in the fall of 1881.

44   Peta Mitchell, "'The Stratified record on which we set our feet': The Spatial Turn and the Multilayering of History, Geography and Geology," in Dear et al., *Geohumanities*, 81.

45   David Bodenhamer, "History and GIS: Implications for the Discipline," in Knowles, *Placing History*, 230.

46   Douglas Richardson, "Spatial Histories: Geohistories," in Dear et al., *Geohumanities*, 210.

47   Massey, "Space-Time," 34.

48   Mikhail Bakhtin, "Forms of Time and of the Chronotope in the Novel; Notes towards a Historical Poetics," in *The Dialogic Imagination*, M. Olquist, ed. (Austin: University of Texas Press, 1981), 250.

2

Mapping the Welland Canals and the St. Lawrence Seaway with Google Earth

Colleen Beard, Daniel Macfarlane, and Jim Clifford

INTRODUCTION

Google Earth is a desktop virtual globe and mapping application that utilizes geospatial data held on the web to visualize the earth in three dimensions. Google Earth also allows users to document, incorporate, and map out their own data and other information onto the virtual globe.

For researchers and local history enthusiasts, the ability to incorporate historical documents into a digital geographic interface is especially compelling. Moreover, this easy-to-use free software considerably decreases the obstacles that might prevent a researcher from employing GIS.

The process of creating custom maps in Google Maps and Google Earth uses many of the same methods as more advanced researchers use with GIS software. It provides the option to create custom lines, polygons, and points. By consulting historical maps, we can add lines representing roads or railways, polygons for the outlines of building or agricultural fields, and points to identify particular places or to locate events. Google Earth also allows users to attach descriptions (using text and HTML code) to these lines, polygons, and points. Although Google Earth cannot link data to an attribute table or query a database in the same manner as GIS, the user can incorporate dates and build a basic spatially referenced digital historic map of a landscape.

Google is now constantly adding historical imagery to their web application, a development that will be well received by environmental historians and geographers. For example, aerial photographs are available for London, England, from 1945 on, along with the more recent series of satellite imagery. However, this information is more limited for Canada, and in the case of the St. Lawrence Valley the images only go back to 1995. The situation is better in the Niagara Region, in part because of aerial photographs uploaded to Google Earth by Sharon Janzen and Colleen Beard, one of the co-authors of this paper. Hopefully this example will lead other map librarians, geographers, and historians to share historical imagery through Google Earth. The "Layers" icon in Google Earth provides access to another huge collection of data, including features such as labels, places, roads, and pictures, but also third-party data from organizations such as NASA and National Geographic. These layers are provided by Google and partner organizations.

Among the layers currently available, the most useful for historical research and teaching is the David Rumsey Historical Maps collection (found under the gallery tab), which includes dozens of historical maps "pinned" or georeferenced onto their location on the digital globe. Working with layers approximates one of the central methods in HGIS, where researchers create and analyze layers of scanned and georeferenced historical maps. While there

are a great number of georeferenced maps for the United States, fewer exist for Canada. It is possible, though not particularly advisable, to add more layers of historical imagery, found for example on the Library and Archives Canada website or from a university map library. The georeferencing tool provides a very basic method for comparing layers of historical maps with more recent satellite images. It should be noted, however, that the georeferencing functions in GIS software are considerably more functional than in Google Earth, and if you find yourself with a collection of scanned historical maps, it is probably best to invest the time and learn the more complicated GIS software.

Even after a researcher learns to work with Quantum GIS or ArcGIS, these Google products remain valuable for presenting digital maps. Many GIS programs allow researchers to export their layers as Google Earth files, making them accessible to collaborators and the public. Moreover, the Movie Maker function allows researchers to create dynamic presentations using Google satellite imagery. This free software and the associated web-based Google Maps is an essential digital mapping tool for learning, presenting, and teaching. University academics have free access to the Pro version of Google Earth, which expands the capacity to work with GIS files and provides higher resolution images for presentations, making it even more useful for historical researchers and teachers. Whether one uses Google Earth in conjunction with GIS software or on its own, it provides a low-budget and easily mastered option for including and presenting digital mapping in historical research. After starting digital mapping education on Google Earth, researchers are well prepared to then learn how to use ArcGIS; Google Earth is an extremely useful stepping stone.

The remainder of this chapter brings together two projects that have used Google Earth to explore the spatial history of the Welland Canals and the St. Lawrence Seaway. Historian Daniel Macfarlane works with Google Maps and Google Earth to facilitate his research on the history of the St. Lawrence Seaway and Power Project, with assistance from Jim Clifford. Brock University Map Librarian Colleen Beard sets her focus on local landscape change, employing Google Earth to explore the three historic Welland Canals that traversed St. Catharines, Ontario, in the nineteenth century.

1. MAPPING THE ST. LAWRENCE SEAWAY AND POWER PROJECT[1]

The St. Lawrence Seaway and Power Project, built between 1954 and 1959, combined a major hydroelectric development in conjunction with a deep water navigation scheme. It was one of the greatest megaprojects of the twentieth century and is now the second largest transborder water control endeavour of its kind. The construction entailed a massive reshaping of the St. Lawrence environment. Over 40,000 acres of land were flooded to create Lake St. Lawrence, which served as the power pool for the international hydro dam between Cornwall and Massena. Since the St. Lawrence River forms the border between Ontario and New York (and the United States and Canada) before flowing through Quebec and emptying

into the Atlantic Ocean, modifications of river water levels necessitated bilateral cooperation between Canada and the United States. Half a century of complicated negotiations and failed compacts passed before the two countries reached a final agreement in the 1950s, a process that was extremely revealing for not only the history of Canadian–American relations and nationalisms but also North American environmental and technological history.

Over the course of my doctoral research on the Seaway, I encountered a great number of maps and other geographically referenced documents concerning the St. Lawrence project. These included blueprints and engineering plans ranging from the late nineteenth century to the early Cold War, from hand-drawn schemes to precise engineering documents. Blueprints of the International Rapids Section of the St. Lawrence River (Fig. 2.1), which was turned into a lake, were particularly fascinating because they showed the ways that the St. Lawrence landscape had been imagined in numbers and measurements. One of my major interests is the ways that North American societies, states, and experts in the mid-third of the twentieth century conceived of the engineering possibilities of riverine environments. This wealth of inherently spatial information presented opportunities to enrich my analysis.

Like many people, I had experience with Google Maps for generic purposes and had even made a few custom maps, but I wasn't sure that the type of multilayered and interactive mapping I envisioned was within my capabilities. Given my lack of time and skills, taking on a large GIS project seemed impossible, but after I had completed my dissertation and started turning it into a book, Google Earth provided a free and accessible alternative that

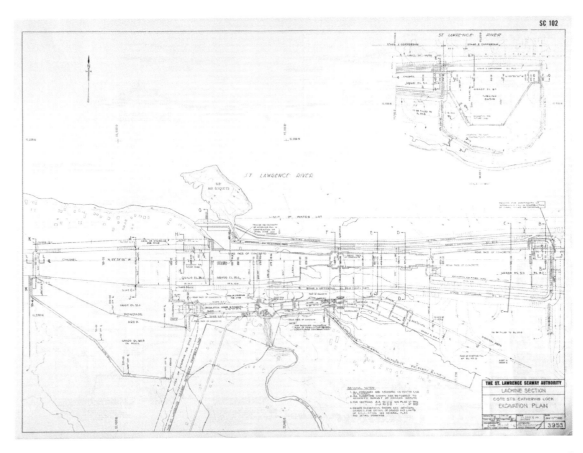

Fig. 2.1. St. Lawrence Seaway Authority blueprint of Lachine section of St. Lawrence Seaway at Montreal © Government of Canada. Reproduced with the permission of the Minister of Public Works and Government Services Canada (2012). (Source: Library and Archives Canada/Department of Transport fonds/RG 12/vol. 5565/Mikan 1254527.)

could be mastered relatively quickly. I sought out further training and education,[2] and began exploring the custom maps feature of Google Maps. Soon I had marked off the main Seaway channel and found that I could use anchors and signs to designate key points of interest and different features. Based on other maps and blueprints, I began adding text, maps, and pictures to the various points of interest, then started tracking the lines of the older canals that the Seaway had replaced. In just a few combined hours, I had essentially produced the map displayed in this chapter. While this did

not result in a comprehensive HGIS database, it did create a digital map with layers of spatial and descriptive information. This sparked my interest in digital mapping and started me on a path towards learning GIS. It also provided me with a map that could be exported into Google Earth, which uses technology that is closely related to GIS.[3]

Google Earth is a virtual globe that allows one to view the earth in three dimensions. For example, the tilt function can be used to look at the earth from an angle, rather than from straight above (this is similar to Google Maps,

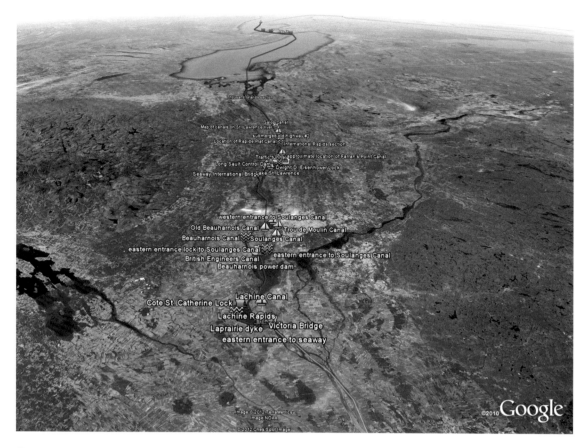

Fig. 2.2. Map of St. Lawrence Seaway route. (Created in Google Earth. © 2012 Google; © 2012 TerraMetrics.)

where the "Street View" option simulates the map's appearance from the perspective of someone standing in front of a landmark, or even adjusts for time of day). This was a particularly effective way of following the Seaway route, as it permitted me to start at one end and track along. Using another option on the same tool bar, a custom tour can be recorded for easy playback.

I used Google Earth to highlight the navigation path of the actual Seaway, as well as ports and features of interest (Fig. 2.2). This provided a useful and potentially powerful visual and geographic representation for future readers but also allowed me to undertake spatial analysis in

a way I could not have otherwise. Using the overlay feature, which will be explained in more detail below, I compared the map of the flooded International Rapids section of the St. Lawrence project (from approximately Prescott to Cornwall) with a historical picture to better understand how much land was flooded and where. The Seaway was digitally viewable from alternative angles, heights, and directions, providing unique perspectives on the subject.

Google Earth also allowed me to map the previous canal systems that predated the Seaway, including the Welland Canal. By adding points, lines, and polygons, I was able to create a custom map of the previous canal routes and

Fig. 2.3. Overlay of village of Aultsville. (Created in Google Earth. © 2012 Google; © 2012 USDA Farm Service Agency.)

the contemporary St. Lawrence Seaway, using satellite imagery to isolate and highlight the features that were pertinent to my research.

The flooding in the International Rapids sections of the St. Lawrence power project required the relocation of a number of riverine communities – the Lost Villages. Archival records, maps, and photographs indicate the placement of individual properties in the various Lost Villages. This information can be represented in Google Earth to reconstruct the layout of the village and, using different features, can be analyzed in order to determine patterns. For example, I was interested in showing how the government hoped to reorient the newly created communities towards

the freeway rather than the river by comparing the distance distribution of residential areas in relation to schools, business centres and other amenities, the river, and the highway.

Even though the historical satellite imagery function of Google Earth does not yet permit me to view images from the time period before the St. Lawrence project was constructed, the fact that it did allow me to go back a decade was nonetheless helpful. The water clarity of the St. Lawrence has varied over the years due to different factors and the sliding historical imagery timeline enabled me to select images from years when the water (and the images themselves) was comparatively clear. Under such conditions, one can view the foundations and remains of many parts of the Lost Villages and transportation networks, even though they are under water, and trace the route of the now-inundated Highway 2. Combining the remains of infrastructure viewable in Google Earth with sufficient archival information makes it possible to create a map showing land covered by water before and after the flooding. To achieve this in Google Earth, I did an image overlay, putting a digitized version of an old map of the flooded village of Aultsville over top of a Google Earth satellite image of its present location (Fig. 2.3).

This was a relatively simple operation. Using the Google Earth overlay function only requires hitting the button along the top border, and selecting the file one wants to upload. Opaqueness – the amount one can see through the image being overlaid – is a key attribute of overlays and can easily be controlled. Using the lines of the main north–south street and old east–west Highway 2 as reference points, I lined up the old maps and the Google Earth image – a simplified form of the more sophisticated georeferencing that can be done with more advanced GIS software.

With this Aultsville overlay, the portions of the area and infrastructure that were submerged, as well as those that remained dry, are obvious. Images such as this have allowed me to discover and explore (if done with remote sensing technology, this would be called "groundtruthing") the actual area depicted on the satellite imagery; this can also be used as a pedagogical tool, as I have taken several classes on field trips to this site. Because the water is relatively shallow here, one can walk the former streets of the town and find building foundations. In fact, when the water is low, one can walk close to the old river shoreline. One

can easily count the number of properties that existed (and were flooded) and discover geographical features that influenced the pattern of the flooding. For example, the main streets and highways remain visible, while flooding was more prominent along the watersheds of tributary creeks and streams. Aided by other information, such as property registries or interviews with former residents, the spatial aspects of the community can be recovered.

2. VISUALIZING THE HISTORIC WELLAND CANALS

In 2010, Brock University Map Librarian Colleen Beard set out to create an interactive visualization tool, or "mashup," to explore the spatial history of the three historic Welland Canals. Inspired by the interest that was generated by the Welland Canal Summit, a public meeting held in 2009 by the Niagara regional government to discuss the heritage designation of the historic canal corridor, the Historic Welland Canals project aimed to increase awareness of the region's canal heritage while showcasing the rich collections of the Brock University Map Library and other local collections. The project brought together hundreds of historic air photos, maps, plans, photographs, and audio interviews held at Brock University and other local collections. These documents were then scanned and embedded into the Google Earth framework to provide a geographic reference for exploring the many features of the three canals.

Fig. 2.4. The Welland Canal System within the Niagara Region. (Created in Google Earth Pro, 2011.)

The Welland Canal is a major route for ships navigating between Lake Ontario and Lake Erie to circumvent Niagara Falls (Fig. 2.4). The canal was first established in 1829 by local businessmen to stimulate local and regional trade, but the government of Upper Canada soon saw its potential as a crucial artery for British North American trade. Over a period of one hundred years, the canal was rebuilt three times. All three routes used Port Dalhousie

as the access point to the canals, with slight variations. The canals climbed the escarpment at Thorold, and then progressed to Lake Erie. The first and second canals had similar routes, each of which followed natural waterways through what is now downtown St. Catharines. The third canal, completed in 1881, deviated from Port Dalhousie and cut a straight diagonal path bypassing St. Catharines. It survives as a scar in the landscape visible primarily from the air. Some of its twenty-six stone-cut locks are still intact and visible where the canal intersects the current fourth canal at its eastern extent.

This project aims to increase the awareness and importance of the canals among the general public and government authorities, particularly within the context of the regional government's application to Parks Canada for historic site designation. It will also serve as a significant resource for planners in making informed decisions associated with cultural, tourism, and environmental planning. For example, historical air photos are currently being used to examine the area where the fourth canal, currently in operation, intersects with the historic third canal, in order to determine zones of concern where tourism may interfere with the daily operation of the canal. Figure 2.5 illustrates the extent of the project content, where each point symbol on the map represents the location of a canal lock, feature, or a point of interest. Each placemark links to photographs, maps, or air photos that annotate and illustrate its history. Navigating the resources using Google Earth provides a virtual tour of canal history.[4]

Historical maps and air photos provide useful sources for exploring and documenting features and landscape change along the historic canal routes. When converted to a

Fig. 2.5. A sample of the project placemarks within the Google Earth framework, displaying the Port Dalhousie area and north terminus of the historic Welland canals. (Created in Google Earth Pro, 2011.)

digital format, these images can be imported into Google Earth and georectified, or adjusted to overlay and match the underlying Google Earth imagery. This technology can reveal some startling landscape changes over time. Geotagging historical photographs to their exact location, for example, allows viewers with handheld devices to stand in the same location as the photo was taken and "look back in time." Audio interviews with employees of shipyards and local businesses, as well as local residents from the time period, provide first-hand accounts and stories of canal life. These historic documents, when placed in a geographic

context, provide a powerful means of relating little-known discoveries about the canals.

Canal Discoveries

The Piers of Old Lock One

The location of the submerged wooden piers of the very first canal in 1829 has been a subject of great interest for local canal enthusiasts. Many area residents were unaware that the north access to the first canal at Lake Ontario cut a path through what is now popular Lakeside Park in Port Dalhousie. The 1855 Welland

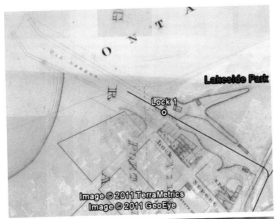

Fig. 2.6. Top: *Welland Canal 1855 survey map, showing the location of the piers and Lock 1 of the first canal, overlayed in Google Earth shown faintly underneath. (Digital image: Brock University Archives, St. Lawrence Seaway Authority [Canada] Fonds, Second Welland Canal Survey Map 1, ca. 1855; reproduced in Google Earth Pro, 2011.)* Bottom: *1934 air photo showing the submerged shadows of the piers. (Digital image reproduction: Brock University Map Library; NAPL photo A4700 #33.)*

Canal Survey (Fig. 2.6) illustrates the position of these piers that extend beyond Lock One into Lake Ontario.

By overlaying this digital image over the Google Earth imagery, I was able to determine the exact geographical location of the piers in today's landscape. Google Earth provides a tool that displays an overlay at gradual transparency levels. As the top layer becomes transparent, it reveals the underlying current imagery. The shoreline of the Lakeside Park area, for example, reveals especially dramatic landscape changes. The shadows of the submerged pier remains, clearly visible from the 1934 air photo taken one hundred years after the canal was in operation, provided another exciting discovery. These shadows are also discernible on the current imagery in Google Earth. Such discoveries are not merely academic; they are of great interest to community members and canal enthusiasts as well. For example, when these shadows were revealed in a community publication,[5] an advocate for the canals hopped in his kayak, equipped with snorkel and mask, determined to locate the underwater remains (unfortunately with little success)!

The 1855 survey map, the 1934 air photo, and one other 1839 map of Port Dalhousie provided the geographical information for accurately locating the buried Lock One for a recent excavation.[6] In 2008, an archaeological assessment was initiated by the City of St. Catharines to determine the significance of buried canal features within the north entrance channel. The study established with certainty the precise location of these features within Lakeside Park, and determined that much of Lock One's wooden structure remained intact.

Shipwreck at Lock 21

This area of the third canal is currently used as a reservoir for regulating water for the Thorold flight locks 4, 5, and 6 of the fourth and current canal. When drained in the winter, the skeletal remains of a sunken vessel can be seen outside the north gate at Lock 21. Although many are aware of these remains, their history is a mystery. Amazingly, the digitized 1934 air photo, when viewed at a zoomed level, shows this vessel in nearly the same position, but afloat (Fig. 2.7).[7] This boat is very difficult to detect from the original air photo print. Not until this image was enhanced through digital processes was it noticeable. These enhancements offer a few more details that can help unravel the mystery. For example, the boat would have had to have been abandoned sometime after 1932 when the canal was no longer in use, since it obstructs the entrance to the lock. Although the boat's mast is not visible, the details in the shadow cast by the mast indicate a vessel larger than a tug boat, as some residents suspect. A 1948 air photo of the same area from the Map Library collection shows no evidence of the boat. It must, therefore, have sunk before then. The history of the ship remains a mystery, but the evidence that the 1934 digital imagery provides has inspired great curiosity among canal enthusiasts and Seaway personnel.

The Railway Tunnel Entrance

A final story unravels the mystery of the railroad tunnel. Not far from Lock 21 on the third canal, between Lock 18 and 19, is the Grand Trunk Railway tunnel. Built in 1887, the tunnel redirected trains under the canal instead of having them pause at the surface rail bridge to make way for ship traffic. Only 220 yards in

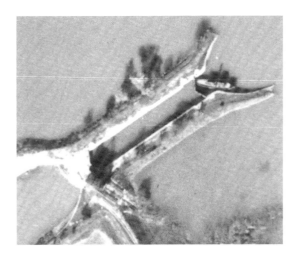

Fig. 2.7. This 1934 air photo shows a boat afloat outside Lock 21 of the third canal. The 1934 imagery is also accessible through the Google Earth Historical Imagery option. (Digital image: Brock University Map Library; NAPL photo A4873 #28.)

length, it is locally known as the Blue Ghost Tunnel due to its spine-chilling effect for those that dare to explore – with a good flashlight and companions in tow! The tunnel is accessible on foot from the west entrance. However, the last thirty yards through to the opposite end are impassable (without hip waders and a hard hat) due to the deterioration of the railroad and ceiling. Although the opening is visible and seems within reach, it is nowhere in sight from the road above. Previous searches for the eastern entrance by canal hikers had met with no success.

A closer look at the 1934 air photo of the area clearly shows the path of the tunnel with both the west and east entrances visible. When overlayed onto the Google Earth framework (Fig. 2.8), the precise latitude and longitude of the east entrance was recorded. We entered these coordinates into a Global Positioning System unit and set out on foot.

Fig. 2.8. Google Earth is used with a 1934 photo overlay to determine the exact position of the east entrance of the Grant Trunk Railroad tunnel under the third canal. (Created in Google Earth Pro, 2011.)

Fig. 2.9. East entrance of the Grand Trunk Railroad tunnel. (Photo: Rene Ressler, May 2009.)

The east side of the canal bank in this area is not easily accessible. After a half-hour trek through deep mud and dense brush, with weak GPS signals, the GPS finally beeped, warning us that we were within a few metres. Several steps further into the dense brush, and there it was – a large concrete ledge. A short descent down the bank revealed the tunnel entrance. To our delight, the concrete structure was still in excellent condition. Protected by underbrush, it had been spared the popularity of the graffiti-covered and garbage-strewn western entrance (see Fig. 2.9).

The Talking Map

Alex Bennett was a pipefitter for Muir Bros Drydock Company in Port Dalhousie during the 1940s (Fig. 2.10). Although the northern terminus of the canal had moved east to Port Weller in 1932, the Drydock Company continued to operate as an important shipyard industry for the region. An audio clip of Bennett's interview with two local historians can be heard from a link within the project's web application.[8] The project can also be accessed on a mobile device, thus making possible a walking tour of the historic canal area. While standing at the location of Muir Bros, now Rennie Park, one can link to the audio and listen to details describing Alex's working experience, at the same time viewing accompanying historical photographs. This application is also referred to as a "talking map." Bennett speaks passionately about coming over from Scotland and being down to his last twenty dollars and about his experience working on a ship called the *Makewelli*. He describes the layout of the Muir building almost to the point where one could recreate the design. Although the Muir Bros Drydock Company had no monopoly over the shipbuilding industry, it managed to operate longer than its competitors, until 1954. This area became the site of the Canadian Canal Society in 1982.

Fig. 2.10. Muir Bros Drydock Company, 1882. (Photo: Brock University Special Collections and Archives.)

The Power of Photography

There are numerous photographs and sketches of the canals. Displaying photographs in the Google Earth framework, however, provides an exciting enhancement to the canal experience. Several books provide historical accounts of the canal with photographs and maps but lack geographic context. To date, 175 photographs or sketches have been added to the project and "geotagged" with their geographical identification. This is a simple, but very effective, procedure. Using photographs in this way can help to reconstruct the historical landscape. Historical maps are used first to identify the names and locations of buildings, lock features, or industries. Once these maps are converted to a digital form and georectified in Google Earth, they become important resources in identifying locations in today's landscape. Photographs of these features are geotagged to the location they once occupied. Several industries along the canal corridor were identified using this method.

Adding old photos of the landscape in the Google Earth context is a very effective method of placing one at the scene. A geotagged photograph can include annotation about a building or historic event to augment the experience of "being there." For example, atop the Glenridge bridge near the corner of St. Paul Street and Westchester in downtown St. Catharines, a current view reveals a neglected landscape – the backend of a series of rundown buildings on St. Paul Street. Standing near this spot in 1871, the view would have been much different! Figure 2.11 illustrates this striking comparison. The 1871 sketch shows a view of the same area where the second canal, now a parking lot and future site of a sports complex, meandered through quite a different landscape. Historical maps were used to identify some of the industry that lined the shores during this time.

Another interesting use of geotagged photos is to document the history and progressive use of buildings over time. The Beaver Cotton Mills, located in Merritton by Lock 16 of the second canal, is documented chronologically in this project through several photos. The mill was situated at the base of the escarpment to take advantage of the water power associated with the series of locks that carried ships up

Fig. 2.11. Top: *A current view of the area once occupied by the second canal, downtown St. Catharines. December 2011. (Photo: Brock University Map Library.)* Bottom: *a sketch-up view of the area in 1871 showing the second canal and surrounding landscape. (Photo: Canadian Illustrated News.)*

the steep incline, also referred to as "Neptune's staircase" – one of the few spots along the corridor that is currently designated as a local historic site. Early sketches of the mill illustrate a flurry of activity in the years before its demise. Current photos offer a good example of a historic mill site repurposed as a restaurant.

Libraries and Digitization

To date, this project has generated interest from the Region of Niagara, as they continue to pursue a heritage designation for the canals. Teachers have expressed interest in using the project as a learning tool by imbedding it into their local history curriculum. It is currently part of the focus for a university geography assignment in a course titled "Digital Cities." The discoveries that have been unveiled through historical maps have sparked the curiosity of local historians and canal enthusiasts and have led to further exploration.

A number of years ago, Brock's Map Library staff embarked on their first digitization project of several hundred 1934 air photos. The enhancements were remarkable: detail could be seen that wasn't apparent from the contact printed version. Since then, digitizing our collections to make them more widely accessible to our researchers has become a priority. Google Earth has provided us with a virtual stage to showcase our digital collections. The Brock Map Library is one of few university libraries to have partnered with Google to have their digital imagery actually incorporated into their Historical Imagery database. By connecting historical documents with geographic locations in the present, it provides a sense of "place" with which we can describe and display the region's history – something that books and other printed materials cannot alone achieve. It is time for librarians and archivists to dust off their historical collections and bring them back to life. Historical GIS has provided many exciting ways to do this. Freely accessible and easily mastered, Google Earth is a good place to start.

CONCLUSION

These two projects each worked with Google Earth in different ways to document and display different aspects of Great Lakes–St. Lawrence canal infrastructure. Both present Google Earth as a flexible, accessible, and readily mastered technology for getting started with historical GIS, or for sharing the results of more comprehensive spatial history projects with a wider audience. Through these examples, we have introduced readers to methods of locating information, building a basic spatially referenced digital historic map, and presenting the results. For example, we have shown how aerial photography, specifically, can be used and georeferenced within Google Earth to generate new information. In doing so, we have demonstrated that Google Earth can benefit those looking to utilize digital mapping technology, be they professional academics, public historians, researchers, librarians, archivists, or public history enthusiasts. Google Earth also provides an ideal platform for teaching historical digital mapping and introducing the basic approaches of HGIS at the undergraduate level.

We have shown how digital technologies have allowed libraries to easily transform their valued historical collections from traditional print format to digital spatial information that reveals new knowledge. The enhancements achieved through digital processes, coupled with the viewing technologies that Google Earth provides, unleash a wealth of information about the history of Canadian landscapes and do so in a way that can be easily presented, disseminated, and absorbed by both academic researchers and the public.

NOTES

1   Daniel Macfarlane, "To the Heart of the Continent: Canada and the Negotiation of the St. Lawrence Seaway and Power Project, 1921–1954," PhD dissertation, University of Ottawa, 2010. An expanded version of this dissertation will be published as a monograph with UBC Press in 2014. For a shorter summary see: Daniel Macfarlane, "Rapid Changes: Canada and the St. Lawrence and Seaway Project," Research Paper, Program on Water Issues (POWI), Munk School of Global Affairs (University of Toronto), 2010, http://www.powi.ca/pdfs/other/Macfarlane-POWI%20paper.pdf.

2   This included a series of blog posts on HGIS for beginners on the NiCHE website: http://niche-canada.org/taxonomy/term/772. Daniel is especially grateful to Jim Clifford for his tutelage in developing GIS skills.

3   Converting a custom Google Map to Google Earth is very easy: on the top of a custom map, just click on "View in Google Earth." A prompt will ask whether you want to open or download the file. Unlike a custom map, which remains stored and saved online, a Google Earth map can be downloaded to your desktop (download a free copy of Google Earth first). Alternatively, you can start directly with Google Earth and skip Maps.

4   This project is accessible from the Brock University Map Library website: http://www.brocku.ca/library/collections/maplibrary. The kml files can be saved and opened in Google Earth.

5   Colleen Beard, "The Three 'Old' Lock Ones – A history in maps," *Dalhousie Peer – Port Dalhousie's Community News Magazine* 10 (2000): 6–8.

6   J. K. Jouppien, "The Location of the North Entrance Channel of the 1st Welland Canal, 1824–1829 and a Stage 1-3 Archaeological Assessment of Select Areas at Lakeside Park, City of St. Catharines, Regional Municipality of Niagara, AhGt-20 CIF/PIF #P119-001-2008" (Niagara Falls, ON: J. K. Jouppien Heritage Resource Consultant Inc., 2008).

7   The entire 1934 air photo imagery of the Niagara area has been digitized by Map Library staff and recently donated to Google. It is now included in the Google Earth Historical Imagery database.

8   The audio interview with Alex Bennett was conducted by Christine Robertson and David Serafino, 2000. http://www.brocku.ca/maplibrary/WellandCanal/Bennett_audio.mp3

3

Reinventing the Map Library: The Don Valley Historical Mapping Project

Jennifer Bonnell and Marcel Fortin

In June 1888, Toronto City Engineer Charles Sproatt stood up to his knees in the muck of the former river channel. Things were not going as planned. The deadline to complete the Don River Improvement Project was fast approaching, and everywhere around him, so much still to be done. The Mayor and Council were losing patience. At every turn, it seemed, more setbacks. Problems with contractors, disputes with landowners, and protracted negotiations with the Canadian Pacific Railway had delayed project progress and sucked up dwindling funds. The previous winter, attempts to use the project as a form of unemployment relief had gone awry when thick frosts slowed the work of cutting into the river banks and resulted in extravagant labour costs. Now, the land itself was revolting. Sproatt's dredges had run up against dense shale deposits on the course of the new river bed north of Queen Street. To remove the shale would consume the remainder of his budget and leave much unfinished; to leave it would be admitting defeat. A river only eight feet deep, rather than the planned twelve, would scuttle plans for a navigable channel north to Gerrard Street, one of the original impetuses for the project. As Sproatt would find in the years to come, costs and time overruns would continue to mount, and the Don River Improvement Project, originally hailed as the fix to turn a languishing district into a thriving industrial hub, would be remembered most of all for its dubious results.[1]

Toronto's Don River has a history of failing to cooperate. Despite the river's small size – just thirty-eight kilometres from its headwaters north of the city to its mouth in the Toronto harbour – it has long carried great capacity for destruction, from seasonal floods to harbour-clogging silt deposits, to the threat of disease outbreak from water-borne pollution. As ship captain (and later harbour master), Hugh Richardson lamented in 1834, the river was a "monster of ingratitude," whose "destructive mouths" threatened to turn the entire harbour into a "marshy delta."[2] Troublesome landscapes tend to attract improvers, and, as a result, the Lower Don River and the area around its mouth have since the 1870s been a landscape subject to rapid and dramatic change. The Don Improvement Project and the associated rail corridor of the 1880s saw the lower river straightened and canalized south of Winchester Street. In the 1910s and 20s, the draining of Ashbridge's Bay Marsh and the creation of the Port Industrial District became one of the largest megaprojects on the continent, converting some 1,200 acres of lacustrine marsh to ship slips and new revenue-generating land for the city.[3] Forty years later, the construction of the Don Valley Parkway along the valley bottom radically altered the river landscape once again, cementing its function as a transportation corridor.

A place subject to so much change has a rich spatial history that makes it especially compelling for mapmakers. And for urban environmental historians, this spatial history presents a great store of evidence with which to address the fundamental questions of our field. How did urban ecosystems, and human relationships with those ecosystems, change over time? What were the effects of these changes upon human health, economic prosperity, class relations, and ecological integrity? How did developments here differ from other parts of the city, and why?

This chapter explores the ways that historical Geographic Information Systems (GIS), or HGIS, can be applied to historians' understanding of the environmental history of Toronto's Don River Valley. It follows the experiences of a small team of researchers (an historian, a map and GIS librarian and a few research assistants) in navigating the complexities of building a historical mapping project using GIS technology. Through the example of the resulting Don River Historical Mapping Project, we discuss the challenges of accessing and working with historical source materials, the uncertainties inherent to historical GIS, and the difficulties of making resources and research findings publicly accessible. Unlike most HGIS projects, which take their origin from a research problem or question and produce specific datasets with which to address that question, this project aimed from the outset to produce a body of data that would be accessible to a broad range of researchers from different disciplines for use in a variety of different projects. While a research project on the Don River Valley was the impetus for the project, the outcomes of the project were in the end much broader, shaped by the mandate of the library to promote its collections and provide open access to data to facilitate research. This chapter, as a result, is as much about the changing role of the academic library as a partner in academic research as it is about mapping the history of the Don Valley.

As noted in the introduction to this volume, Canadian historians, like their counterparts in other parts of the world, have to date only tentatively explored the possibilities of working with historical GIS. Access to digitized historical data and technical support are among the obstacles that historians face in adopting HGIS methods in their research. Here the academic library has much to offer. More than simply a repository for collections, university libraries have over the past twenty years increasingly stepped into the role of service providers. Particularly in specialized libraries such as map and data libraries, these services go beyond traditional supports such as reference and circulation to encompass teaching, in-depth research consultation, and even partnership in research. Increasingly, library staff are being included in the research process because of their expertise with the source material housed in libraries, with technical issues, and with the process of academic research. As much as historians stand to benefit from digitized historical data, librarians are looking for ways to give little-used paper historical collections a new life as digital data. Digitized and georeferenced to real-world coordinates, the information contained within historical maps can be extracted, collated, queried, and compared to produce new knowledge about the past.

A collaborative endeavour, this paper makes a case for collaboration: fruitful partnerships, we demonstrate, can emerge between historians and academic librarians, not only in finding and accessing data, but also in learning from each other in order to work more effectively with the often unusual problems (and promises) that HGIS research presents to the historian.

INTERPRETING ENVIRONMENTAL CHANGE IN THE LOWER DON VALLEY

Toronto's Don River winds through the most urbanized watershed in Canada. Like many of the rivers on the north shore of Lake Ontario, this small waterway follows a generally south-westerly course as it moves from the porous moraine lands at its headwaters to its outfall in Toronto harbour. Two main branches, the East and West Don, join to form a single stream (the Lower Don River) at the forks about seven kilometres north of Lake Ontario. A third tributary, Taylor-Massey Creek, flows into the forks from the east (Figure 3.1).

The environmental history of the Don River valley is in large part a story of the relationship between the developing city and the river valley at its eastern periphery. This history of the two-hundred-year relationship between a small urban river and what is today Canada's largest city revolves around the idea of the river valley and the city as mutually constitutive – each shaping the development of the other. As the city grew, it radically altered the physical and ecological composition of the valley, denuding slopes, polluting waterways, filling wetlands, and levelling hills. Just as the city transformed the valley, the valley presented certain possibilities and foreclosed others as the city expanded. From the mosquito-infested marsh at its mouth to the occasionally devastating floods it wrought upon valley landowners, to the large quantities of silt and debris it washed into Toronto harbour, the river was an active participant in the city's development.

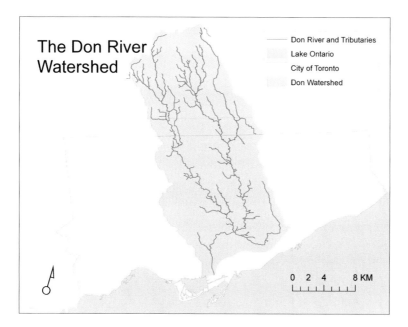

Fig. 3.1. Don River watershed. (Sources: Don Watershed boundary: Toronto Region Conservation Authority 2008; current shoreline: DMTI CanMap Postal Geography FSA Boundaries, v2008; City of Toronto Boundary: City Wards Boundary File, City of Toronto Open Data Catalogue.)

The valley's geography, with its steep ravine walls and wide plateaus, was even more influential, at once a formidable barrier to the eastward expansion of the city and an enabling corridor for transportation and urban growth. Until the completion of the Prince Edward (Bloor Street) Viaduct in 1918, no bridges existed across the wide valley expanse north of Gerrard Street, and travellers were forced to route south to Winchester Street or north to Pottery Road to access communities east of the river. Those bridges that did exist were precarious structures prone to washout during seasonal floods, further constraining access for landholders and industrialists east of the Don. As much as the lower valley posed a barrier to east–west communication, it invited movement north–south. Rail development happened first, in association with the Don River improvement plan of the 1880s and 90s. Seventy years later, the Don Valley Parkway took the valley's corridor function to its full potential, carving six lanes of highway from the Gardiner Expressway near the lake shore to Highway 401 north of the city. Reconfigured as a metropolitan corridor, the valley facilitated suburban development along its length, stimulating the growth of the city.[4]

The river valley has also served as a different kind of corridor, laying a swath of green space through the heart of the city. Through the nineteenth century, its steep, corrugated ravines resisted agricultural and residential development, surviving as pockets of woodland within an increasingly deforested landscape. Parkway construction in the mid-twentieth century capitalized on valley woodlands as an aesthetically pleasing backdrop to the curving ribbon of road, and a site for roadside parkland and recreational areas. In the aftermath of Hurricane Hazel in 1954, valley parklands served a secondary function as development-free drainage corridors. For 1940s-era conservationists and twenty-first-century urban explorers, valley green spaces provided, and continue to provide, a welcome respite from the monotony of the urban grid, a place to restore body and mind within easy distance of the city core.

Corridors of movement for urban wildlife, producers of oxygen and sinks for carbon, these green spaces also serve important ecological functions. Once feared as a harbour for gangsters and social deviants, today's valley lands are appreciated for their role in "wilding the city."[5] The valley is, however, as it always has been, an ambiguous space, subject to multiple uses and divergent ideas about its future: busy recreational trails expose the ramshackle tents of the homeless; the burble of a blackbird at a restored wetland site challenges the hum of traffic on the parkway; at the river's edge, hardy riparian grasses push through the metal grid of a discarded shopping cart.

The river valley has also claimed an important place in the history of ideas about the city and its future, its landscapes conceived by different groups in different periods as verdant wilderness, picturesque countryside, polluted periphery, predestined industrial district, restorative retreat, vital refuge, dangerous underworld. Over the course of the river's relationship with the city, a series of improvement schemes, from major channel reconstruction to highway construction and parkland acquisition, have mobilized these competing ideas, harnessing the river and its valley as a transformative force in building a prosperous and productive future metropolis. The relative success of these plans, and the effects they had upon valley ecologies, upon individual lives, and upon the life of the city, have served as important catalysts for change in the historical relationship between Toronto residents and the natural environment upon which they depend.

Much of this two-hundred-year history of the relationship between a city and its iconic river valley can be explored using existing sources and historiographical methods. Evidence of the valley's social history, for example, and changing cultural perceptions of valley landscapes, can be gleaned from a close reading of newspaper articles, municipal reports and correspondence, city council minutes, and other sources. A project focussed so fundamentally on landscape and environmental change, however, requires a comprehensive understanding of the area's spatial history. Issues surrounding historical land use are especially difficult to puzzle out. What kinds of land uses did the river valley attract in different locations and periods? How did the river channel change over time, and in which periods were these changes most pronounced? How did the spatial representation of these changes align with contemporary planning documents?

The rich legacy of historical maps, fire insurance plans, engineering drawings, and aerial photographs that document the city's development contains the evidence needed to address these questions. Extracting this evidence to make comparisons across space and time, however, is not an easy task. To begin with, the large number of sources available for the river valley makes conflicting representations inevitable – a challenge common to all forms of historical analysis. The nature of the sources themselves also complicates the process of analysis. Historical maps are drawn at different scales and with varying degrees of accuracy. The large format of many of these sources, furthermore, makes them difficult to work with. Jennifer Bonnell, the historian in this partnership, recalls attempting to document the river's industrial history by photocopying unwieldy fire insurance plans and taping them together to create a giant visual mosaic of building outlines along the lower river. Tiled together, each paper collage stretched the height of her office walls; each represented just one year in the history of the river.

Fortunately, the challenge of building and interpreting eight-foot map mosaics led Bonnell to seek assistance from staff at the University of Toronto Map and Data Library. There, conversations with map and GIS librarian Marcel Fortin revealed the potential for a collaborative project that would both digitize and pool together existing historical sources for this iconic Toronto landscape and use this information to build something new: a comprehensive geospatial database for the watershed as a whole. For Bonnell, the prospect of such a database removed months of tedious work with paper maps from her analysis and opened up exciting possibilities for historical insight. For the map library, the project provided an opportunity to showcase the rich resources of the university library's map collection and the dramatic interpretive and presentation capabilities of GIS. Seed funding from the Network in Canadian History and Environment (NiCHE) allowed us to hire a research assistant to begin the work of scanning and georeferencing historical maps and building geospatial datasets of the area, and, within a few months, the Don Valley Historical Mapping Project was underway.

Over the next two years, with some additional support from NiCHE and the University of Toronto Map and Data Library, we produced a series of geospatial datasets for Toronto's Don River watershed between the years of 1858 and 1950. In keeping with the core themes in the valley's environmental history and with available source materials, we produced data in four main categories: 1) industrial development in the lower valley from 1858 to 1950; 2) changes to the river channel, tributaries, and the Lake Ontario shoreline near the river mouth from 1858 to 1931; 3) land ownership in the watershed in 1860 and 1878; and 4) historical points of interest throughout the watershed. The project extracted information from a wide range of source materials, including topographical maps, detailed city maps, fire insurance plans and atlases, city directories, county atlases, and planning and conservation reports. Nearly two hundred maps were digitized over the course of the project, most of which have been made available to the public on the project website.[6]

BUILDING THE DON VALLEY HISTORICAL GIS

For librarians, involvement in a project like this engages new strengths in research and technical services and partnerships. But projects like this also draw upon librarians' long-standing expertise in creating and promoting free and open access to information. Sharing and disseminating information flows naturally from their mandate to serve the public. To the academic, the library may not only act as the technical arm of a project but also relieve the burden of personally archiving or disseminating resulting data.

Although GIS has become an obvious choice for historical geographic research, access to historical data sets continues to limit researchers' ability to use and apply the technology. Like most digital collections, for example, the Map and Data Library's digital geospatial data holdings contain thousands of datasets, but its historical datasets are still few in number. Because of these limitations in the availability of scalable and expandable data for original research, data creation is more often than not a significant component of these projects.

The first step in the creation of any historical dataset is the assembly of historical source materials. Here again, Canadian researchers face access challenges unique to the Canadian regulatory environment. The Don Valley Historical Mapping Project, for example, relied heavily on fire insurance plans and atlases to assemble information on the valley's industrial history. Fire insurance plans are richly detailed documents that provide scaled renderings of street grids, building outlines and construction materials, and other structural land uses (oil tanks, coal and lumber storage, number of boilers, etc.); they were produced for all parts of the old city beginning in the 1880s, and revised frequently. The larger-format and smaller-scale fire insurance atlases also proved valuable in places and periods where the more-detailed fire insurance plans were unavailable or inaccessible. Valuable as these documents are, they are also very difficult to access in digital form. As we discussed in the introduction to this volume, fear of copyright infringement has led many Canadian libraries and archives to restrict and in some cases eliminate the duplication of all fire insurance plans by researchers. These restrictions constrained the breadth and accuracy of our work. In some cases, rather than digital colour reproductions of the plans, we had to make do with black and white microfiche reproductions from the University of Toronto map collection. The legibility of these images posed problems in a few cases. Issues of quality arose in part from the process of reproduction: microfiche scanning is a complex process that

involves numerous experiments with resolution, image size, scanner surface placement, and automation. The nature of the source documents themselves also contributed to the quality of these reproductions. Fire insurance plans were often revised incrementally by pasting revised drawings over small portions of the original map. The result in many urban areas are layers of revisions and annotations covering several years. Held to the light, the original notations are often visible beneath the revisions. When photographed for reproduction, the revised areas of these plans are often difficult to decipher.

The nature of historical maps and the circumstances in which they were typically created also pose challenges for historical GIS projects. In creating and later interpreting historical datasets such as ours, it is important to bear in mind that the process of mapmaking in nineteenth-century Canada differed markedly from the process today. Maps from this period often took years to create, incorporating the process of surveying the sites, compiling information from each site or building, drawing the maps based on the information accumulated, and printing the final (dated) map. Once published, paper maps were not easily altered, as digitized maps are today, to reflect subsequent changes in landscape features. The potential for error compounded as maps were frequently used to create other maps. Features and information were copied from one map to another, and errors and changes to features were often long to be registered across a generation of maps. Within this context, historical maps and the data they contain are best understood as representing a date-range of several years, rather than the specific year of the map's publication.

Geographic extent is another issue that one must grapple with in working with historical sources. Reflecting the jurisdictional realities of the time in which they were produced, maps often present information in different "containers" than we are familiar with today: municipal boundaries differ; other boundaries, such as watersheds, are not apparent. This can have benefits and disadvantages in HGIS. In mapping changes to the river channel over time, for example, we found that many of our sources (city maps in particular) drew the river only as far as the forks, omitting the upper valleys. Our datasets reflect the sources available to us: many, especially for the nineteenth and early twentieth century, comprise information for the lower river alone. At the same time, the limited coverage of these city maps often resulted in more detailed representations of the river channel. Because the Lower Don formed the eastern boundary of the city for much of the nineteenth century, depictions of the bends and oxbows of its lower reaches are especially detailed on city maps from this period.

Finally, as in all historical accounts, the presence of competing representations of the same place and time adds complexity to the process of interpretation. Historical representations of the river channel are a case in point. Multiple and varying representations of "what was water" and "what was land" made it difficult to delineate the changing course of the river in different periods. Certainly, part of the uncertainty stemmed from the landscape itself: the river, especially in its lower reaches, was a dynamic landscape subject to transformation, not only from year to year, but from season to season. Seasonal meteorological events such as spring snowmelt, summer rainstorms, periods of drought, and ice jams in winter worked to

continually alter the land/water interface, as spring freshets submerged marshy wetlands and summer heat waves made previously water-logged channels passable by foot. Mapping valley lands, as a result, was not a simple assignment, and the maps that were produced, even within a short time period of each other, often differed in substantive ways. The individual objectives of mapmakers also influenced the representations they produced. Depending on the surveyor's assignment – to assess the boundaries of the river, for example, or to determine the extent of traversable, marketable land – what constituted "water" and what constituted "land" were represented differently.

The accuracy of our work depended in large part, therefore, on the assessments of early mapmakers. While we could judge the relative "soundness" of the map – for example, whether its features could be georeferenced with current landscape features, such as buildings and street intersections – we could not determine what was the "best" representation of the landscape that the mapmaker had before him. Take for instance the two maps in Figures 3.2 and 3.3. Both are from the late 1850s: the one on the left from 1857, and the one on the right from 1858. Both maps georeferenced relatively well to current features, but both present the course of the river differently. Had the course of the river changed in the space of one year, or was one map a more "accurate" representation of the landscape than the other? We can take a guess at the answer but we cannot know with certainty. Several years can pass between the time a particular place was surveyed and the time the resulting map is printed. Individual maps can also fluctuate in the accuracy of their representations across the landscape: historical and current features may, for example, align

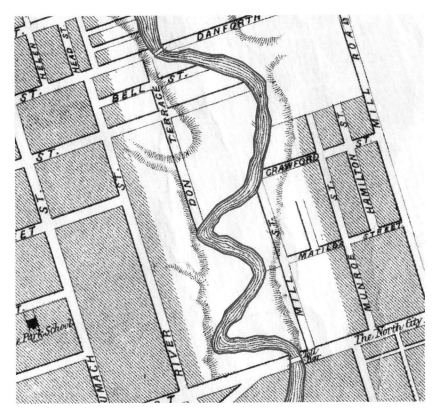

Figs. 3.2 and 3.3. Two
representations of the river
channel, 1857 and 1858.
(Map Sources: Toronto,
Canada West, Waterlow
& Sons. Lith. London,
1857; Boulton, W.S. Atlas
of the City of Toronto and
Vicinity, 1858 [courtesy of
Toronto Public Library].)

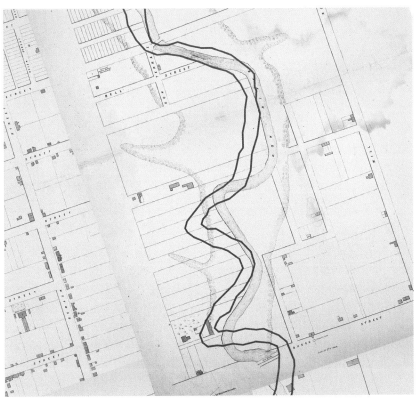

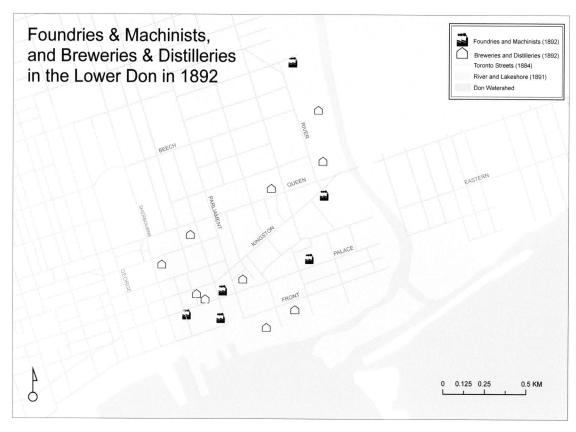

Fig. 3.4. Foundries and machinists, and breweries and distilleries in the Lower Don in 1892. (Sources: Don Valley Historical Mapping Project Database; Don Watershed boundary: Toronto Region Conservation Authority 2008; Toronto Historic Streets, University of Toronto Map and Data Library 2011.)

well in one section of the map, but not as accurately in others. To avoid making unsupportable statements of accuracy about these maps and their resulting data, we elected to include multiple representations in our database. Our data represent, therefore, not a definitive portrayal of the river in a particular year, but instead the information extracted from a certain map printed on a certain date.

In addition to the issues that arose from the nature of our source materials, we also faced decisions around how best to use GIS technology to represent past landscapes. The database we developed for the valley's industrial history

is a case in point. The database incorporated industrial sites visible in maps published between 1858 and 1950, identifying for each "point" on the map the industry's address, ownership, industrial category,[7] and the source map and year. Once complete, the database contained the potential to display industries not only by location but by industry type and map year. We could, for example, query the database to show the number of slaughterhouses represented on maps in 1891 or the number and location of breweries along the river in different periods (Fig. 3.4).

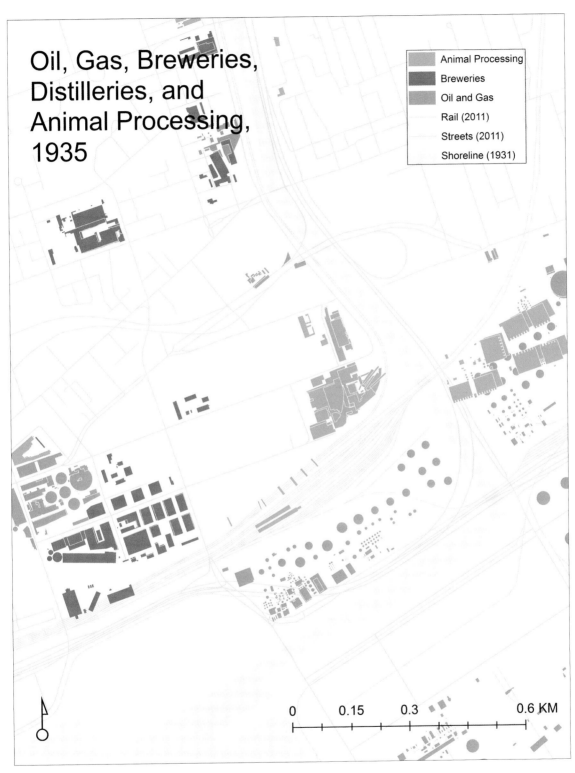

Oil, Gas, Breweries, Distilleries, and Animal Processing, 1935

	Animal Processing
	Breweries
	Oil and Gas
	Rail (2011)
	Streets (2011)
	Shoreline (1931)

0   0.15   0.3   0.6 KM

Fig. 3.5. Oil and gas, breweries, and animal processing, 1935. (Roads and Railroads from DMTI Spatial Inc. CanMap RouteLogistics 2011.3.)

While our point layer of industrial sites allowed for the efficient display of a large number of sites in different periods, it failed to take into account the variance in the "footprints" of different establishments. A mammoth site such as the Gooderham & Worts distillery, for example, which occupied several hectares near the river mouth, was represented with the same small point as a small tanning operation upriver. The solution to this problem lay in creating a "polygon" layer that traced the building outlines of industrial establishments. Fire insurance plans and atlases provide rich evidence for this work, in their documentation of surveyed building outlines and lots, and a range of specific information pertaining to industrial operations on these sites such as the volume of toxic liquid contained in tanks or the amount of coal or lumber stockpiled on site. As the example below shows, these industries displayed considerable variability in the size and relative land consumption of establishments. Once built, polygon layers provide a wealth of accessible data for researchers seeking to address a diverse range of research questions. By building polygons of oil tanks, for instance, including information from the fire insurance plans on the capacity of these tanks, models could be built to investigate the potential impact of pollutants on various industrial sites. Patterns of dispersal of hazardous material could also be investigated for specific known toxic sites.

Polygons also proved more appropriate than a simple line layer in mapping the river channel. We began by drawing the course of the river as a line layer for the year 1857. We soon realized, however, that, while easier to build, single lines failed to represent the variable width of the river and its tributaries in a useful way. We could draw lines to depict both shores of the river, but we could not shade these in to depict hydrography. Building the layers as both polygons and lines solved the problem. Displaying the river using polygons, however, brought its own complications. As discrete shapes, polygons are not easily used to represent continuity. Rivers, of course, are naturally connective in function. For the hydrographic data we were building to be useful, we needed to represent the Lake Ontario shoreline east and west of the river mouth. These polygons had to be stretched out into the lake and somewhat unnaturally squared off at the east and west extents of the historical maps they were drawn from.

When the database was complete, our coverage of the lakeshore spanned from Humber Bay in the west end of the city to just past Ashbridge's Bay in the east, with slight variations on this extent for years where maps were not available for the full area covered. The river channel was not the only ecological feature to change in this period. The Lake Ontario shoreline also fluctuated dramatically with dredging and land reclamation activities in the harbour and adjacent marsh. By extending our geographic coverage, we were able to document this changing shoreline along with the changing course of the Don River. In the end, we built the river and lakeshore layers using lines and polygons for nine "snapshots" between the years 1857 and 1931. These dates encompassed dramatic changes in the area surrounding the lower river, including the straightening of the river south of Winchester Street in the 1880s, the construction of Keating Channel in the 1910s, and the reclamation of Ashbridge's Bay to create the Port Industrial District in the 1910s and 20s. River layers also included a number of historical tributary creeks, many of which were buried or culverted as the city expanded.

The Don Valley Historical Mapping Project produced new insights on the environmental history of the river valley. New possibilities for interpretation emerged, not so much from our evidence (the sources we worked with were for the most part already familiar to historians), but from the collating and comparative functions of the method itself. The capacity of GIS technology to extract, group, and display data from diverse sources, to alter the scale of analysis, and to alternately highlight and suppress particular groups of data, allowed for detailed comparison of spatial information within and between selected time periods. Nowhere was this more evident than in the spatial history of the river channel.

In the late nineteenth and early twentieth centuries, a series of improvement plans directed at the lower reaches of the river produced dramatic changes to the course and character of the river channel. Among these was the Don River Improvement Plan, proposed by civic politicians in the 1880s with the goal of producing a sanitary and rational river landscape as a basis for prosperity. The plan, which deepened and straightened the river's serpentine lower reaches, aimed to transform flood-prone and polluted valley lands into a hub for industry and a driver of residential development on adjacent

table lands. Unravelling the timing and the extent of the material changes brought about by the 1880s improvements is an important facet of the environmental history of the river valley.

Most prominent among the sources that exist for the 1880s improvement is the 1888 "River Don Straightening Plan" (Fig. 3.6). The document provides a detailed projection of the future envisioned for the Lower Don, depicting the existing course of the river between Winchester and Eastern Streets, the proposed route of the straightened channel, and the lot numbers of properties to be expropriated. It is, nevertheless, a plan for a project that encountered numerous hiccoughs in implementation, both in terms of projected timelines and projected results. What did the river channel actually look like after straightening? What aspects of the project were completed, and what was omitted from the initial plans? How long did the project take? These questions were surprisingly difficult to answer upon reviewing existing source materials. Textual sources, such as newspaper articles and city council minutes, often failed to include precise spatial information. In other cases, accompanying maps had been lost or misplaced. Historical maps and plans presented conflicting information, furthermore, that made it difficult to track what changes had actually occurred on the ground, and when.

GIS technology proved unmatchable in documenting the environmental changes associated with the 1880s improvements. It enabled new insights about the timing of particular project components, and the disconnect that often existed between engineers' plans and timelines and what actually transpired on the ground. The capacity of HGIS to create layers of data for different periods was especially useful in

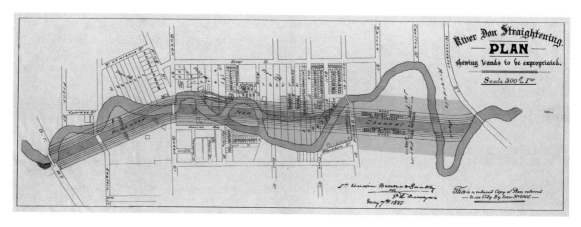

Fig. 3.6. River Don straightening plan. (Unwin, Browne and Sankey, Surveyors. May 7, 1888. City of Toronto Archives, Series 725, File 12.)

interpreting these changes, facilitating as it did the detection of patterns and anomalies over time – something very difficult to achieve in the one-to-one comparison of print or even scanned maps. This increased capacity is evident in Figure 7, which depicts the river's changing course in three different periods. The extent of the lakeshore is also visible in its different stages of reclamation during these periods. Without GIS technology, the overlay of historic data is both more difficult, and it lacks interoperability – the capacity to be used in a variety of software for different purposes – which is crucial for data sharing and reuse.

GIS also created interpretive possibilities in tracking smaller landscape changes unmentioned in the textual sources, and perhaps not readily apparent in a review of paper or digital maps of the period. References to isolated relocations of the river channel in association with the construction of the Don Valley Parkway in the 1950s and 60s, for example, are easily discernible by overlaying polygons of the river channel before and after highway construction and adjusting the scale of the map to view particular reaches of the river.

In other aspects of the project, the capacity to select out certain information for analysis (for example, the locations of oil refineries that established in the lower valley in the early twentieth century) and the ability to adjust the scales of analysis – to zoom out to see the large picture of industrial development in the lower valley, and the concentration of particular types of industries; to zoom in to analyze the outlines of individual buildings and their relationship with the river – provided opportunities for insights into the history of pollution in the river valley and the services the river provided in different places and periods.

A GIS of a river's history is, of course, only as good as the source material upon which it is based. While we would have liked to gain an appreciation, in spatial terms, of historical changes to the condition and character of the river, including depth and flow rates and levels of pollution and sediment, the absence of detailed historical source materials made this impossible. An understanding of the changing sensory experience of the river – its visual appearance, sounds, and smells – was another

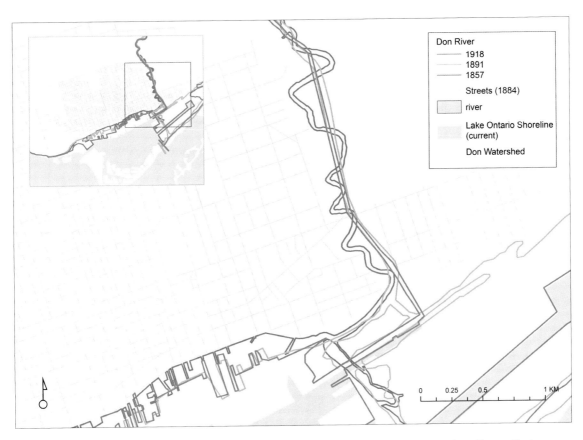

Fig. 3.7. River and Lake Ontario shoreline, 1857, 1891, and 1918. (Sources: Don Watershed boundary: Toronto Region Conservation Authority 2008; current shoreline: DMTI CanMap Postal Geography FSA Boundaries, v2008; Historical Roads [1818–1884]: University of Toronto Map and Data Library 2011.)

aspect of the river's past that would remain dependent upon scattered textual references often lacking in spatial specificity.

CONCLUSION

On average, about two hundred people per month visit the Don Valley Historical Mapping web pages (see maps.library.utoronto.ca/dvhmp/). Presentations on the project to local environmental and citizens' advocacy groups such as Lost Rivers and the Task Force to Bring Back the Don have generated substantial interest; these groups aim to incorporate project data and maps into their own planning and public education initiatives for the valley. People have used the data we created in ways we didn't expect. For example, the scanned map images and the Google Earth files, and not the GIS files, have been by far the most popular downloads. Students and the general public, rather than GIS researchers, have been the main beneficiaries of the project, and, consequently, the "web ready" and open format files such as our Google Earth data are downloaded with much greater frequency than the more stringent and difficult-to-use but highly versatile and powerful shapefiles.[8]

By digitizing hundreds of historical maps and producing over fifty features in GIS and Google Earth format, all freely available for download by the public and other researchers, the project accomplished its primary goal of making library resources more accessible. Its results demonstrate the possibilities opened up through collaboration. From the library's perspective, projects such as these showcase what libraries are all about. They justify the acquisition and maintenance of expensive collections and the hiring of skilled and academically inclined personnel that are dedicated to research and services. For the historian on the team, participation in such a project stemmed from the recognition that, despite the time and energy invested to learn HGIS methods, she was unlikely to become proficient enough to produce her own data and maps expediently. An understanding of how GIS works, however, and the particular challenges of HGIS did prove extremely useful in working with the data, and members of the project team, effectively.

The project presents numerous opportunities for expansion. A wider geographic expanse would provide a striking demonstration of the spatial distribution of industry across the city. More lost streams and rivers could be uncovered, bringing Toronto's history alive cartographically and enhancing, at the same time,

the appreciation of the map collections at the University of Toronto and other libraries. The project might also be expanded to include other important geospatial features, such as tree cover, topography, elevation change, agricultural development, and transportation and municipal infrastructure development.

Projects such as these can also fuel the development of other projects. They can feed into existing ones, such as the Ontario County Map Project currently underway,[9] and stimulate future projects. A commitment to open access to the data produced by projects such as this one is an essential part of this process. Agreements between institutions to share not only aggregated data but source materials such as archival maps and images is another way forward. Many historical GIS projects fail to get off the ground in part due to the prohibitive cost of acquiring digital map reproductions. A large proportion of the material we work with in urban environmental history is in the public domain but restricted from use because of cost-recovery mechanisms in a number of institutions that restrict reproductions without payment. Making this information accessible to researchers is the first step in enabling new historical knowledge. New relationships and new practices within archives and other collecting institutions are vital to the development of HGIS projects in Canada.

NOTES

1   For a detailed history of the Don River Valley and the various infrastructure projects that have altered the landscape of the river valley over time, see Jennifer L. Bonnell, *Reclaiming the Don: An Environmental History of Toronto's Don River Valley* (Toronto: University of Toronto Press,

forthcoming). See also Bonnell, "A Social History of a Changing Environment: The Don River Valley, 1910–1931," in *Reshaping Toronto's Waterfront*, ed. Gene Desfor and Jennefer Laidley (Toronto: University of Toronto Press, 2011), 123–50; Gene Desfor and Jennifer Bonnell, "Socio-ecological

Change in the Nineteenth and Twenty-first Centuries: The Lower Don River," in *Reshaping Toronto's Waterfront*, 305–25; and Bonnell, "An Intimate Understanding of Place: Charles Sauriol and Toronto's Don River Valley, 1927–1989," *Canadian Historical Review* 92, no. 4 (2011): 607–36.

2　H. Richardson, W. Chisholm, and J. G. Chewett, "Report of the Select Committee on the Improvement of the Harbour of York," in *Memorandum with Accompanying Plans and Documents Relative to the Past and Present State of the Harbour of Toronto* (Ottawa: Department of Public Works, 1881), Appendix, pp. 1–3.

3　For an excellent history of the reclamation of Ashbridge's Bay Marsh and the creation of the Port Industrial District, see Gene Desfor, "Planning Urban Waterfront Industrial Districts: Toronto's Ashbridge's Bay, 1889–1910," *Urban History Review* 17, no. 2 (1988): 77–91; and Gene Desfor and Jennefer Laidley, eds., *Reshaping Toronto's Waterfront* (Toronto: University of Toronto Press, 2011).

4　For a good overview of the history of Toronto and its development, see J.M.S. Careless, *Toronto to 1918: An Illustrated History* (Toronto: James Lorimer, 1984); James Lemon, *Toronto since 1918: An Illustrated History* (Toronto: James Lorimer, 1985). Bonnell details the history of the Don Valley Parkway and floodplain acquisition in the valley in *Reclaiming Toronto's Don River Valley*. See also Stephen Bocking, "Constructing Urban Expertise: Professional and Political Authority in Toronto, 1940–1970," *Journal of Urban History* 33, no. 1 (2006): 51–76; Wayne Reeves, "From Acquisition to Restoration: A History of Protecting Toronto's Natural Places," in *Special Places: The Changing Ecosystems of the Toronto Region*, ed. Betty I. Roots, Donald A. Chant, and Conrad E. Heidenreich (Vancouver: UBC Press, 1999), 229–41; Richard W. White, *Urban Infrastructure and Urban Growth in the Toronto Region: 1950s to the 1990s* (Toronto: Neptis Foundation, 2003).

5　On the history of the urban ecology movement in Toronto and the significance of valley green spaces, see Gene Desfor and Roger Keil, "Every River Tells a Story: The Don River (Toronto) and the Los Angeles River (Los Angeles) as Articulating Landscapes," *Journal of Environmental Policy and Planning* 2, no. 1 (2000): 5–23; Gene Desfor and Roger Keil, *Nature and the City: Making Environmental Policy in Toronto and Los Angeles* (Tucson: University of Arizona Press, 2004).

6　Several maps still fall under copyright and cannot be made available as digital copy. Other maps were obtained through licensing, which restricts their availability to the public. The Don Valley Historical Mapping Project can be accessed at maps.library.utoronto.ca/dvhmp/.

7　Although detailed classification codes exist for industries in the present (the North American Industrial Classification System [NAICS] and the Standard Industrial Classifications [SIC] are two examples), they are not easily applied to historical industry types, a number of which no longer exist, or are difficult to distill into a single category in the present. With this in mind, we devised our own classification system based on the industries represented in our source materials, and the themes that emerged from the valley's environmental history. These categories included: saw mills; paper mills; grist mills; breweries, distilleries, and their suppliers; foundries and machinists; oil and gas refineries and paint manufacturers; soap works; textile manufacturers and carding mills; other food production; other light manufacturing; fuel storage; building materials producers and suppliers; agricultural suppliers; general suppliers and warehousing; transportation; utilities and public works; chemical producers and suppliers; printers and lithographers; and animal processing.

8　Not widely known outside the GIS community, shapefiles are the most commonly used GIS format around the world. While the format is proprietary and owned by ESRI, most GIS software packages, including open source software, can read and write this format.

9　The Ontario County Map Project can be accessed at: http://maps.library.utoronto.ca/hgis/countymaps/. The project's goal is to compile land occupancy, cultural and physical information in the form of geospatial data, from nineteenth-century Ontario County Maps. The ongoing collaborative project is a partnership between the University of Toronto, the University of Guelph, Western University, and McGill University.

4

The Best Seat in the House: Using Historical GIS to Explore Religion and Ethnicity in Late-Nineteenth-Century Toronto

Andrew Hinson, Jennifer Marvin, and Cameron Metcalf

In 1881 Toronto's Knox Presbyterian Church underwent major renovations to the inside of the building. The pulpit was lowered, the gallery front changed to iron, the pews comfortably upholstered, the entire floor carpeted, and the ceiling repainted and decorated. Most significantly, the seating arrangement was changed from the traditional straight-benched pews in formal order to a modified amphitheatrical layout with semi-circles forming around the pulpit. The realignment of the church's pews necessitated a reshuffling of where congregants were seated. Yet if contemporary reports were true that the refurbished auditorium was "virtually a new room" and in terms of artistic arrangement and taste "second ... to no church in Toronto,"[1] this congregant reshuffling was only a minor inconvenience.

In choosing a new pew, church members had to decide not only where they wished to be seated but also how much they were willing to pay for the privilege. Pew rents were a long-established mechanism for generating church income and enabled parishioners to contribute an amount in keeping with their means. From 1881 Knox had five levels of rental, the amount reflective of the pew's proximity to the pulpit. With the exception of those at its side, which were rented at $1 per quarter, all seats within the body of the church were $1.25. In the gallery all front rows were also $1.25, with the second, third, and fourth tiers along the length of the church being 80¢, 60¢ and 50¢, respectively, and those facing the front $1. Where everyone opted to sit was recorded in the Knox Presbyterian Church Pew Rent Book, 1882–1887.[2]

While surviving pew rent books are relatively rare, the value of documents that provide insight into the lives of "ordinary" people have long been recognized by social historians, including those focusing on religious history who since the 1980s have been primarily concerned with the "view from the pew" rather than a top-down history.[3] Organized by pew number, the book clearly shows who sat next to whom, how much rent each entrant paid, and the duration in which they remained in that place. Arrivals to and departures from the church can easily be determined, as can movement within the church, the pew rent book recording when and where a change in pew occurred. Also preserved are details of those who paid the rent, which in most cases would be the head of family, including their name, occupation, and address. For example, throughout 1882 sitting in pew 44 (a $1.25 seat in the body of the church) was George Noble, a merchant residing at 701 Yonge Street, who sat next to John Ritchie, a plumber living at 189 Jarvis Street. Both paid for three sittings at a cost of $3.75 per quarter, meaning each was accompanied by two companions, possibly a wife and child, or an elder dependant. Although the disappearance of historical blueprints of the church interior make it impossible to determine the exact layout of the church, it can be surmised that both men were surrounded in some fashion by Alexander Cameron, a barrister at pew 42, John Sinclair, another merchant at 43, James Fleming, a seedsman and florist at 45, and A. M. Smith, a wholesale grocer at pew 46. As such, the data from the pew rent book can be used to shed considerable light on the Knox congregation, fitting alongside the work of other historians and social scientists in Canada who have used church memberships and similar types of records.[4]

Yet as well as being located within a few seats of each other, these men shared something else in common. Although ethnicity was not recorded in the pew rent book, by linking them to the 1881 Canadian census we discover that they were each Scottish, a finding consistent with another study, which showed a high proportion of Scots among church elders and managers from Toronto Presbyterian churches.[5] This is not particularly surprising considering Presbyterianism originated primarily in Scotland, and Knox Church itself was established after members broke away from Toronto's St. Andrew's Church following a major schism in the Church of Scotland in 1843. The church also had a connection with the city's Scottish Gaelic community, playing host to the Toronto Gaelic Society's bible classes and holding occasional Gaelic sermons. Although not all Scots were Presbyterian, and not all Presbyterians living in Toronto were Scottish, Knox, along with most Presbyterian churches in the city, did maintain a strong Scottish character and, as with the relationship between other ethnic groups and their religious buildings, there is much evidence to suggest that these churches formed the core of Toronto's Scottish community.[6] Indeed the correlation between the Scots and the Presbyterian Church is arguably of greater significance than similar connections with other groups owing to the fact that, in almost every other respect, Scots blended into the wider Toronto population.[7] This was particularly so regarding their geographic concentration. Whereas Little Italies and Chinatowns were readily identifiable ethnic enclaves, there was no such common equivalent among Scots in Toronto. While ethnic trappings such as cafes, grocery stores and restaurants, travel agencies, and other services were apparent in other ethnic neighbourhoods, the same does not appear to have existed in any one particular area for the Scots.[8] Although there were Scottish clubs and societies throughout the city, none enjoyed the same level of membership nor were any as pivotal to Scottish identity as the Presbyterian Church.

The pew rent book is therefore of value to both religious and ethnic history and, by using traditional historical methods, could help to inform in both these areas. It is, however, by combing the data from the pew rent book with GIS technology that genuinely significant steps can be taken in providing new insights into previously unexplored relationships between where congregants lived and their place of worship. With the use of GIS, residential patterns can be examined, which not only inform about the geographic dynamics of Toronto's decentralized Scottish community but enable spatial questions of a Canadian church congregation that otherwise could not be answered. Whereas the drive for a "bottom-up" history of religion in Canada has led to considerable advances in our understanding of church demographics, the role of family and gender in worship, and the cultural history of religion, to date there are few detailed geographic analyses. Even where studies of religion and ethnicity intertwine, these have yet to take advantage of the potential of GIS.

The research in this chapter shows how GIS can make a significant contribution to religious and ethnic history. In presenting our findings, we have two aims. The first is to provide a general spatial analysis of the Knox congregation using the address data for the period between 1882 and 1887. What we find is a church that drew its congregants from across the city. Congregants demonstrated considerable loyalty to the Knox community, choosing Knox over other Presbyterian churches closer to home. This study also reveals a dynamic social space, both within the church, where congregants changed pews regularly, and outside it, as a large number of congregants moved house, some more than once, within the period of study. The second objective is more specific and pertains to the unusual insight offered by the pew rent book into where congregants in the church were seated. Recall the merchant George Noble, who was seated next to John Ritchie, a plumber. On the one hand, it may be surprising to find two individuals from different socioeconomic classes seated next to each other; on the other, the church was a setting where, at least in theory, social divisions did not matter. Using GIS, we explore the socioeconomic dynamics within Knox and show how they manifested themselves outside the church building.

METHODOLOGY

Detailing all quarterly payments for seats occupied between 1882 and 1887 along with those who made them, the pew rent book provides a fascinating insight into a church community in this period. What is not given are the names and details of family members for whom payments were made. In order to capture this information, Communion Rolls were used as a second, complementary data source. Published in the church's annual reports, these lists include the name of each church member, their address, the church district in which they lived, and whether or not they were receiving communion. As well as providing details of the excluded family members, the primary advantage of introducing this second source is the inclusion of addresses for all church members throughout the six-year period, which unlike those in the pew rent book, appear to have been kept up-to-date.

The data from both sources were entered into spreadsheets and then imported into a relational database, which was required to link the two datasets. Having the data stored in a database also enabled anomalies to be corrected, including variations in first names, where, for example, "Wm" and "William" might both appear as separate entries for the same individual. The completed database contains a total of 785 persons in the pew rent data set, 2,736 in the Communion Rolls, with 389 linkages being made between the two sources, a valuable analytical tool in its own right. It also provided a secure repository for the data, protecting the investment of time and effort associated with data collection, revision, and consolidation for the project. Finally, it ensured compatibility with GIS software, enabling the creation of tables that could in turn be imported into a GIS environment.

To accurately reflect Toronto in the late nineteenth century, a modern street network dataset was backdated to 1884 using contemporary fire insurance plans as a reference, by altering the physical road layout, street names,

and address ranges. The addresses from the database were then geocoded, a process by which addresses are plotted onto a map surface, with a match rate of 87 per cent (7,383) matched, 7 per cent (594) tied or possible match, and only 6 per cent (509) unmatched. While the database administration and the historic roads layer preparation were both time-consuming pursuits, the high match rate meant that any results produced through spatial analysis could be regarded with considerable confidence. Throughout the project various GIS techniques were used. All the maps were generated using GIS and several different tools were employed to analyze the spatial data that contribute to our findings. The project utilized a diverse range of skill sets brought to it by a project team that consisted of a database administrator, a GIS librarian, and a historian.

A COMMUNITY OF THE MIND

Having grown steadily since its establishment in 1843, from the 1880s Knox underwent unprecedented growth. These changes reflected Toronto's dramatic rise in population in the last decades of the nineteenth century. From 86,000 residents in 1881, Toronto's population more than doubled to 181,000 by 1891. Although this was in part due to annexations of some of the city's neighbouring districts, the onset of industrialization precipitated the arrival of many newcomers in search of employment opportunities. An increasing number of Toronto residents, new and established, chose Knox Presbyterian Church as their place of worship.

During the six-year period covered by the pew rent book, the church's population climbed from 1,093 in 1882 to 1,624 by 1887. That so many new members could be physically accommodated was due to the increased seating capacity brought about by the church's timely renovations. Certain administrative changes did, however, have to be made. Most notably the number of church elders and deacons who were responsible for the care and oversight of the congregation had to be increased. All congregants were assigned to a church district, each with their own elder and deacon. The boundaries of these districts were printed alongside the Communion Rolls and can be recreated in a GIS to visualize how these administrative units changed over time (Fig. 4.1). As well as revealing the creation of three new districts over the six years, the map can also be used to show the proportion of congregants living in each district. Comparisons of the before (1882) and after (1887) images demonstrate that in redrawing the boundaries and establishing new districts, the church's administrators sought a more equal distribution of population among them. This is clearly illustrated by examining district 12, which in 1882 contained an above-average proportion of the church's population (between 9 and 12 per cent) but was subsequently split in two (with the creation of district 17), both of which by 1887 contained less than 6 per cent of church members. What is also evident is that, along with a growth in congregation size, there was a considerable increase in the church's geographical catchment area. Whereas, in 1882, the districts' northern boundaries stopped at Bloor Street and did not go much east of Berkeley Street, within five years they incorporated parts of the Rosedale and Yorkville neighbourhoods

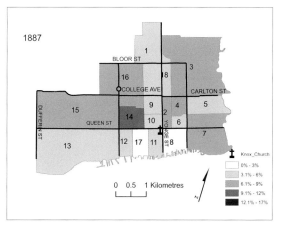

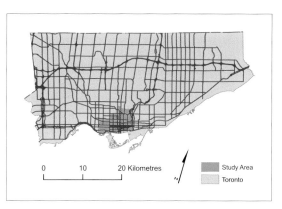

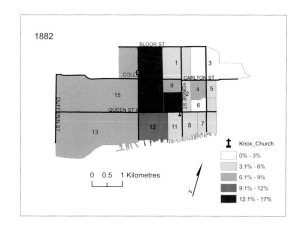

Fig. 4.1. Percentage of Knox congregants by church district. (Sources: 1882 and 1887 Church Districts: created using description of districts in Knox Presbyterian Congregation Rolls 1882–1887; Knox Presbyterian Church Congregants: Knox Presbyterian Congregation Rolls 1882–1887.)

in the north, and stretched to the Don River in the east. In total the church districts went from covering an area of fifteen square kilometres to twenty.

The specific whereabouts of the congregation by 1887, as detailed in the Communion Rolls, is shown in Figure 4.2. Confirming the extent to which the membership was dispersed across the city, it is apparent that Knox cannot be considered a neighbourhood church. This is contrary to what may be assumed in an age when the primary means of getting to church was on foot. Although some may have relied on private carriages or hired hacks (taxi drivers), streetcars were the city's primary form of transport and were utilized by all classes of the population but would not be introduced on Sundays for at least another decade.[9] GIS can be used to measure distances, and by doing so it can be calculated that on average parishioners travelled 2.4 kilometres to and from church, and for those living in the city's outer limits, it could involve a round trip of up to ten kilometres.

In other respects, however, it is not unexpected that the Knox congregation mostly came from outside the church's immediate proximity. When the eighty-three members broke away from St. Andrew's to form Knox Presbyterian Church, it was because of differences they had with its parent body, the Church of Scotland. As the city's only representative of the Free Church of Scotland, its congregation was probably never confined to its immediate geographic locality. The significance of this, however, should have lessened after 1861 when the Free Church Synod in Canada and the United Presbyterian Church (formed following an earlier Church of Scotland schism) merged to form the Canadian Presbyterian

Church, and even more so after 1875, when the Church of Scotland in Canada was brought into the fold to form the Presbyterian Church in Canada. In theory these changes eliminated any of the denominational differences that had previously existed, leaving no theological reason to prevent Toronto's Presbyterians from attending their local branch. Yet this appears not to have happened. Within its historical records, the only real reference to Knox's immediate surroundings around this time pertains to its missionary activities. In a sermon to the Knox congregation to mark the Rev. H. M. Parsons' tenth anniversary at the pulpit, the minister made an appeal for a greater voluntary effort from his parishioners, stating: "The field around this church building is more needy than ever.… The city between Queen Street and the Bay, from Sherbourne to York, is our field with no one else to till it."[10] He made reference to the good work being carried out by the church's mission on Duchess Street but appealed for another on York Street, where a special effort was needed within what he described as a "leprous portion of the city." While the church may have accepted responsibility for this part of the city, few of the church's congregants actually lived there.

As well as displaying the distances congregants lived from the church, Figure 4.2 shows that, in making their way to Knox, many had to pass other Presbyterian churches on route. By using GIS to calculate which Presbyterian church parishioners resided nearest, it is found that only 145 members (14 per cent) lived closer to Knox than another Presbyterian church. Considering the close proximity of some of the churches, a degree of membership crossover would be expected, but the extent to which this occurred reaffirms that any relationship

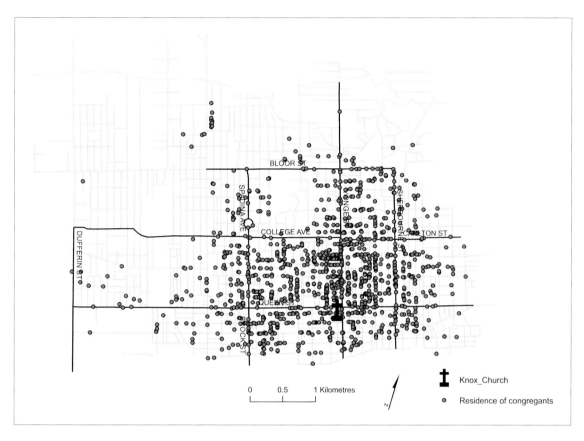

Fig. 4.2. Residential distribution of Knox congregants, 1887. (Sources: 1884 Toronto Streets: Adaptation of DMTI rte 2010; Knox Presbyterian Church Congregant Addresses: Knox Presbyterian Congregation Rolls 1882–1887.)

between Knox and its immediate geographical community was minimal.

In *Streets of Glory: Church and Community in a Black Urban Neighbourhood*, American sociologist Omar McRoberts has shown a similar lack of attachment between parishioners and the surrounding area of their church building in Boston at the turn of the twenty-first century.[11] Rather than being connected through geography, he argues congregants shared something else in common, such as ethnicity, class, lifestyle, or political orientation. This helps to explain another recent observation by religious commentators, that of "shopping for Faith," where worshippers find where they

are most comfortable by trying out not only different branches of the same denomination but different religions altogether.[12] While this may be recognized among social scientists as a current trend, this study indicates that it is not a new one. Hannah Lane's demographic analysis of small-town New Brunswick in the mid-nineteenth century shows that there was a high degree of religious fluidity with worshippers moving frequently among the major and minor Protestant denominations, but what we see here is that churches of the same Protestant denomination were also theoretically in competition with each other for members.[13]

The question that arises is what specifically attracted worshippers to Knox.

Of the possible commonalities between church members listed by McRoberts, several may have been applicable to Knox. Ethnicity was undoubtedly one factor, and in the case of Knox's nearest Presbyterian neighbour, Cooke's Church, it can be used to explain why one church was chosen over the other. Cooke's was established in 1851 after a group of Irish members left Knox to form their own congregation. What led to this break remains a mystery, but, from the outset, ethnicity was an important part of the breakaway church's identity, its name being taken from a key figure in the development of Irish Protestantism, and the membership and ministers being mostly of Irish origin.[14] But while Cooke's may have been the city's only Irish Presbyterian Church, Knox was certainly not alone in being of Scottish character. Regionalism offers another possibility. With Knox playing host to city's Gaelic Society bible classes and occasional Gaelic sermons, there could have been a link to Scotland's Highland community, but beyond the Gaelic connection there is little evidence to support this.

Although there are no data to support an analysis of political orientation (another possible commonality), John Moir, author of the most authoritative history of Presbyterianism in Canada, argues that members of the Established Church of Scotland were more likely to support the Conservative party in politics, whereas the Free Church Secession groups usually could be counted as Liberals.[15] Separate denominations within the Presbyterian Church in Canada no longer existed, but it is quite possible that erstwhile traditions lived on.

The inclusion of occupational data in the pew rent book makes it possible to analyze the socioeconomic characteristics of the church. Occupations are given in the pew rent book for 270 congregants, which have subsequently been divided into occupational classifications.[16] As Table 4.1 demonstrates, occupations that fell into the skilled non-manual category were greatest in number, followed by skilled manual, and then professional. Although there were members in the unskilled and semi-skilled categories, these groups were underrepresented with a combined proportion of only 12 per cent. There are several possibilities as to why this was. First, only 34 per cent of entrants in the pew rent book have their occupation listed next to them, and it is probable that those with higher status occupations would be more inclined to share this information than those at the opposite end. In Scotland, where lower attendance has also been found among the poorer classes, several explanations have been offered, including alienation through the implementation of church discipline based on middle-class values, or simply being too poor to attend.[17] As well as not possessing "Sunday best" attire, the demand for pew rents was given by the urban poor as a reason for not attending church. A Royal Commission into Religious Instruction in Scotland carried out in 1836 stated:

> The dislike of the people to occupy low priced or gratuitous sittings, avowedly set apart for the poor, which in general such as to make those who occupy them marked and distinguished from the rest of the Congregation, and the inferior nature of the accommodation provided for them, operate in preventing attendance.... [W]hile it may

not be difficult to supply himself with a sitting, a poor man is frequently unable to pay for adequate accommodation for himself and his family.[18]

While exceptions exist, such as the case of Mrs. Adams of 18 Ord Street whose fees were waived because of her inability to pay, Knox did not assign general free seats. The church did of course have variations in pew rents, and it is possible that there was a stigma attached to sitting in the cheaper seats. Rosalyn Trigger's study of Protestant churches in Montreal discusses moves to abolish pew rents at certain churches in the late nineteenth century specifically because they made distinctions on the basis of wealth, and, at Knox, some of the movements between pews that took place from 1882 to 1887 suggest that ability to pay did influence where people sat.[19] Mrs. Hunter of 149 Sherbourne Street, for example, started out at a $1.25 seat (pew 123) but moved to a $1 pew (pew 183) when the number of sittings she was paying for increased from one to three. William McFarlane, on the other hand, when paying for three sittings, was located at a $1 pew (pew 11) but moved to a $1.25 seat (pew 107) when the number of sittings he was paying for reduced to two. Money does not however appear to have been the only factor influencing why people moved seats. One of the most striking features of the original pew rent book is the extent of transiency between pews. Overall, 132 separate rent payers changed pew between 1882 and 1887, many of whom moved more than once. A clerk named James Donaldson was recorded as moving pew no fewer than five times, shifting inconsistently between the $1 and $1.25 seats. With a skilled non-manual occupation and no

Table 4.1:
Occupational Categorization of Knox Congregants, 1882–87.

	No.	%
Professional	59	21.9
Skilled non-manual	95	35.2
Skilled manual	84	31.1
Semi-skilled	20	7.4
Unskilled	12	4.4
Total	270	100

Table 4.1: Occupational Categorization of Knox Congregants, 1882–87. (Sources: occupation: Knox Pew Rent Books 1882–1887.)

apparent dependents, it is unlikely that cost was an influencing factor in where he sat.

Just as it would involve little more than guesswork to provide a reason as to what prompted James Donaldson to move around the church with such regularity, explanations of what brought the Knox community together are equally speculative. Individuals may have preferred the church for reasons previously mentioned; they may, too, have made their choices based on much less quantifiable reasons, such as a shared preference for the Rev. Parsons' preaching style or the comforts of sitting on a cushioned as opposed to a hardback wooden pew. Indeed, one of the purposes behind the Knox renovations was to boost church membership, an acknowledgment that, even in 1882, aesthetics (and comfort) mattered.[20] Another localized issue that may have affected membership was the use of organ music. A hotly debated topic among the Presbyterian Church in Canada's General Assembly, the issue was passed down to individual congregations to decide for themselves. At Knox, the

Fig. 4.3. Number of Knox congregants by address, 1887. (Sources: 1884 Toronto Streets: Adaptation of DMTI rte 2010; Knox Presbyterian Church Congregant Addresses: Knox Presbyterian Congregation Rolls 1882–1887.)

use of organ music was first raised in 1873 but was consistently voted against until 1878, after which a further three years passed before an organ was finally installed. That the controversy was long to abate is suggested by a *Telegram* reporter who, after recounting the offering being taken after the service "in silence," commented: "no doubt this unusual custom is a concession to those who yet, *in spite*, oppose the use of the organ."[21] While some clearly opposed the bringing of music into the church, for others it may have been what enticed them to Knox over other places of worship.

Finally, one of the most likely reasons that people chose Knox over other places of worship was because of family. Although neither the pew rent book or Communion Rolls indicate family relationships, Figure 4.3 gives an indication of the multiple-person households among the congregation. Not all, but probably most, of these households were made up of families. Furthermore, not all family members lived in the same household. Scottish households in Toronto consisted of mostly nuclear families, and it is quite possible that many of the congregants were part of extended families who attended the church. More than just a place of Sunday worship, church for Scottish Presbyterians formed an important part of their lives. As such it makes sense that families attended the same church and more generally that, in picking where to become a member, parishioners did not simply choose the church closest to home.

A TRANSIENT COMMUNITY

In spite of the considerable growth Knox experienced between 1882 and 1887, the church also lost many members during this time. The fluidity of church membership can be seen in the statistical tables included in most of its annual reports. In 1883, for example, although 138 new communicants were added to the roll during the year, the net increase was only 56.[22] That it was not higher was due to nine deaths, thirty-one members being placed on the retired list due to absence, and forty-two being removed by certificate to other churches. Over the entire period between 1882 and 1887, of the 311 heads of household recorded in the pew rent books at the beginning, less than half (189) were still there in 1887. According to the Communion Rolls, 433 joined the church in this period, and 250 left. The destinations of those who left the church are largely unknown. An 1885 annual report for Knox, however, grants us a small insight in stating the destinations of those parishioners who were granted certificates to join another church. Some were leaving the city, such as Alice Brodie, who was destined for Edinburgh, Scotland; a further seven were headed to the United States and

four to other parts of Canada. Most, however, (twenty-four) were to remain in Toronto, splitting themselves between Charles Street, West, Cooke's, Leslieville, St. Andrew's, Old St. Andrew's, Parkdale, College Street, and Central Churches. In at least three cases, the moves seem to have been the result of a marriage; for others, the reasons for leaving are as debatable as those that brought congregants to Knox in the first place. That no one church was favoured would suggest that individuals considered a range of factors in making their decisions. This level of transiency was not unique to Toronto and, as Peter Hillis' study of church membership in Glasgow would suggest, was part of a much wider phenomenon. Of the 2,481 members recorded on the Barony of Glasgow Communion Roll between 1879 and 1883, 668 people joined the church in this period while 713 left.[23] As with Knox, some of the members who left did so for destinations far afield while others moved on a more local scale, remaining within Glasgow.

Those who remained in Glasgow, Hillis argues, left the church because they moved house. While this may account for some of the membership turnover at Knox, as has been shown, close geographical proximity to the church building was not of great priority to its members. Furthermore, it can be seen from the Knox Communion Rolls that many congregants who did move house continued to worship at Knox. In total 759 parishioners (28 per cent) moved house and remained at Knox, and many did so several times in the period of study. This high level of mobility is consistent with other North American studies of urban centres. Howard Chudacoff's examination of residential mobility in Omaha found that, between 1880 and 1920, only 3 per cent of his case

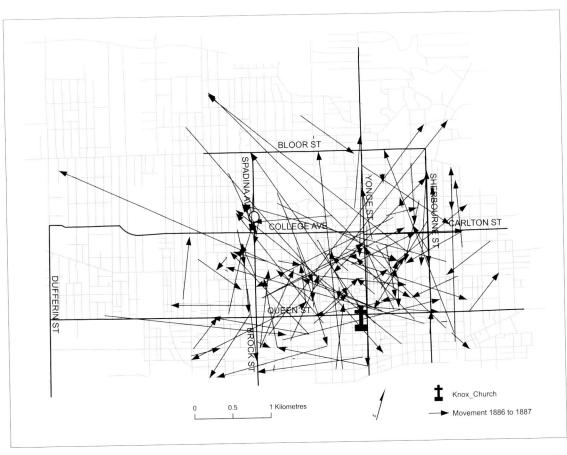

Fig. 4.4. Tracking residential movement of Knox congregants, 1886–87. (Sources: 1884 Toronto Streets: Adaptation of DMTI rte 2010; Knox Presbyterian Church Congregant Addresses: Knox Presbyterian Congregation Rolls 1882–1887.)

study lived in the same place for as long as two decades. Slightly under half moved away from the city within five years and over two-thirds left within twenty. Of those who remained in Omaha for twenty years, the overwhelming majority occupied three or more homes.[24] Closer to Toronto, Michael Katz's study of Hamilton, Ontario, has shown a similarly high turnover in population. Looking at the decades between 1851 and 1871 in his study, only 6 per cent of those living in Hamilton at the end of this period had been there twenty years earlier. Of those who were there at the beginning of each decade, about two-thirds of the entire population and over one-half of household heads left Hamilton in the ten years that followed.[25] What Katz's study does not take into consideration, however, is the levels of transiency amongst those living within the city.

By separately plotting the addresses from each of the annual Communion Rolls, GIS makes it possible to track the movements of parishioners who changed residences from year to year. Figure 4.4 shows the relocations between 1886 and 1887, the year in which there was greatest movement. Not only does this reinforce the extent of transiency, but the map gives a sense of where people were moving.

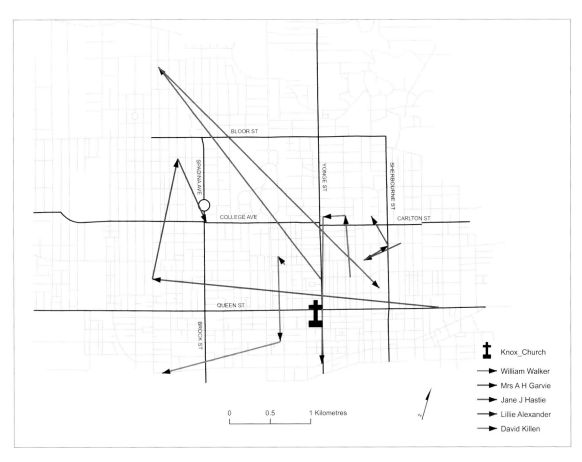

Fig. 4.5. Residential movement of selected Knox congregants, 1882–87. (Sources: 1884 Toronto Streets: Adaptation of DMTI rte 2010; Knox Presbyterian Church Congregant Addresses: Knox Presbyterian Congregation Rolls 1882–1887.)

What is striking is the lack of discernible pattern, there being no consistency to the direction or distances moved. This is not unique to this year, and when the movements for each consecutive year are compared (not shown), the only notable trends are the general increase in people moving, which is most likely a reflection of the congregation's growth, and a greater average distance being moved each year.

Even more intriguing are those congregants who moved more than once. Chudacoff's study of Omaha recognizes that multiple moves took place over two decades. Some Knox congregants had different addresses for each of the six years under examination, suggesting that the levels of transiency within Toronto may have been even greater. This should be of particular interest to urban historians and is a good illustration of how the findings of a narrowly focussed case study such as this can inform more widely. Focussing on the individual movements of several of those parishioners who relocated more than once between 1882 and 1887, Figure 4.5 gives a sense of their contrasting relocation patterns. Of those shown, the shortest overall distance moved was by Mrs. A. H. Garvie: although she relocated on three occasions over those six years, the accumulated distance between houses was little over 1.2 kilometres. Clearly in moving house, Mrs. Garvie made a conscious effort to remain in the same area. Elizabeth Platt, on the other hand, moved a total of seven kilometres in only two moves. As can be seen though, having moved to the outskirts of the city, where she stayed for two years, she subsequently returned to within a few hundred metres of where she originally resided. Both David Killan and William Walker can be seen making two relatively localized moves before a more significant shift, in the case of Killan considerably to the east of the downtown core. Conversely Lillie Alexander begins with a significant move across the city, before two further localized moves.

Why were people moving in these ways? Michael Katz warned that "the search for tidy reasons to explain why some men moved and others did not will never succeed." Even still, economic reasons are generally seen as being at the heart of the decision to move.[26] On the one hand, this could involve moving to be nearer a place of work, or as a recent study by William Jenkins on the Irish in Buffalo argues, it could reflect a change in occupational status.[27] Using a sample of Irish heads of household, Jenkins traces them at five-year intervals from 1881 to 1911 using city directories. Consistent with the two previously mentioned studies, half of these were not traceable in 1881, and by 1911 there were details for only seventy-four households. For those who could be traced, however, Jenkins's analysis shows a spreading out of families from a distinctively Irish and working-class area of the city to those with mixed class and ethnicity. Although he cautions against overdrawn conclusions of a blooming Irish American "middle class," he does suggest that Buffalo's west side emerged as the "choice destination" for the city's Irish working class.[28] While the overall inconsistency of the movement among the Knox parishioners makes it difficult to identify a similar pattern, it is possible to explore movement among those living in one of the city's poorer areas. We tracked the movement of sixty-one parishioners who either moved to, from, or within Knox's missionary area. Of these, only seven moved from one part of the zone to another, while a total of sixteen moved in and thirty-eight moved out. As can be seen in Figure 4.6, although there is no consistent destination, most made significant moves away from the city's downtown core. While this could be an indication of upward social mobility, the apparent randomness of their destinations suggests that other factors were at work. In his study of Omaha, Chudacoff found that the most important factors affecting the decision to move were household and environmental needs. As a family passed through its lifecycle, those needs changed, and the inflexibility of any one home in meeting these needs produced residential turnover. Pointing to a mid-twentieth-century survey of American urban dwellers, Chudacoff notes that the prime complaint against a former residence was lack of closets and lack of open space. "In other words, when a man made the decision to move, he had in mind a series of optimal specifications concerning the quality of and amount of space in his new and still unchosen residence."[29] The inconsistency in the pattern of residential mobility among Knox parishioners would suggest that the decision to move was an independent choice based on factors pertinent to each individual.

More research is required to determine the specific reasons as to why people moved when and where they did. While GIS cannot alone

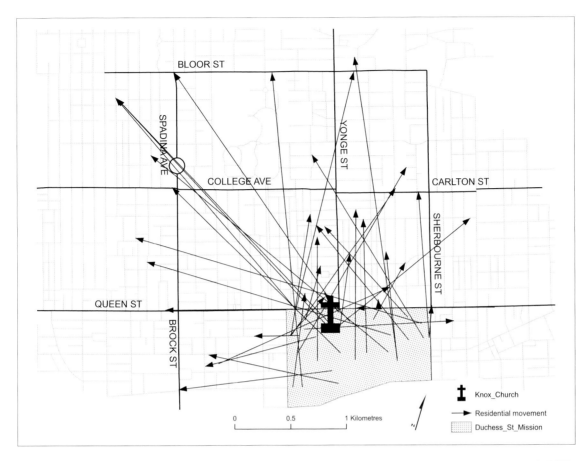

Fig. 4.6. Residential movement from Knox missionary zone, 1882–87. (Sources: 1884 Toronto Streets: Adaptation of DMTI rte 2010; Knox Presbyterian Church Congregant Addresses: Knox Presbyterian Congregation Rolls 1882 –1887.)

answer these questions, it is because of GIS that our understanding of transiency among the Knox congregation is considerably greater than simply knowing its extent. By being able to project onto a map the destinations to which individuals moved, we have gained a significant insight into, not only the Knox community in the late nineteenth century, but Toronto urban history more generally.

AN EGALITARIAN COMMUNITY?

Although upward social mobility does not appear to have been the primary factor behind moving, it does return us to the issue of social status. As we have shown, one of the values of the pew rent book is its insight into the seating arrangements of the church, which, paired with its accompanying occupational information, gives the potential for examining how the socioeconomic dynamics of the congregation manifested themselves within the church

Table 4.2: Occupational Categorization Cross-tabulated with Quarterly Pew Amount, 1882–87.

	Quarterly pew amount					Total
	50¢	60¢	80¢	$1	$1.25	
Professional	0	2	0	8	115	125
Skilled non-manual	4	3	3	22	91	123
Skilled manual	3	7	18	61	84	172
Semi-skilled	0	0	0	8	19	27
Unskilled	0	0	0	3	9	12
Total	6	12	21	102	318	459

Table 4.2: Occupational Categorization Cross-tabulated with Quarterly Pew Amount, 1882–87. (Sources: Pew rent amount and occupation: Knox Pew Rent Books 1882–1887.)

walls. It has already been shown that there were socioeconomic differences among parishioners, but as our example of the plumber seated next to the merchant suggests, these may have mattered little. Further scrutiny using the pew rent data (Table 4.2) reveals that, while those in the top occupational category (professional) were predominantly seated in $1.25 seats, the converse was not true for those in the lower levels. Most congregants with skilled non-manual occupations were also in $1.25 value pews, as were almost half of those in the skilled manual category. Clearly socioeconomic status had some bearing on where one sat but not to the extent that it could be considered a barrier to being seated next to a higher socioeconomic grouped parishioner, and certainly not to enjoying a good view of the alter. Before concluding that Knox Presbyterian Church was a place where socioeconomic realities could be left outside, however, it is important to determine if this was something that occurred only inside the church or if it was part of a more general phenomenon. That professionals and manual workers sat alongside each other could in fact

be a reflection of Toronto's Scottish community or Toronto as a whole. It must therefore be ascertained if there was any evidence of class "segregation" outside the church.

Unlike Buffalo, Toronto did not have a clear frontier dividing social classes, but as with all Victorian cities the extremes of rich and poor were all too evident. Charles Pelham Mulvany's *Toronto: Past and Present*, published in 1884, describes some of the city's main arteries. Among the most elite addresses were the "sumptuous private residences" of Rosedale, closely followed by Jarvis and Sherboure Streets, both of which were lined on either side "by the mansions of the upper ten." In contrast were Elizabeth Street with its "unsavoury appearance and repute," Centre Street, "another slum," and worse still York Street, which according to Mulvany was "occupied by dingy and rotten wooden shanties."[30] By mapping the pew rent payers based on their occupational groups, it is possible to determine the extent to which these correlate to Mulvany's evaluation.

The GIS techniques used so far have mostly involved producing maps that can

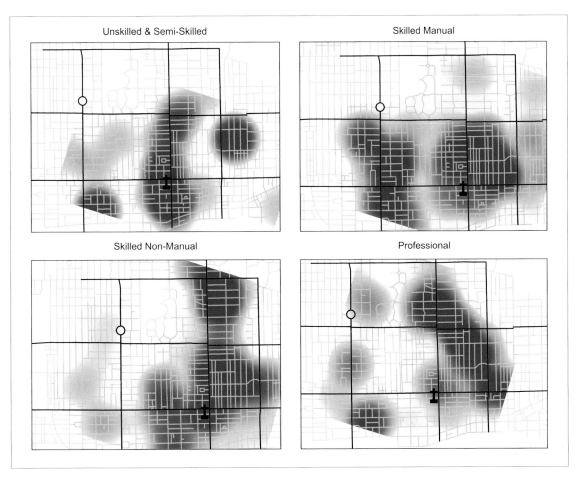

Fig. 4.7. Residential density of Knox congregants by occupational categorization, 1882–87. (Sources: 1884 Toronto Streets: Adaptation of DMTI rte 2010; Knox Presbyterian Church Congregant Addresses: Knox Presbyterian Congregation Rolls 1882–1887; Pew rent amount and occupation: Knox Pew Rent Books 1882–1887.)

subsequently be analyzed through observation. While the GIS maps allow us to visually compare the household locations of parishioners of different occupational levels, spatial statistics enable heavily populated areas to be more easily compared, and with greater accuracy. We used a kernel density technique, which aggregates address points together within a specified search radius and creates a smooth, continuous surface representing the density of members from a particular group. In Figure 4.7, the parameters have been adjusted to show clearly defined hotspots which are useful for comparative purposes but statistically less reliable than using a smaller search radius. As can be seen, the greatest concentration of unskilled and semi-skilled congregants resided within a few blocks west of Yonge, between College Avenue and Lake Ontario. Skilled manual category members were concentrated between Yonge and Sherbourne, and College and Queen, as well as several blocks either side of Spadina and Brock, between College Avenue and the lake. Many of the skilled non-manual congregants lived between Yonge and Sherbourne and College and Queen, but also to the east of Yonge, several blocks north of College Avenue. The professional category members overlap both of the skilled non-manual areas of concentration but also the area in between. While there was clearly some crossover in where those classified in different occupational levels lived, there is a definite contrast in where the lowest and highest level congregants lived, which roughly corresponds with Mulvaney's observations.

The confirmation that late-nineteenth-century Toronto can be divided into areas in which congregants of different socioeconomic status lived is not in itself a major research finding. It is not surprising that doctors are found living separately from labourers; factory workers from clerks; or merchants from plumbers. Yet taken in the context of what we have previously established about the seating arrangements of Knox Church, it is highly significant. Knox was evidently a place where congregants could come where socioeconomic status was of little consequence. The absence of more unskilled and semi-skilled workers should caution us against making sweeping claims of a truly egalitarian community, but that the class divisions so evident outside the church walls were even somewhat reduced is a significant breakthrough in our understanding of this religious and ethnic community.

CONCLUSION

GIS has the potential to answer specific research questions or to be used as a general investigative tool. It offers a valuable complement to traditional methods, which, as has been demonstrated, can lead to significant and in some cases unexpected findings. In this project

GIS was used most fundamentally to plot onto a contemporary map where the parishioners of Knox Presbyterian Church lived. In doing this, we found that Knox cannot be considered a neighbourhood church, and, while the factors that drew the congregation together remain open to speculation, GIS methods allowed us to look for reasons beyond geography. Furthermore, we know that these factors were strong enough to keep people worshipping at Knox, even when most worshippers lived closer to another church of the same denomination. One of the most striking aspects of the pew rent book was the extent of transiency, both inside and outside the church. For those who moved house, GIS can be used to track their movements and determine patterns that could not be detected from the written records alone. Regarding the Knox congregants, it is the lack of consistency to these movements that is interesting, which, together with the exceptionally high levels of transiency, make this a clear area of future research. Another area that has wider implications for urban historians is the identification of distinct residential areas in relation to socioeconomic backgrounds. With further investigation, this could lead to a much deeper understanding of class dynamics in Toronto. Here, it is enough to confirm that, among the Knox congregation, unlike inside the church, socioeconomic differences did manifest themselves quite clearly. By incorporating HGIS methods into our study, we came to see Knox as a place where people from, not only different parts of town, but also from very different backgrounds, could comfortably mix. The question that arises, and which will only be answered when data from similar sources are analyzed in a GIS, is whether or not our findings extend to other houses of religious worship beyond Knox.

NOTES

1 J. Ross Robertson, *Robertson's Landmarks of Toronto: A Collection of Historical Sketches of York from 1793 to 1837 and of Toronto from 1834 to 1904* (Toronto: J. R. Robertson, 1904), 215.

2 Deacon's Court and Board of Managers Seat Rents, 1882–1887, Knox Presbyterian Church Toronto collection, Presbyterian Church in Canada Archives.

3 Mark McGowan, "Coming Out of the Cloister: Some Reflections on the Developments in the Study of Religion in Canada, 1980–1990," *International Journal of Canadian Studies* 1–2 (1990): 175–202.

4 See for example Hannah M. Lane, who uses church census records in "Tribalism, Proselytism, and Pluralism: Protestant, Family, and Denominational Identity in Mid-Nineteenth-Century St Stephen, New Brunswick," in Nancy Christie, ed., *Households of Faith: Family, Gender, and Community in Canada, 1760–1969* (Montreal and Kingston: McGill-Queen's University Press, 2002); Rosalyn Trigger, "God's Mobile Mansions: Protestant Church Relocation and Extension in Montreal, 1850–1914," PhD thesis, McGill University, 2004; and Jordan Stanger-Ross, who uses marriage records in "An Inviting Parish: Community without Locality in Postwar Italian Toronto," *Canadian Historical Review* 87, no. 3 (2006): 381–407.

5 Andrew Hinson, "A Hub of Community: The Presbyterian Church in Toronto and its Role among the City's Scots," in Tanja Bueltmann, Andrew Hinson, and Graeme Morton, eds., *Ties of Bluid Kin and Countrie: Scottish Associational Culture in the Diaspora* (Guelph, ON: Centre for Scottish Studies, 2009).

6 Several examples of the importance of religious buildings to ethnic communities in Toronto are given in Robert Harney, ed., *Gathering Place: Peoples and Neighbourhoods of Toronto, 1834–1945* (Toronto: Multicultural History Society of Ontario, 1985).

7 Andrew Hinson, "Migrant Scots in a British City: Toronto's Scottish Community, 1881–1911," PhD thesis, University of Guelph, 2010.

8 John Zucchi, *A History of Ethnic Enclaves in Canada* (Ottawa: Canadian Historical Association, 2007), 2.

9 Christopher Armstrong and H. V. Nelles, *Revenge of the Methodist Streetcar Company: Sunday Streetcars and Municipal Reform, 1888–1897* (Toronto: P. Martin Associated, 1977), 32.

10 Rev. H. M. Parsons, sermon preached in Knox Church, 20 April 1890, reprinted in H. M. Parsons, *Biographical Sketches and Review, First Presbyterian Church in Toronto and Knox Church, 1820–1890* (Toronto: Oxford Press, 1890).

11 Omar M. McRoberts, *Streets of Glory: Church and Community in a Black Urban Neighbourhood* (Chicago: University of Chicago Press, 2003).

12 Richard Cimino and Don Lattin, *Shopping for Faith: American Religion in the New Millennium* (San Francisco: Jossey-Bass, 1998).

13 Lane, "Tribalism, Proselytism, and Pluralism."

14 Hinson, "A Hub of Community," 123.

15 John Moir, *Enduring Witness: A History of the Presbyterian Church in Canada* (Toronto: Presbyterian Publications, 1974), 136.

16 The occupational classification scheme used is the Social Power (SOCPO) scheme. The scheme distinguishes five levels of social class. Lower-class subgroups are SP (social power) level 1 (mainly unskilled workers), SP level 2 (mainly semiskilled workers), and SP level 3 (mainly skilled manual workers). SP level 4 is mainly composed of skilled non-manual workers and SP level 5 comprises white-collar and/or professional specialists (e.g., lawyers), wholesale dealers, factory owners and the like. Full details are available in Bart Van de Putte and Andrew Miles, "A Social Classification Scheme for Historical Occupational Data," *Historical Methods* 38, no. 2 (2005): 61–92.

17 Allan A. McLaren, *Religion and Social Class: The Disruption Years in Aberdeen* (London: Routledge and Kegan Paul, 1974) and Peter Hillis, *The Barony of Glasgow: A Window into Church and People in Nineteenth-Century Scotland* (Edinburgh: Dunedin Academic Press, 2007).

18 Hillis, *The Barony of Glasgow*, 143.

19 Trigger, "God's Mobile Mansions," 87.

20 Deacon's Court Minute Book, The Presbyterian Church in Canada Archive.

21 William Fitch, *Knox Church Toronto: Avant-garde Evangelical Advancing* (Toronto: John Deyell, 1971), 28.

22 *Annual Report of the Trustees and Deacons Court of Knox Church for Congregational Year 1883* (Toronto: Globe Printing, 1884).

23 Hillis, *The Barony of Glasgow*, 109.

24 Howard Chudacoff, *Mobile Americans: Residential and Social Mobility in Omaha, 1880–1920* (New York: Oxford University Press, 1972), 150, 151.

25 Michael B. Katz, Michael J. Doucet, and Mark J. Stern, "Population Persistence and Early Industrialization in a Canadian City: Hamilton, Ontario, 1851–1871," *Social Science History* 2, no. 2 (1978): 220.

26 Michale Katz, *The People of Hamilton, Canada West: Family and Class in a Mid-Nineteenth-Century City* (Cambridge, MA: Harvard University Press, 1975), 104.

27 William Jenkins, "In Search of the Lace Curtain: Residential Mobility, Class Transformation, and Everyday Practise among Buffalo's Irish, 1880–1910," *Journal of Urban History* 35, no. 7 (2009): 970–97.

28 Jenkins, "In Search of the Lace Curtain," 982.

29 Chudacoff, *Mobile Americans*, 158.

30 Charles Pelham Mulvany, *Toronto: Past and Present* (Toronto: W. E. Caiger, 1884), 43.

5

Stories of People, Land, and Water: Using Spatial Technologies to Explore Regional Environmental History

Stephen Bocking and Barbara Znamirowski

INTRODUCTION

South-central Ontario is rich in the stories of Canadian environmental history. Extending north from Lake Ontario, the region encompasses diverse landscapes. Some are the products of glacial history: fertile plains, rolling hills, and lakes and rivers that became essential to transportation, settlement, and industry. The Canadian Shield imposes its own character on the region's north. These landscapes formed the setting for central themes in Canadian environmental history: survey, settlement, forest clearing and agriculture, formation of transportation networks, industrial development. There have also been efforts to make sense of this landscape, debate its appropriate use, and resolve conflicts between diverse interests. These themes played out in ways that were specific

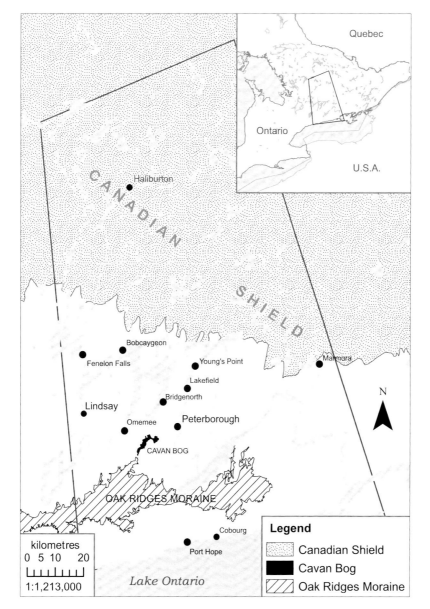

Fig. 5.1. Regional Environmental History Atlas (REHA): Project location and boundary. (Ontario. Ministry of Natural Resources, Structured Data. 2009–2011. Software: ESRI Inc. ArcGIS Desktop 10, Adobe CS4 Illustrator.)

to this regional landscape, and therefore study of its environmental history must relate to local geography and environmental features.

In this chapter we discuss the value of Historical Geographic Information Systems (HGIS) as a tool for the study of the environmental history of a region. We focus on South-central Ontario: an area extending from Lake Ontario in the south to Haliburton in the north, and from Oshawa in the west to Belleville in the east. This region, encompassing several counties and the Trent River watershed, is centred on Peterborough and Trent University, the home of our Regional Environmental History Atlas project (REHA) (Fig. 5.1).

This project responds to the opportunities and challenges inherent in diverse research resources that relate to environmental history.

These resources are closely tied to specific land-scapes, and so it is appropriate to assemble them in a format that can make these ties immediately evident, enabling analysis of places and spatial patterns (including the novel patterns produced by industrialization, new forms of agricultural production, and new markets), and presenting the possibility of new interpretations of historical change. Beyond the display of geographically situated historical data, HGIS can support the telling and analysis of stories of Canadian environmental history, such as the expansion of agricultural settlement, the rise and decline of resource industries, and the emergence of perceptions of landscapes.[1] In doing so, they can serve as tools for research, communication, and teaching and as a foundation for collaboration between historians and others who share an interest in history. The contributions explored in this project also illustrate the potential for regional HGIS projects elsewhere in Canada.

South-central Ontario presents several advantages for a project of this kind. Numerous themes relevant to Canadian environmental history are present here. The region is also small enough to support the long-term goal of assembling a reasonably comprehensive collection of knowledge of its environmental history. And finally, this region is of the right size to enable affordable collaboration amongst the small but diverse group of individuals (environmental historians, historical geographers, map and data librarians, local historians, conservationists, and others) who share an interest in its history and heritage.

Before discussing the project itself, it is appropriate to outline its regional context. Glacial forces shaped this landscape and many of its features: the Oak Ridges Moraine, eskers, and countless drumlins; lakes that line the edge of the Shield; the transition zone between the Shield and the St. Lawrence Lowlands that is known as The Land Between; rivers (such as the Trent, Otonabee, and Moira) that channelled melt waters; and waterlogged areas (such as the Cavan Bog) created by glacial drainage and ice dams. Suitability for agriculture generally declines as one travels north. Although there is much fertile land, the land is often rough and stony, better suited to pasture than crops. Drumlins and other glacial remains are interspersed with poorly drained plains underlain by clay, creating areas of marsh and bog that posed obstacles to agriculture. On the Shield, thin soils discouraged permanent settlement.[2]

Our focus is on the period since European settlement. Aboriginal history is an essential and complex aspect of the history of this region. However, it also requires knowledge and experience that are beyond the expertise of the project team. We hope that future collaboration will enable the project to pay appropriate regard to the region's Aboriginal history.

In the 1800s, here as elsewhere in Ontario, the land was surveyed, as part of the Colonial Office's effort to impose an orderly system of land allocation. This occurred usually, but not always, prior to settlement. Areas near Peterborough were surveyed in the 1810s and 1820s; the northern region, including Haliburton, was surveyed in the late 1840s and 1850s. Surveys imposed a rigid grid on the landscape, rendering the region amenable to ownership and settlement. They also neglected its ecological complexity, ignoring local features such as soil, slope, streams, and drainage patterns. This would become especially apparent where, for example, roads, property lines, and farm fields cross steep-sloped drumlins. These surveys were only the beginning of mapping: county maps followed, and in the twentieth century several agencies produced topographical map series, as well as other, more specialized maps.

Settlement took place in several stages. Beginning in the 1790s clearing and settlement near Lake Ontario proceeded unevenly, with much of the land allocated to absentee owners.[3] Soon, however, the region shared in Upper Canada's population growth (from 158,000 in 1825 to 952,000 in 1851), with settlement of lakeshore townships accelerating after the 1820s, while inland immigration initiatives focussed on the Peterborough area (including the settlement in 1825 of 2,000 Irish, arranged by Peter Robinson).[4] By the 1840s a predominantly agricultural landscape, made up mainly of family farms, had been established south of the Shield. A brief spasm of efforts to settle the Shield began in the late 1850s, and in the 1860s and 1870s the Canadian Land and Emigration Company sponsored mostly unsuccessful attempts to settle the Haliburton region. Rural population peaked around the 1880s, before declining, particularly in the north.[5] These trends reflected factors specific to the region, including abandonment of marginal areas, as well as wider patterns of migration: from rural areas to cities, and from Ontario to western Canada. Together these factors created the patchwork of settlements, bounded by rock,

frost, and borders, that became characteristic of early Canada.[6]

Settlements often developed in response to environmental features. For example, Port Hope had fertile land, a sheltered harbour, and power from the river. Peterborough was a promising site for water power, and at the head of river transport. Many communities were established at sites for grist or saw mills, including Omemee, Lindsay, Bobcaygeon, and Fenelon Falls. In Smith Township, several settlements owed their origins to the timber trade: Bridgenorth, Young's Point, and Lakefield all grew up around sawmills.[7] Figure 5.2 exhibits the distribution of mills in this region, and the historical periods when they were established. The map demonstrates numerous features of their history, including their concentration in Peterborough, and their distribution in the region south of the Shield; by the 1850s the Shield had become the major source of timber, linked to mill sites by rivers and other waterways.

Transportation networks were essential to settlement and transformation. Natural waterways – the Trent and Otonabee Rivers, lakes, and portage routes – played an early role in moving people, goods, and raw materials.[8] From the 1830s to the 1920s, the Trent-Severn Waterway moved fitfully towards completion. The waterway's history was closely tied to regional economic and political developments and furnishes a local instance of the transformation of a river system to serve human purposes.[9] Early roads were also important, albeit of widely varying quality and comfort. They generally followed survey lines, respecting property rights while ignoring topography. Colonization roads were pushed north into the Shield in the 1850s and 1860s; they were stimulated by the *1853 Land Act*, which encouraged

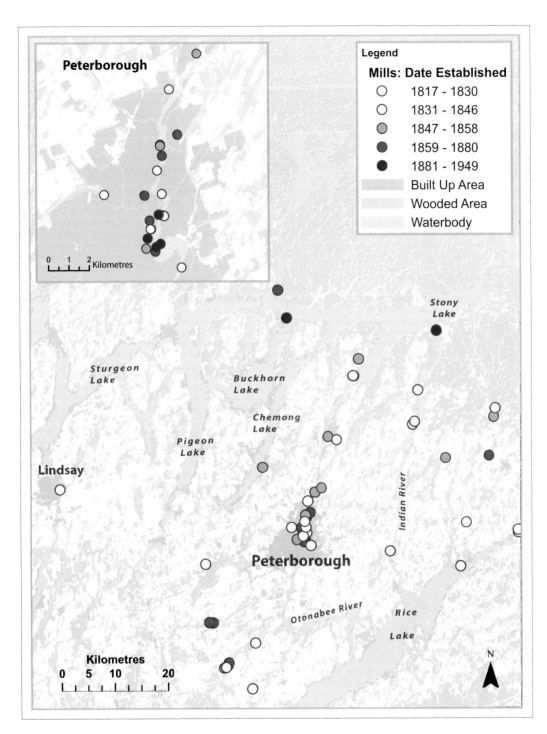

Fig. 5.2. Historic mill sites with dates. (Base Map: Ontario. Ministry of Natural Resources, Structured Data. 2009–2011. Software: ESRI Inc. ArcGIS Desktop 10, Adobe CS4 Illustrator and Photoshop.)

settlement by authorizing free grants of up to a hundred acres.[10] Beginning in the 1850s, railroads became essential to moving resources and people.[11] Railroads also rearranged regional trade relations (lines built at right angles to Lake Ontario both exhibited and reinforced trade ties to the United States) and encouraged concentration of industry and population. In the case of Haliburton in the 1870s, they encouraged depopulation of northern townships, when a new rail line made it less expensive for logging camps to import their supplies, thereby eliminating a major market for local farmers who had been making, at best, only a marginal living.[12] HGIS can enable analysis of the relations between transportation, settlement, and land clearing and aspects of the landscape, such as agricultural potential.

Initially, the regional economy was tied to the local environment. Industries produced essential goods such as leather, furniture, and food, at cheese factories, meat packers, distilleries, and breweries. Most industries were widely distributed across the region, relying on local materials from forest or field, and on local sources of energy, mainly waterpower, while producing goods for the local market. As transportation improved and shipping costs declined, manufacturing became gradually concentrated in fewer population centres.

In the case of the timber industry, external factors were important, such as British demand for squared timber and trade agreements with the United States, which accelerated exports of sawn lumber. The industry quickly transformed from meeting local needs (such as for building materials) using timber from areas cleared for agriculture into a commercial operation focussed on export. Lumber production developed rapidly after the 1830s, and the

timber industry became a significant economic actor. This transformation was accompanied by a geographic shift: from small mills located near Lake Ontario to larger mills located to the north.[13] In 1860, for example, Peterborough County had thirty-seven mills, employing 637 workers. For its part, Port Hope became a busy export centre: in 1879, 50 million board feet of lumber passed through, and 68 million the following year.[14] Dependency on export (and on a resource that could be driven to exhaustion) also encouraged a boom and bust economy of the kind typically associated with the staples trade.[15] Maps can aid in the analysis of the development and transportation of this industry in relation to other factors, such as modes of transport (waterways and railroads) and the availability of timber.

Agriculture played essential roles in regional environmental history. Production evolved from subsistence crops (potatoes, turnips, corn) in the early years of settlement, to, by 1850, wheat for export and modest mixed agriculture for local consumption. There was also a substantial early presence of livestock: in 1851–52, more than one third of farm land in Cavan, Emily, Douro, and Ops townships was used for pasturing; in three other townships more than one quarter of the land was used for pasturing. In the 1870s American wheat displaced Ontario farmers from that market, encouraging farmers to switch to mixtures of field crops and livestock. Overall, the nineteenth-century agricultural landscape formed a complex mosaic of wheat cultivation, livestock and other mixed cultivation, and woodlots.[16] Much marginal land has also been abandoned. In South Victoria, for example, area under crops decreased from 196,603 acres in 1920 to 121,844 in 1964.[17] HGIS can aid in understanding in more detail the relation between agricultural potential (particularly soil quality) and patterns of clearing, cultivation, and abandonment.

Throughout this history, landscape features were manipulated to serve human purposes. Rivers were transformed to generate power, to create the Trent-Severn Waterway (raising water levels and flooding shorelines), and to transport timber. These transformations could also be indirect, as when land clearing changed the hydrological regime, making floods more common, as occurred on the Ganaraska River. Much of the forest cover was destroyed to make way for agriculture, and beginning in the early 1860s there was a rush to secure and harvest the pine forests of the Shield. With habitat transformation, several species declined, including salmon (virtually eliminated by stream obstructions such as mill dams and by damage to spawning habitat) and passenger pigeons, among others.

Throughout the history of the region, a variety of forms of knowledge have been significant. One was farmers' skills and knowledge: essential to adapting to this landscape, guiding land uses such as raising potatoes and pasturing livestock. Some skills were brought from Ireland (where the landscape and glacial features were somewhat similar to those in this region), and farmers obtained others from the Mississauga people (particularly relating to the use of fire to clear land) and from their own observations in Canada.[18]

Evolving attitudes and perceptions of the landscape also influenced its transformation. The nineteenth-century aspiration to "conquer" and clear the landscape was dominant; but other attitudes, including admiration for the scenery, were also present, particularly among those with the wealth and leisure to enjoy it.[19] The Shield was the object of diverse and shifting views: at first a potential new farming frontier, it became (in Al Purdy's words) a "country of defeat" for farmers, as well as a valuable source of timber, a recreational playground, and an iconic landscape, representing Canadian identity.[20] Conflicts on the Shield between settlement and the timber industry also provoked discussion of the need to define distinct areas for each activity. By 1913 the region had become central to Canadian discussions about resource conservation.[21]

Evolving uses and perceptions of the landscape were also reflected in the development of tourism and recreation. Tourist facilities appeared, such as hotels and resorts, as well as transportation facilities, including railroads and steamship lines.[22] Landscapes that once produced timber or agricultural crops became redefined as landscapes of consumption; today, the contemporary landscape of cottages and resorts still reflects the influence of the timber industry: its railways, dams, canals, locks, and boarding houses were reconfigured to exploit tourists rather than trees.[23] These evolving land uses and perceptions can be mapped in order to better understand how they related to economic activities and to specific features of the landscape.

These various trends and transformations: immigration and settlement, land clearing and agriculture, industrial development (particularly the timber industry), formation of transportation routes, and new uses of the landscape, including for recreation, together constituted, as J. David Wood has suggested, an ecological revolution.[24] As such, the history of this region parallels landscape transformations elsewhere that have received more attention from environmental historians.[25] And, as this brief review suggests, a conventional historical narrative can provide a solid basis for understanding this revolution.

DEVELOPING AN ATLAS OF REGIONAL ENVIRONMENTAL HISTORY

Much can be added to this narrative, however, by integrating it with spatial technologies. Historians have demonstrated the contribution of conventional mapping to understanding the historical geography and environmental history of southern Ontario; they have used maps to illustrate patterns in agriculture, population growth, railroads, timber cutting and export, and industry.[26] HGIS and other spatial technologies can enhance these advantages of mapping. Their benefits include the capacity to assemble and display diverse forms of information – documents, maps, photos, environmental information – to enable exploration of spatial patterns. Visualizing places and the relations between them can make evident patterns that an historian may only otherwise vaguely sense. Specific questions can also be explored, including those that concern the causes and effects of historical change, such as the relation between transportation routes and industrial development. With its potential to extend spatial analysis across several scales, HGIS can also assist in the telling of complex stories of social and environmental change.[27] Most fundamentally, these technologies present the possibility of using maps and related materials, not just as the end products of research, but as research tools.

The aim of our project is to explore these benefits by constructing an online atlas of regional environmental history that juxtaposes the products of historical research with materials for further study. We see this atlas as encompassing four types of resources and tools for research:

1. An online HGIS, with historical sites identified and linked to relevant information, to support dynamic generation of maps, as well as queries and classification. Clicking on a point of the map would provide access to a range of information about that site, and would, eventually, permit users to add their own content.

2. Presentation of historical information in spatial form, including changes over time. This would include, for example, the formation and discontinuation of railroad lines, and the construction of the Trent-Severn Waterway.

3. Georeferenced historical materials, presented through overlay of the base map with primary sources, such as topographic maps, historic maps, and aerial photos, to enable comparison of past environmental features and land uses (such as forest cover) with current conditions.

4. Linking of the HGIS with diverse other materials: historical, geographical, literary and scientific documents, legislative debates and acts (these are sometimes specific

to a region), photos, oral histories, bibliographies, and landscape observations. Where possible, these materials would be georeferenced, perhaps to county or township.

This design reflects the two chief purposes of the atlas. First, because it is organized in terms of the landscape itself, it will present environmental history information in a more accessible and intuitive format than conventional bibliographies or other research tools. This will expand the possibilities for communication between environmental historians and their audiences. Second, it will provide a point of departure for further study of the region's environmental history. The geographic presentation of historical information can itself become a research instrument, by enabling study of spatial patterns and relationships, making evident otherwise obscure aspects of historical change.

To fulfill these purposes, our project uses spatial technologies to assemble and juxtapose historical materials and to explore how they relate to each other and how they can illuminate our understanding of the region's environmental history. As practitioners often note, HGIS projects can require several years before they begin to be useful in scholarly or practical contexts.[28] They require major investment and collaboration amongst individuals with academic qualifications and technical expertise. This has also been our experience. Our project is intended to be a long-term undertaking – a "living" project, always open to additions and revisions and to collaborative partnerships.

Our work on this project began in late 2008, with seed funding from the Network for Canadian History and Environment (NiCHE).

Since then, the project has developed through collaboration with several institutions and individuals. Additional funding and in-kind support has been obtained from various sources, including Trent University and the GEOIDE (GEOmatics for Informed Decisions) Network. Major work on assembly and processing of historical resources and spatial technologies has been accomplished by the Maps, Data, and Government Information (MaDGIC) Unit of Trent University Library. The emerging roles of academic map libraries as spatial and statistical data centres and also as research collaborators make them natural partners in HGIS projects. This collaboration also aligns with the strategic priority of many university libraries – including that of Trent University – to become more involved in academic research. Collaboration with regional organizations has demonstrated the benefits of working with the local heritage community; these relationships take time to develop but also enrich and help to sustain projects.

Since its origins, the project has evolved considerably. This evolution and the numerous decisions that have been made regarding technical design reflect how the project has been a learning process for its developers, as challenges have been encountered and overcome. Project development has comprised several distinct activities: developing and applying software, assembling and processing historical materials and information, relating these to geographic locations, and exploring their application to environmental history.

A major focus of effort has been the development of spatial technologies, operating on multiple platforms. This includes a range of proprietary and open source technologies for desktop GIS, publishing web map services (for serving and consuming dynamic maps over the web), database development, and spatial processing. With this broadening of technologies, it has become appropriate to describe this project in terms of the application not just of GIS but of spatial technologies more generally.

Development of these technologies has been a complex process – one not yet completed. This complexity reflects a process of learning and experimentation and also of change in the GIS web mapping technologies themselves. For example, when our project started, we used proprietary software to publish maps over the web following a process that was relatively compatible with how we developed spatial content in our GIS desktop production. The end-product, however, was fairly "out of the box," with only limited customization.

With the emergence of the next generation of web mapping software – ESRI's ArcGIS for Server – we accordingly began migrating components of the project into it. In addition, and in tandem with our work with ArcGIS for Server, we experimented with the use of the Google Maps API (Application Programming Interface) to present time-series of maps: initially with historic topographic maps, but eventually expanding to include other historic materials, including aerial photos. Our evolving use of these spatial technologies provided considerable opportunity to explore their potential. New web mapping technologies have rapidly affected the potential of HGIS: we have gone from using technology with largely limited functionality – mostly displaying spatial information that can be turned "on" or "off" – to combining GIS with web programming scripts to support interactive processing by the user. For example, users can change opacity and overlay features to support their

own time-series analysis. However, this has also required frequent rethinking of technical design and writing of new or upgraded code. For example, the upgrade to ArcGIS for Server involved a complete change in how we publish maps and a very different set of programming skills. The outcome in terms of presentation of historical materials has certainly been worth it, but the work has been time-consuming and technically demanding.

Another significant change has been the manner in which spatial content itself has become available. For example, the emergence of a variety of online base maps makes creation of base maps for web projects no longer essential. Our site, for example, now uses ESRI's ArcGIS Online Base Maps for parts of the project. As in Google, users can directly select what base map they wish to show, enabling use of topographic, imagery, relief, or street maps as background. This has many advantages, as the maps are already created and do not require local hosting. However, these maps function as "backdrops" only, and they cannot replace the detailed information provided through provincial layers. For example, we have included Ontario provincial soils information and hydro network surface water information, which show streams, rapids, dams, locks, and obstacles in water. Overall, this development work has provided the basis for relating the environment, including land forms, soil types, water bodies, and other geographic features, to human activities such as agriculture, settlement, and transportation. This environmental information, when mapped against land use, can also provide clues as to how, historically, people have evaluated the landscape's potential and limitations.[29]

A wide range of texts, maps, photos, fire insurance maps, and other historical resources have been assembled for the project. They include maps at different scales and dates (such as nineteenth-century county maps and twentieth-century topographical maps), as well as historical aerial photographs from the Trent map collection. A major effort has been devoted to scanning and georeferencing these materials, so as to make them available in digital form and to relate them to the region's geography. Depending on the age and quality of the image, this process can be time-consuming and requires skill and patience. We began this effort by selecting, scanning, and georeferencing aerial photography held by the Trent University Library. We were able to identify approximately 4,400 aerial photographs in the collection from our project area. Air photos range from 1928 to 1977 (federal photographs purchased from the National Air Photo Library (NAPL) and from 1977 to 1993 (photography purchased from the Ontario Ministry of Natural Resources (OMNR). Aerial photos were scanned to meet archival standards, with scanning and georeferencing largely completed by student assistants who received training as well as written instructions. The NAPL also assisted: our library provided NAPL with information about the air photo holdings (including year, the "A" flight line number, and picture numbers); NAPL compared this information against their database and supplied spreadsheets that included a variety of information about each photo, including, for example, spatial references for all corners and the centre, the altitude, NTS sheet number location, camera details, season, scale and precise date. This information considerably reduced the work

required for georeferencing and provided valuable metadata for our geodatabase.

The project has also drawn on Canada's rich history of topographic mapping. We have drawn on maps prepared by several agencies, including the Geographical Section, Department of National Defence, and Canada Surveys and Mapping Branch, Department of Energy, Mines and Resources.[30] For historical GIS, awareness of how maps differ over time and what different series show is essential. For example, only early military maps distinguish coniferous and deciduous trees (using hand-drawn symbols), and they are also unique in distinguishing masonry and wood buildings, which was done until 1927. Both topographic maps and aerial photos are highly relevant to environmental history: they reveal changes in land cover and land use, as well as the expansion of settlements and transportation routes, including roads and railways. By superimposing these materials and adjusting their transparency, it is possible to compare different aspects of these historical changes.

Starting with 1:50,000 scale maps (and their predecessor, the 1:63,360 [one inch to one mile] map) our objective was to archive and make digitally available all editions of historic maps at this scale going back to the first edition (the editions we are working with span from 1929 to 1985). Several challenges have been encountered in locating, digitizing, and georeferencing these maps, and developing associated databases. Our first challenge was to determine what was available for our area, as there is no central record of map availability. While Canada's Map and Chart Depository Program has ensured that maps are readily available in libraries, early maps are often kept in storage and must be requested, and not all universities (including our own institution) have been able to acquire editions that predate the institution. A fiche inventory does exist for 1:50,000 and 1:250,000 maps produced by Energy, Mines and Resources, Surveys and Mapping Branch. Another challenge related to copyright, which can restrict digitization and presentation. For example, many archives prohibit the scanning of fire insurance plans. We have also found that the policies regarding digital pictures of even small areas of a fire insurance map differed significantly between institutions: some allow (even encourage) this, while others view even pictures of sections as a violation of copyright.

We have been interested in seeing how Google Maps and Google Earth could be used for HGIS. At this time, we use Google Maps to display early topographic maps. Several steps were required to post maps on Google. Maps were scanned and saved as uncompressed tiff files, at a resolution of 600 dpi. By testing scans at different resolutions, we found that the improved quality of detail justified this high resolution. Our choice of this resolution was also influenced by the fact that we are archiving early maps and imagery collected for this project; as the main university library in our region, this has become an important consideration in terms of local heritage preservation. Although there is the risk of slowing down the visualization of maps when working at higher resolutions (because of the size of these files – the scanned maps ranged from 414 MB to 1.04 GB each), the use of tiling and map caches (in which maps are pre-rendered into thousands of small image tiles) has made this less of an issue. Tiling was done using Map Cruncher, a freeware program designed by Microsoft Research for Virtual Earth (now Bing Maps) and also useful in Google Maps.[31]

One example of the results of this effort is displayed in Figure 5.3. This displays an excerpt from the atlas: a georeferenced 1933 topographic map superimposed on Google Maps. Below this (left side) is a detail of this map (the black box), focused on the west end of Rice Lake. To its right is another excerpt from the atlas: a detail of a georeferenced 1973 topographic map of the same area. These (and other topographic maps) overlay each other in our atlas; by adjusting their transparency, the history of environmental change in this area – including, in this case, changes in the distribution of wetlands and forest, road construction, and growth of the town of Bewdley – can be easily displayed.

Another focus of the project was on mapping the thousands of features that are relevant to the region's environmental history, using information drawn from documents, historic maps, and other sources. As a first step towards developing mapping strategies, we decided to focus on a specific category of features: mills (including saw mills and grist mills used for grinding corn, wheat, and other crops). Historically, these were an essential feature of the Ontario landscape, important to early settlement, agriculture, and industry. By 1836, there were 350 grist mills in Upper Canada, and, by 1840, 1,000 saw mills, growing to 1,600 in 1848.[32] Saw mills relate to the distribution of land clearing and settlement activities, the timber industry, and appropriate sites on watercourses. The appearance and disappearance of grist mills relates to several factors, including the distribution of settlements and the agricultural economy.

This pilot project involved several steps: i) tracking down the locations of mills, using published and archival sources, such as old maps,

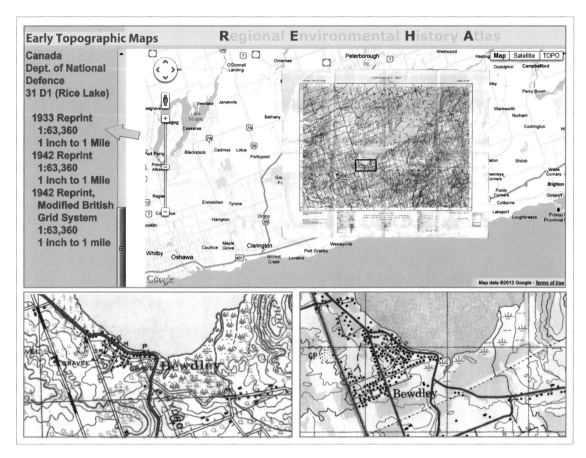

Fig. 5.3. Using Google to display historic topographic maps. (Topographic Map: Canada. Geographical Section, Department of National Defence. Topographic Map Ontario Rice Lake Sheet, 31 D1, 1 inch to 1 mile, 1:63,360, 1932, Reprinted 1933.) (Topographic Map: Canada. Surveys and Mapping Branch, Department of Energy, Mines and Resources, Rice Lake, Ontario, 1:50,000, edition 4, produced 1970, printed 1973. Base Map: Google Maps, January 2012. Software: Google Maps API, MSR MapCruncher, Adobe CS 4 Photoshop.)

historic county atlases, and fire insurance maps (no trace remains today of most mills at their original sites);[33] ii) constructing a database of attribute data (descriptions of the mills), spatial data (locations), and temporal data (when the mills were constructed, changed purpose or location, or ceased to exist); iii) presenting mills information using HGIS. To date, the project has identified 310 unique mills, and this number is expected to increase. Of the 310, 171 are located on unique sites, a different set of 171 have known dates of establishment,

66 have known closing dates, and 170 have text annotations. As much as possible, we tried to maintain a consistent approach to location and attribute information. This could be a challenge, particularly for those mills whose location could be determined only approximately. A spreadsheet template was devised for research assistants to collect spatial and attribute information. This template, which formed the basis for our geodatabase, was designed with community partners in mind to enable sharing and integration of database information.[34]

Some of the functions of the atlas can be demonstrated using information gathered and mapped as part of this pilot project on mills. All figures are excerpted from the atlas and are available online. Figure 5.2 displays the distribution of mills in the central portion of the study region and the time periods in which they were established. Mapping this distribution of mills provides a useful foundation for examining the geographic distribution over time of this industry in relation to land-clearing and settlement. This distribution can also be compared with published maps of wood production in the region.[35] Figure 5.4 (left side) displays the distribution of mill sites in the city of Peterborough, with an overlay of an 1878 map of the city. This figure thus illustrates the atlas's ability to integrate historical data with georeferenced historical materials. On the right of Figure 5.4 are two close-ups of this base map, with part of a georeferenced fire insurance plan; these exhibit the use of these plans to identify the locations of mills. Figure 5.5 demonstrates the atlas's query function: as described earlier in this chapter, clicking on a point of the online atlas provides access to a range of relevant information. In this case, the information includes text and a photo relating to a mill in Millbrook.

The mills project thus served as an opportunity to experiment with assembly, manipulation, and presentation of data and to explore their relevance to environmental history. The project also demonstrated several of the challenges involved in relating historical information to geographic locations. Information regarding the location of mills is often uncertain, contradictory, or non-existent. Mills themselves, and particularly saw mills, were usually transitory: located close to where trees were

being cut, they were often moved elsewhere after the timber supply had been exhausted. Others were carried away by floods or were simply abandoned.[36] Conservation authorities have pulled down many abandoned mills, dams, and foundations. Other challenges related to the construction of the database, such as ensuring compatibility between Excel files and ArcGIS. Although compatibility problems encountered were overcome, they prompted us to evaluate more robust relational database management systems (RDBMS). Advantages of these systems would include more efficient storage and linkages between database content, multi-user editing, and the capacity for a greater number of users to query the database without conflict.

Another component of our project involved reading text sources (such as township histories), identifying information relevant to regional environmental history, and georeferencing this information. Our purpose was to "harvest" the knowledge accumulated by local historians across various thematic categories, including conservation, fisheries, industry, mills, natural heritage, settlement, timber industry, tourism, transportation, and water development. Information relating to approximately 300 locations of historical interest was extracted from these sources. While this exercise was valuable, the extraction of geographic information from textual sources also proved to be very time-consuming. As one author has noted, while this approach can be described as "data mining," it is, at best, mining using only a pick and shovel. These sources also typically provided only vague, incomplete, or ambiguous geographic information.[37]

Finally, we note that mapping census information relating to changing populations and activities in townships can provide a basis

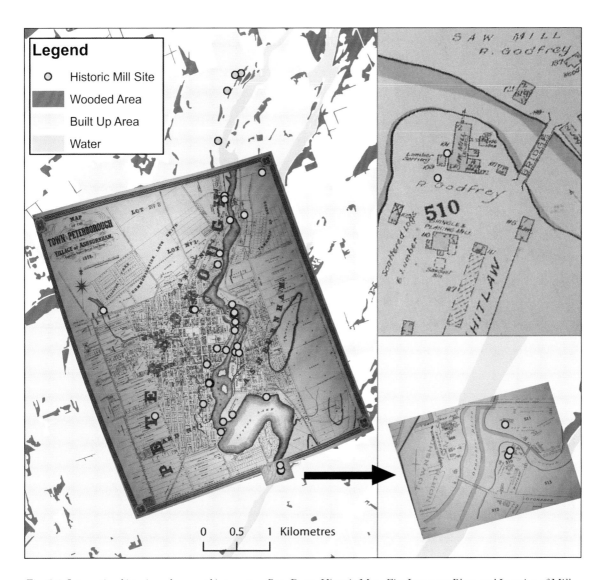

Fig. 5.4. Integrating historic and geographic resources: Base Data, Historic Map, Fire Insurance Plans and Location of Mills. (Left Panel of Figure: Historic Map: Map of the Town of Peterborough and Village of Ashburnham; compiled from registered plans and actual surveys, The Burland Desbarats Lith. Co. Montreal, 1878, in: H.C. Miles & Co. Toronto (Ont.), The New Topographical Atlas of the Province of Ontario, Canada: Compiled from the Latest Official and General Maps and Surveys, and Corrected to Date from the most Reliable Public and Private Sources of Information, Comprising an Official Railway, Postal and Distance Map of the Whole Province and a Correct and Complete Series of Separate County Maps on a Large Scale ... also a Series of Recently Issued Maps Showing the Whole Dominion of Canada and the United States. Toronto: Miles, 1879, p. 60. Base Map: Ontario. Ministry of Natural Resources, Structured Data. 2009–2011. All panels: Insurance Plan of the City of Peterborough Ontario, Toronto and Montreal: Underwriters Survey Bureau Limited 1929, p. 31. Software: ESRI Inc. ArcGIS Desktop 10, Adobe CS4 Illustrator and Photoshop.)

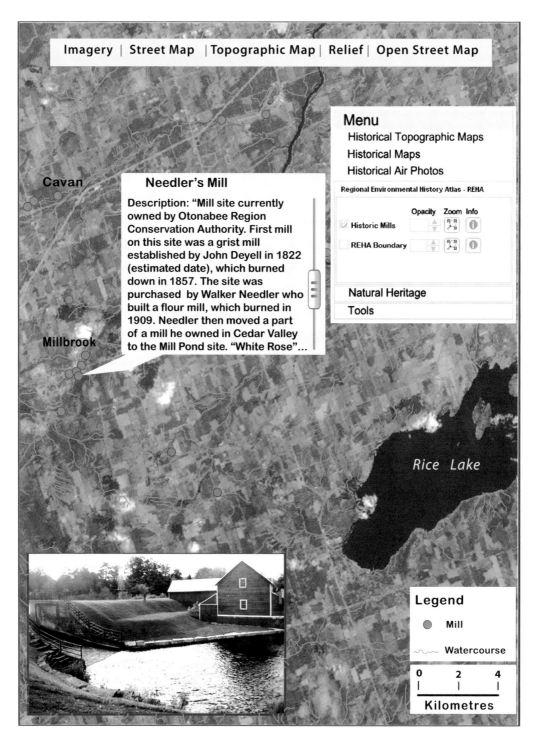

Fig. 5.5. Demonstrating REHA Web Site Query Function: Needler's Mill, Millbrook. (Photo of Needler's Mill, Millbrook. Courtesy of Stephen Bocking, 2010. Water layer: Ontario. Ministry of Natural Resources. Ontario Hydro Network. 2011. Satellite Imagery Base Map: ESRI Inc. ArcGIS Online ESRI World Imagery Base Map, January 2012. Software: ESRI Inc. ArcGIS for Server 10, Adobe CS 4 Photoshop, Adobe CS4 Illustrator.)

for understanding spatial patterns of land uses and economic activities in relation to environmental features. This component of the project awaits development.

The spatial technologies applied in this atlas have several applications for environmental history. These include the presentation and comparison of historical materials and information, enabling exploration of questions of interest to historians and opening up the possibility of visualizing the spatial dimensions of history in novel and productive ways. Another potential use is as a basis for collaboration between historians and other interested parties. There is a substantial local community of interest: amateur historians and naturalists, academic researchers (retired professors have extensive knowledge of the region's natural history and historical geography), conservationists, and local agencies (including the Trent-Severn Waterway, the OMNR [headquartered in Peterborough], and Otonabee Conservation). An atlas of environmental history can expand and consolidate the community that is interested in the region's history.

In this chapter, however, we wish to examine in more detail how spatial technologies can contribute to telling the stories of environmental history. Places hold multiple meanings and identities: they "gather" ideas, activities, and practices. Several recent environmental histories have emphasized this aspect of various regions such as Georgian Bay, the Chilcotin, and the Trent Valley.[38] Our region has also gathered diverse meanings and activities. Many stories can therefore be told of its environmental history, which together illustrate how environmental features have shaped human activities, while being, in turn, transformed. Spatial technologies cannot on their own tell these stories.[39] But what they can do is support conventional narrative text by illustrating spatial patterns and processes. Three brief examples and a more detailed case study of the role that can be played by HGIS in telling stories of environmental history are presented below.

Our first example relates to the history of land clearing and agriculture. This is, most obviously, a story of expansion and retreat, and of transformation of the regional environment. However, spatial technologies also present an opportunity to tell a more subtle and complex story of the encounter between cultivation and environment. External factors were important, such as market demands and trading relationships. Factors within the region must also be considered, each of which had distinctive spatial patterns.[40] Agricultural potential varied greatly: on and off the Shield, as well as on the finer scales of drumlins and other glacial relics, patterns of drainage, bodies of water, and other features. The role of potential in influencing agricultural development can be difficult to determine: farmers had varying levels of awareness (tree cover was not always a reliable indicator of soil fertility), and other attributes such as drainage and road access could be more important when they evaluated a plot.[41] Roads, railways, and other transportation networks also influenced agricultural patterns: by the 1860s improved transportation and declining shipping costs were making agriculture more sensitive to local environments and land costs, encouraging local specialization.[42] These patterns can be represented spatially. Using spatial technologies, a variety of questions can also be asked. How did patterns of mixed farming relate to agricultural potential at different scales? How were land clearing and choice of crops tied to the formation of transportation networks? Through these and other questions, possibilities for interpreting subtle patterns in the encounter between agriculture and the environment can emerge.

The story of industry in this region has been one of evolving relations with the landscape: from the use of local inputs such as waterpower, wood, leather, and food to supply local markets, to industries disengaging from these inputs (particularly as steam power replaced water power), and, thanks to the railroads, exporting to markets elsewhere. Local features shaped this story. For example, while the Marmora iron works were situated close to waterpower and steam power, there was no railway nearby. Accordingly, neither raw nor refined materials could be carried economically to or away from the site, and hence the works survived for only a brief period.[43] On the other hand, Peterborough was a promising site for water-powered industry because the Otonabee River had a significant drop at that location. Mapping mills and other industrial sites on the landscape, relating their distribution to environmental factors such as sites of energy production (especially waterpower and local electricity generation) and transportation networks, can add depth and complexity to the story of shifts in materials and markets and of concentration of industry in larger communities.

Our third example concerns the history of ideas about the value and purpose of this landscape. In many instances, these ideas were related to specific features of the landscape. Among the earliest and most well-known European perceptions were those of Catherine Parr Traill and Susanna Moodie. In *The Backwoods of Canada*, Traill recorded her observations as she travelled in the 1830s north from Lake Ontario to the Peterborough area, and these can be at least roughly situated. A more recent chapter in the history of landscape perception relates to tourism and to the role of this region as a resort, sometimes in opposition to the perceived ill-health and crowded conditions of the city. As early as the 1850s, certain sites, such as the shores of Rice Lake, had gained status as vacation resorts. By the late 1800s, the lakes lining the Shield, with their rocky shorelines and islands, had begun to attract tourists.[44] Resorts such as the Viamede Hotel and the Mount Julian Hotel began to host tourists in the 1870s. In 1883, the American Canoe Association held a meet on Stoney Lake, and this also encouraged tourism. Sites such as Sturgeon Point and Pleasant Point became noteworthy resorts. Many children's camps were established. Recreational activity north of the lakes was slower to develop, especially in areas distant from railway lines. Algonquin Park began its shift from lumber centre to camping resort. Some cottage lots or camping took place near colonization roads, such as on Big Cedar Lake and Long Lake. In the 1930s, new or improved roads, such as the Burleigh Road, encouraged tourism.[45] However, relatively isolated cottages on the Shield also encouraged perceptions of remote wilderness.

This history of perceptions and activities can be interpreted in terms of Canadian art and literature – through the work, for example, of the local poet Isabella Valancy Crawford, who in the 1880s expressed disgust for the crowded, noisy city and a preference for the healthy countryside. This history can also be placed in the context of the history of nature tourism.[46] The history of perceptions of landscapes of health and ill-health, and the development of health resorts, is also relevant.[47] However, there are also opportunities to situate these perceptions: to understand how they related, not just to general views of the countryside, but to specific environmental features, such as lake coastlines, or communities with distinctive social and economic features. This presents its own challenges: statements regarding beauty, recreation, or health benefits were often not tied to specific places. But they often were, too: the benefits of hunting, fishing, and canoeing in the lake country were extolled by Victorian health movements, and the 1888 Peterborough Directory referred to Stoney Lake as "Peterborough's supplementary lungs."[48] We can also consider the significance to Crawford of her life in the Kawarthas, or the placement of resorts on Stoney Lake and elsewhere, and the movement of tourists through the railways and steamship lines that carried tourists into the region, thereby relating the material evidence of situated activities to landscape perceptions. Several relevant questions can also be examined using spatial technologies. For example: What landscapes were particularly important in inspiring aesthetic appreciation and recreation? What were the links between perceptions and how landscapes were used – for example, the shift from timber exploitation to tourism? How did landscape features encourage this shift from production to consumption? The extensive history of landscape art in this region can provide much geographically situated information regarding perceptions and uses of the land. For example, A. J. Casson painted extensively in this region; his paintings of Lakes Kushog and Kashagawigamog in the 1920s illustrate not only their appearance ninety years ago but how access to these lakes have changed: today largely privatized through cottage development, they are off-limits to all but a select few.[49]

A CASE STUDY: THE TIMBER INDUSTRY

To continue our exploration of the potential of spatial technologies to assist in telling stories, we will examine in more detail the evolving relationship between the timber industry, settlement, and conservation. This industry has been the basis for a classic narrative. Between 1850 and 1910, the industry experienced rapid northward expansion, followed by contraction as forests were depleted and mills closed. Observers have presented this as a morality tale of a rapacious industry, shortsightedly exhausting its own resources:

> The ravenous sawmills in this pine wilderness are not unlike the huge dragons that used in popular legend to lay waste the country; and like dragons, they die when their prey, the lordly pines, are all devoured.[50]

Later observers agreed (albeit employing a less poetic but more managerial language) and drew implications regarding the waste of a once-valuable resource:

The following report … will serve to exhibit in a precise and detailed manner the consequences of mismanagement.… The slopes, once, for the most part, covered with valuable pine and hardwood forest, had been cut over. A large area, the pinery in particular, had been repeatedly subjected to fires and rendered liable to eventual total destruction … if the present policy of indifference and neglect continues, what might have been a continuous source of wealth will become … a useless waste.[51]

The moral was clear: conservation was an imperative. Ideas of conservation in Canada have been attributed to perceptions of resource waste and the influence of ideas from the United States. Evolving views of forests were also important: from being merely temporary obstructions to farmers to renewable resources meriting their own place on the landscape. South-central Ontario is at the centre of this narrative as one of the most noteworthy studies of the Canadian Commission of Conservation (1909–21) focussed its attention here.[52]

But a more complex story is also possible, of evolving relations between the timber industry, settlement, and the landscape. Using spatial technologies, the factors affecting the timber industry can be examined and related to each other. This would include economic factors and policies that influenced the formation of the timber industry: trading patterns and agreements, colonization policies, and elite conservation discourse. These factors could be related to circumstances within the region itself: the distribution of forests, settlements, agricultural areas, transportation routes (colonization

roads, railroads, rivers, and the Trent-Severn Waterway), protected areas such as Algonquin Park, land ownership, and timber company initiatives.

Industry initiatives, including acquisition of timber rights, and cutting operations, were also important. In broad terms, the industry swept northwards, beginning in the 1840s with the "Pine Land Grab" in the middle and northern reaches of the Ganaraska River, followed by exploitation of the Otonabee Region, the Pigeon Lake area, and, beginning in the 1860s, the Haliburton region. In the 1860s, Peterborough declined as a sawmilling centre because of rising transport costs (as cutting went further north) and the use of portable steam sawmills.[53] Major timber operators (such as the Mossom Boyd Company) typically dispersed their activities across several townships to adapt to unreliable spring stream flows that were needed to move logs downstream.

Mapping transportation routes in relation to industry activity can contribute to understanding how these were linked, and, thus, how the availability of timber – that is, the redefinition of forests as resources – was partly a question of access, with transportation playing a role in the transition from cutting for local purposes to commercial export. Logs were usually supplied to mills in the spring by streams, but movement of milled lumber required navigable waterways or railways.[54] Considerable information is available regarding these routes. Between the 1840s and the 1860s, the Trent, Indian, Otonabee, and possibly Cavanville Creek were being used for running timber.[55] In Dummer Township, canals were built to supplement the Indian River as a transportation route for lumber. In one day in 1864, 280 cribs of timber came down the Otonabee

River from Lakefield.[56] The role of railways was related to the shifting importance of the ports of Cobourg and Port Hope, which was in turn tied to increasing timber exports to the United States after the 1840s. The significance of the Trent-Severn Waterway can also be explored: an early priority in building the waterway and dams on upstream lakes and rivers to control stream flow was to ease the passage of timber. In 1844–45, timber slides were built at Healey's Falls, Middle Falls, and Chisholm Rapids, on the section of the waterway between Rice Lake and Trenton.

The conversion of forests into resources can also be understood by mapping the relationship between settlement and industry. Timber-cutting and settlement co-existed across the region. However, the relationship varied in different areas: in the south clearing for settlement occurred without industrial development, while elsewhere clearing fed the industry; on the Shield, timber exploitation took place alongside small-scale settlement. One factor shaping this relationship was views of the suitability of particular areas for either activity. In the predominantly agricultural landscape of the south, some areas were set aside as woodlots, and their distribution may reflect, among other factors, local variations in soil, slope, and other conditions. On the Shield, timber interests were dominant and were able to impose their view of agriculture as inappropriate. Timber companies alleged that settlers took land only to cut timber and that they started fires and filled rivers with rubbish, impeding log runs. However, at a finer scale, the relationship appeared to combine antagonism and co-existence. Timber companies established depot farms to lessen their dependence on local farmers. Yet the industry also provided a market for local

farmers who, given the poor state of the roads, had few other options.[57] In some cases, colonization roads served both interests. In fact, this had been one motivation for the roads: industry would cut the timber, and settlers would occupy the cleared land. The roads sometimes provided access to good timber-cutting sites as well as pockets of soil suitable for agriculture. But given the poor quality of the land, such outcomes were likely exceptional.

Divergent views of the Shield's agricultural potential complicated this interaction. In the 1820s, the surveyor Alexander Shirreff expressed an optimistic view of this potential – a conclusion apparently based on the abundant forests.[58] Other observers agreed: in 1847, the first survey evaluated the region as suitable for settlement, as did the 1856 report of the Commissioner of Crown Lands. In the 1860s, this view was apparently widespread: an 1862 ad in the *Peterborough Examiner* noted the region's "fertile soil." The English Land Company apparently assumed the wealth of Haliburton lay primarily in its soil, not its timber.[59] Such views were consistent with the general assumption that the progress of Upper Canada depended on continued northwards expansion of settlement, especially given concerns that settlers were draining away to the United States.[60] However, other surveyors expressed skepticism regarding this optimistic view. These divergent evaluations – either optimistic or skeptical – were often expressed in relation to specific townships, raising the possibility that they were related to specific places – perhaps, for example, to surveyors' routes. To what extent were contrasting impressions of potential based on different preconceptions, attitudes, or territories? The specific landscapes that inspired contradictory impressions could be assessed by examining surveyors' original records and relating these to the territories covered.

Mapping the timber industry and settlement can assist in situating their relationship and in understanding how it could be both antagonistic and complementary. More generally, with the centre of the industry shifting over time, the evolving relations between the industry and other activities reflected an evolving geography of land use: different landscapes, with different associated values, were at stake at different times. Several questions are amenable to spatial analysis. For example, to what extent was co-existence or antagonism between industry and settlement organized geographically: co-existence where agriculture was on suitable soils and could serve industry needs, but antagonism elsewhere? To what extent was the conflict between industry and settlement instigated by the industry's move onto the Shield country?

The emerging conflict between timber and settlement became most evident in the last decades of the nineteenth century, coinciding with the period of most rapid exploitation of the northern forests.[61] This conflict, as well as associated anxieties regarding future timber supplies, eventually encouraged interest in planning and conservation. Solutions to the conflict tended to be framed in spatial terms, including recommendations that areas suitable for settlement or for timber be identified and kept separate. In 1866, the Commissioner of Crown Lands advocated distinguishing between land that was and was not suitable for agriculture, with settlement prohibited on the latter. The 1868 *Ontario Free Grant and Homestead Act* imposed this compromise, effectively acknowledging that the forests were themselves a valuable industrial resource.[62] Algonquin

Park was another, and the most prominent, effort to impose this kind of spatial solution. A series of similar recommendations, often based on the argument that settlement would destroy lands better suited for forestry, culminated in the 1913 Commission of Conservation report on the Trent Watershed.[63] Other approaches to conservation also became evident in this region. In the early 1900s, Edmund Zavitz, an influential forester and conservationist, expressed concern about the state of the land and streams of Northumberland County adjacent to Lake Ontario (and especially on the Oak Ridges Moraine). He also stressed the financial benefits of forests and urged reforestation.[64] In the 1920s, county forests were established in Northumberland and Durham counties.[65] These activities illustrated how conservation had different meanings in different environments: in the south, tree planting and rehabilitation of degraded landscapes; in the north, separation of forests and settlement and protection of forests against fire. A spatial analysis can provide the basis for exploring these divergent meanings and implications of conservation.

A key issue is understanding how elite discussions regarding conservation played out in the local context, as expressed through government and industry initiatives, with a variety of consequences for the environment and for other land uses in the region. Conversely, and acknowledging that this region was a key terrain for Canadian conservation, it is worth examining how local circumstances framed national conservation perspectives.[66]

Finally, these analytical approaches regarding timber exploitation, conflicts with other land uses, and efforts to manage these conflicts through conservation can also be applied to other contexts. These include other resource development conflicts in the region, such as that stemming from the transformation of the region's rivers into transportation conduits and disposal sites for sawdust and other industrial wastes.[67] They also present the potential of comparing this region's history to the history of other regions, such as New England, that have experienced potentially similar episodes of resource conflict and that have been studied more thoroughly by historians.[68]

CONCLUSIONS

Spatial technologies have much to contribute to the practice of environmental history. These contributions range from presenting primary sources and information in ways that enable visualization of relationships between historical phenomena to supporting the telling of complex stories about the historical relation between people and the land. Three roles of HGIS are especially evident: i) organizing historical sources in relation to the landscape to make these available to anyone with an interest in the history of the region; ii) visualizing the geographic implications of information from these sources, including visualization of past landscapes; and iii) performing spatial analysis of this information to understand the development of patterns of environmental features, land uses, and other human activities. Of particular promise is the potential to identify meaningful relationships between diverse factors: for example, the relation between agricultural expansion and soil fertility; or between industry location, water power potential, and railroads; or, more unexpectedly, between

the nineteenth-century timber industry and twenty-first century recreational landscapes.

Spatial technologies therefore provide intriguing possibilities for generating and communicating new interpretations of historical change: serving not just as products of research but as research tools by enabling the asking of novel questions regarding the spatial arrangements of historical events and processes. They thus carry implications for historical practice, including the need for historians to consider explicitly the geographical dimensions of their research. In teaching contexts, they present opportunities for interactive exploration of historical themes and for online "publishing" of student work. In the community, there are possibilities for collaboration between academic historians and those interested in heritage conservation.

At the same time, there are numerous challenges. These include practical issues relating to the technical skills and work required before useful benefits become evident. These challenges imply a need for collaboration between individuals with historical, geographical, information science, GIS, and other technical expertise and for substantial funding and access to technical facilities. There is also the challenge inherent in translating historical information into geographic formats, given that this information is often not tied to specific locations. Applying GIS to historical work is not just a technical but a social and conceptual challenge – one inherent in the contrast between the precisely defined locations and spatial patterns represented by GIS and the more subtle concepts of place and landscape employed by geographers and historians.[69]

However, as this project demonstrates, it is possible to overcome these challenges while generating novel insights into the historical relation between humans and their environment. The regional scale of this project can also serve as a model for initiatives elsewhere. Indeed, given the specificity of Canada's regions (however they are defined), initiatives at this scale may have a special role to play in relating historical knowledge to Canada's geographical context.

NOTES

1   On these and other themes in environmental history, see: Matthew Evenden and Graeme Wynn, "54, 40 or Fight: Writing within and across Borders in North American Environmental History," in Sverker Sörlin and Paul Warde, eds., *Nature's End: History and the Environment* (New York: Palgrave Macmillan, 2009).

2   Peter Adams and Colin Taylor, *Peterborough and the Kawarthas*, 3rd ed. (Peterborough: Trent University, 2009).

3   See, for example, maps of Hamilton Township, in Graeme Wynn, *Canada and Arctic North America: An Environmental History* (Santa Barbara: ABC-CLIO, 2007), 125.

4   J. David Wood, *Making Ontario: Agricultural Colonization and Landscape Re-creation before the Railway* (Montreal and Kingston: McGill-Queen's University Press, 2000), 39.

5   Alan Brunger, "Early Settlement in Contrasting Areas of Peterborough County, Ontario," in J. David Wood, ed., *Perspectives on Landscape and Settlement in Nineteenth Century Ontario* (Toronto: McClelland & Stewart, 1975), 125.

6   Cole Harris, *The Reluctant Land: Society, Space, and Environment in Canada before Confederation* (Vancouver: UBC Press, 2008), xv.

7   Alan Brunger, "The Cultural Landscape," in Adams and Taylor, *Peterborough and the Kawarthas*, 119–54.

8   Brunger, "Early Settlement"; Peter Ennals, "Cobourg and Port Hope: The Struggle for Control of 'The Back Country'," in Wood, *Perspectives on Landscape*, 182–95.

9   James Angus, *A Respectable Ditch: A History of the Trent-Severn Waterway, 1833–1920* (Montreal and Kingston: McGill-Queen's University Press, 1988); on rivers and their environmental histories, see, for example, Christopher Armstrong, Matthew Evenden, and H. V. Nelles, *The River Returns: An Environmental History of the Bow* (Montreal and Kingston: McGill-Queen's University Press, 2009).

10 Wood, *Making Ontario*; Neil S. Forkey, *Shaping the Upper Canadian Frontier: Environment, Society, and Culture in the Trent Valley* (Calgary: University of Calgary Press, 2003), 77.

11 H. V. Nelles, "Introduction," in T. C. Keefer, *Philosophy of Railroads* (Toronto: University of Toronto Press, 1972), ix–lxiii.

12 H. R. Cummings, *Early Days in Haliburton* (Toronto: Ontario Department of Lands and Forests, 1963), 161–67.

13 C. Grant Head, "An Introduction to Forest Exploitation in Nineteenth Century Ontario," in Wood, *Perspectives on Landscape*, 78–112.

14 Ian Montagnes, *Port Hope: A History* (Port Hope, ON: Ganaraska Press, 2007), 45.

15 A.R.M. Lower, *The North American Assault on the Canadian Forest: A History of the Lumber Trade between Canada and the United States* (Toronto: Ryerson Press, 1938).

16   Kenneth Kelly, "The Impact of Nineteenth Century Agricultural Settlement on the Land," in Wood, *Perspectives on Landscape*, 71–76; Douglas McCalla, *Planting the Province: The Economic History of Upper Canada, 1784–1870* (Toronto: University of Toronto Press, 1993), 46.

17   Watson Kirkconnell, *County of Victoria: Centennial History* (Lindsay, ON: Victoria County Council, 1967), 60.

18   Forkey, *Shaping the Upper Canadian Frontier*, 25–30.

19   Wood, *Making Ontario*.

20   Al Purdy, "The country north of Belleville," in *The Cariboo Horses* (Toronto: McClelland & Stewart, 1965).

21   C. D. Howe and J. H. White, *Trent Watershed Survey* (Toronto: Commission of Conservation, 1913).

22   Christie Bentham and Katharine Hooke, *From Burleigh to Boschink: A Community Called Stony Lake* (Toronto: Natural Heritage Books, 2000); Doug Lavery and Mary Lavery, *Up the Burleigh Road … Beyond the Boulders* (Peterborough: Trent Valley Archives, 2006, 2007).

23   Head, "Introduction to Forest Exploitation."

24   Wood, *Making Ontario*.

25   William Cronon, *Changes in the Land: Indians, Colonists, and the Ecology of New England* (New York: Hill & Wang, 1983); Carolyn Merchant, *Ecological Revolutions: Nature, Gender, and Science in New England* (Chapel Hill: University of North Carolina Press, 1989).

26   Wood, *Making Ontario*; Graeme Wynn, "Timber Production and Trade to 1850," in R. Louis Gentilcore, ed., *Historical Atlas of Canada* (Toronto: University of Toronto Press, 1987), Plate 11; C. Grant Head, "The Forest Industry, 1850–1890," in Gentilcore, *Historical Atlas*, Plate 38; Marvin McInnis, "Ontario Agriculture, 1851–1901: A Cartographic Overview," in Donald H. Akenson, ed., *Canadian Papers in Rural History*, vol. 5 (Gananoque, ON: Langdale Press, 1986), 290–301.

27   Richard White, "Foreword," in Anne Kelly Knowles, ed., *Placing History: How Maps, Spatial Data, and GIS are Changing Historical Scholarship* (Redlands, CA: ESRI Press, 2008), ix–xi.

28   Ian Gregory and Paul Ell, *Historical GIS: Technologies, Methodologies and Scholarship* (Cambridge: Cambridge University Press, 2007); Knowles, *Placing History*.

29   Brian Donahue, "Mapping Husbandry in Concord: GIS as a Tool for Environmental History," in Knowles, *Placing History*, 151–77.

30   N. L. Nicholson and L. M. Sebert, *The Maps of Canada: A Guide to Official Canadian Maps, Charts, Atlases and Gazetteers* (Folkestone, UK: Wm. Dawson; Hamden, CT: Archon Books, 1981).

31   Map Cruncher has two functions that we found essential: it allows one to tile large images into many small tiles, and also to georeference them, so that maps can be rendered at a variety of zoom levels. Elson, Jeremy, Jon Howell, and John R. Douceur, "MapCruncher: Integrating the World's Geographic Information," Microsoft Research Redmond, April 2007 (http://research.microsoft.com/pubs/74210/OSR2007-4b.pdf, accessed February 24, 2011).

32   McCalla, *Planting the Province*, 93–98; Harris, *Reluctant Land*, 337.

33   Diane Robnik, *The Mills of Peterborough County* (Peterborough: Trent Valley Archives, 2006) was a particularly valuable source of information on mills.

34   Spreadsheet template fields included: Object ID, Site ID, Easting, Northing, Mill Name, Township, Type_Summary (this is a collapsed category, bringing types of mills together), Type (such as grist, saw, feed, flour, planing, oat, etc.), Precision (of location), Date Established, Date Closed, Images, Photographs, Location (textual information on how the location was determined), and Descriptions (historic information gathered from texts).

35   See, for example, Head, "Introduction to Forest Exploitation," fig. 6.3, 82.

36   Wood, *Making Ontario*, 109.

37   Knowles, *Placing History*, 13.

38   Claire Campbell, *Shaped by the West Wind: Nature and History in Georgian Bay* (Vancouver: UBC

Press, 2005); Forkey, *Shaping the Upper Canadian Frontier*; William J. Turkel, *The Archive of Place: Unearthing the Pasts of the Chilcotin Plateau* (Vancouver: UBC Press, 2007).

39   Gregory and Ell, *Historical GIS*, 118.

40   H. W. Taylor, J. Clarke, and W. R. Wightman, "Contrasting Land Development Rates in Southern Ontario to 1891," in Akenson, *Canadian Papers in Rural History*, vol. 5, 50–72.

41   Wood, *Making Ontario*, 106–7.

42   Harris, *Reluctant Land*, 365.

43   Wood, *Making Ontario*, 112–13.

44   Clifford Theberge and Elaine Theberge, *At the Edge of the Shield: A History of Smith Township, 1818–1980* (Peterborough: Smith Township Historical Committee, 1982), 139.

45   Lavery and Lavery, *Up the Burleigh Road*.

46   Patricia Jasen, *Wild Things: Nature, Culture, and Tourism in Ontario, 1790–1914* (Toronto: University of Toronto Press, 1995).

47   Gregg Mitman, *Breathing Spaces: How Allergies Shape Our Lives and Landscapes* (New Haven, CT: Yale University Press, 2007).

48   Katharine Hooke, *From Campsite to Cottage: Early Stoney Lake* (Peterborough: Peterborough Historical Society, 1992), 2–5.

49   We thank John Wadland for explaining this point to us.

50   Withrow, quoted in Lower, *North American Assault*, frontispiece.

51   Howe and White, *Trent Watershed Survey*, 1–4.

52   Ibid.

53   A. H. Richardson, *A Report on the Ganaraska Watershed* (Toronto: Dominion and Ontario Governments, 1944), 24; Head, "Introduction to Forest Exploitation"; Forkey, *Shaping the Upper Canadian Frontier*, 79; Ontario, Department of Energy and Resources Management,

Otonabee Region Conservation Report: Summary (Toronto: Department of Energy and Resources Management, 1965), 21–22.

54   Head, "Introduction to Forest Exploitation."

55   Ontario, *Otonabee Region*, 20.

56   Theberge and Theberge, *At the Edge of the Shield*, 109.

57   Brunger, "Early Settlement."

58   Forkey, *Shaping the Upper Canadian Frontier*, 76.

59   Cummings, *Early Days in Haliburton*.

60   Graeme Wynn, "Notes on Society and Environment in Old Ontario," *Journal of Social History* 13, no. 1 (1979): 49–65.

61   Lower, *North American Assault*.

62   Wynn, "Notes on Society and Environment."

63   Forkey, *Shaping the Upper Canadian Frontier*, 89–93; Howe and White, *Trent Watershed Survey*.

64   E. J. Zavitz, *Report on the Reforestation of Waste Lands in Southern Ontario* (Toronto: L. K. Cameron, 1908).

65   Richardson, *Ganaraska Watershed*, 226.

66   H. V. Nelles, *The Politics of Development: Forests, Mines, and Hydro-Electric Power in Ontario, 1849–1941* (Montreal and Kingston: McGill-Queen's University Press, 2005).

67   R. Peter Gillis, "Rivers of Sawdust: The Battle over Industrial Pollution in Canada, 1865–1903." *Journal of Canadian Studies* 21 (1986): 84–103; Jamie Benedickson, *The Culture of Flushing: A Social and Legal History of Sewage* (Vancouver: UBC Press, 2007).

68   Theodore Steinberg, *Nature Incorporated: Industrialization and the Waters of New England* (Amherst: University of Massachusetts Press, 1991).

69   Gregory and Ell, *Historical GIS*.

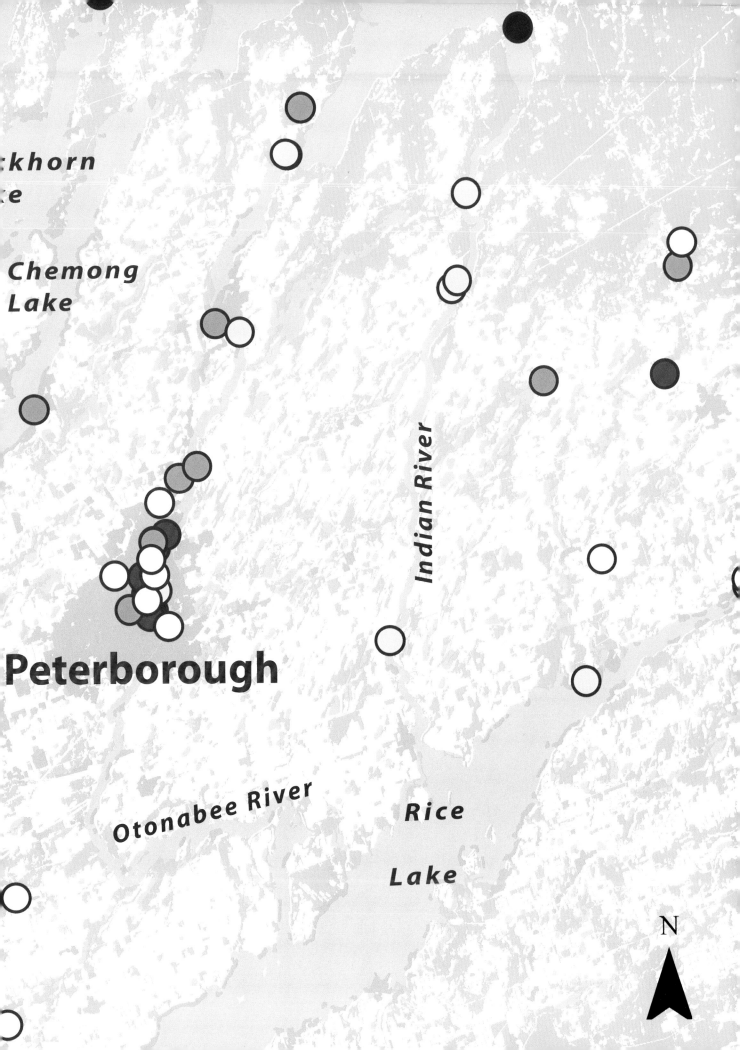

6

Mapping Ottawa's Urban Forest, 1928–2005

Joanna Dean and Jon Pasher

Like many HGIS projects, this research began with a question that could not be answered with traditional historical sources. Municipal records showed dramatic shifts in attitudes to city trees in the mid-twentieth century. We wanted to know whether these cultural shifts reflected changes in the urban forest itself.[1] Were there too many trees in the 1920s, as critics charged? Were there too few in the 1960s, as environmentalists claimed? And how many trees are too many, anyway? The question was of more than academic interest. Urban foresters have established that trees provide significant health and climate benefits by cleaning and cooling city air. Trees can live for decades, and historical analysis can contribute to our understanding of forest growth and improve management in the future.

Our pilot project assessed the use of historical aerial photographs to measure canopy cover on selected neighbourhoods in Ottawa. The techniques for measuring forest canopy cover from aerial photographs and satellite images are well tested, but little historical analysis has been done, and the use of historical aerial photographs complicated the project in a variety of ways.[2] We found that it was possible to measure canopy cover within a reasonable margin of error. It appears that the critics may have been right, that tree cover was dense in central parts of the city in the early twentieth century and that it subsequently declined. It appears that the environmentalists were also right to worry about trees in the 1960s, when the inner-city decline was matched by the deforestation in early suburbs like Alta Vista.

What was interesting, however, was that geospatial analysis took the research beyond the original question: not only could we draw connections between canopy cover and shifts in popular opinion, but we could also correlate canopy cover to social indices such as class, income, or race. Inner-city neighbourhoods were subjected to close scrutiny for urban renewal in the postwar period, and a wealth of statistics are available. Our pilot project had not been designed with this analysis in mind, but even so the correlation of income and canopy cover is clear. The statistics allow us, following Nik Heynen in the United States, to insert the urban forest into an environmental justice framework as an environmental benefit that is socially produced and unevenly distributed.[3]

The project also offered an opportunity to move beyond statistics. Our methodology created maps of canopy cover that can be read visually and stacked in a dynamic time series. This has the potential to alter our perception of the urban forest. Trees appear to humans to be static entities, and their growth is only recognized retrospectively: it is a common trope of memoir writers to observe with surprise that a childhood tree has grown. The dynamic changes of canopy cover visible in a time series reminds us that forests are living agential communities that move and respond to changes in the built environment.

While our own work and experiences are focussed on urban forests, the methods and issues are applicable to other HGIS applications. The observations arise from the different perspectives of project team members. Jon Pasher, who was completing his doctorate in Geography at the time of the project, focussed on the methods, development, and accuracy of the statistics. Joanna Dean, a historian of the urban forest, with very little experience in HGIS, focussed on the applicability of the method for environmental history. The project described here was a pilot project, funded with a small SSHRC grant, and possible only because of the generosity of colleagues in the Geomatics and Landscape Ecology Laboratory at Carleton. We analyzed only five carefully selected areas, and, as in any good pilot project, we learned from our mistakes. We offer the following observations as "lessons learned," intended to be useful to others interested in incorporating historical air photos into their HGIS research.

CONTEXT

Urban forests have been the subject of much recent analysis in Europe and North America. The recognition of the environmental benefits or services provided by city trees led to a growing demand for the quantification and monetization of these benefits, and the developing science has provided the context for our research.[4] The term "urban forest" was coined by Erik Jorgensen, at the University of Toronto Faculty of Forestry, and has been widely adopted. The term is generally understood to include all urban trees: the street trees that line roads, as well as single trees and groups of trees in gardens, yards, cemeteries, parks, and woodlands. Although the urban tree is normally distinguished from shrubs and other vegetation by its trunk and crown (the urban tree is defined as "a woody perennial plant growing in towns and cities, typically having a single stem or trunk – and usually a distinct crown – growing to a considerable height, and bearing lateral branches at some height from the ground,"[5]) most observers include shrubs within an urban forest. For those involved in canopy cover analysis of aerial photographs and remote sensing, it is often difficult to distinguish between trees and shrubs, and, as the ecoservice benefits are similar, the inclusion of shrubs makes sense methodologically. Ecologists sensitive to the interrelationships between plants and animals tend to think more broadly, and the term urban forest is sometimes extended to include grass, and even the related biota in an urban ecosystem.[6]

Urban foresters have established that urban trees provide a wide array of ecosystem services: they moderate air temperature, attenuate storm water flooding, mitigate urban heat island effects, and reduce noise and air pollution. Trees have been shown to have a measurable impact on urban levels of particulate matter, ozone, sulphur dioxide, nitrogen dioxide, and ozone. Woodlands increase urban biodiversity by providing semi-natural habitats to a wide variety of species, and at a broader scale urban trees contribute to carbon sequestration and storage.[7] They also provide social benefits: not only do they make a city aesthetically pleasing, (and raise real estate values), but they have measurable impact on residents' health and sense of wellbeing. Trees growing within sight of a hospital room have been shown to improve recovery rates of patients.[8] At the same time, urban trees also provide disservices: they produce allergenic pollen, volatile organic compounds (contributing to smog), and green waste. The costs and inconveniences of tree management can be high in urban areas, especially during storms, when they do significant damage to infrastructure. Not all city residents value the aesthetics of a treed urban landscape.[9]

But the consensus is that the services outweigh the disservices, or have the potential to do so if well managed, and so the emphasis in the literature is on improved understanding and management of urban trees.[10]

Because of the focus on these ecosystem services, and the interest in assigning monetary values, many urban forest studies adopt quantitative methods: field surveys for smaller areas and GIS, aerial photography and remote sensing for larger areas.[11] The science has developed with advances in remote sensing and spatial analysis. It is often possible to acquire high-resolution satellite images with a spatial resolution of 60 cm and as well many municipalities have acquired low-altitude aerial photographs, which, although often intended for other purposes, can prove very useful for identifying canopy cover. One recent study demonstrated the inequitable distribution of greenspace in Montreal by extracting vegetation indicators from high-resolution satellite images, indicating the proportion of city blocks, streets, alleys, and backyards covered by total vegetation and trees/shrubs. These data were correlated to census data at the level of the dissemination area (400–700 people, roughly a city block in Montreal) to show environmental inequities.[12] One of the authors of this chapter, Jon Pasher, is currently engaged in a national-scale analysis of carbon sequestration within the urban areas of Canada. As a pilot project at Environment Canada, high-resolution imagery is being used to assess urban canopy cover to improve on previous estimates of the contribution to Canada's urban area carbon budget calculations.

Historical studies are limited by the available data: aerial photographs from the mid-twentieth century do not permit this kind of close digital analysis. Historical studies are worth attempting, however, because of the longevity of trees. Unlike most vegetation, trees survive for decades, even in some cases centuries. In a young city like Ottawa, trees can be older than the city itself. Dendrochronology showed that a bur oak, recently cut down for infill housing, was 154 years old, the same age as the City of Ottawa, ninety years older than the house it shaded and at least a hundred years older than the man who felled it.[13] Trees also take many years to die. Analysis of a group of bur oaks in Winnipeg showed that they had been in decline for decades because of changes in water table levels caused by residential construction in the 1940s.[14] In another project, core analysis of urban Norway maples led ecologists to conclude that the trees had been in decline for twenty years. They speculated the decline was related to a combination of drought and sidewalk renovation and noted: "serious symptoms of deterioration in the crown may not occur for many years after the onset of decline."[15] If the purpose of urban forest analysis is to improve management, then historical studies capable of assessing arboreal growth and decline over a number of decades will be essential.

METHODOLOGY

Historical air photos were obtained from the Canadian National Air Photo Library. Each photo was scanned and georeferenced to real-world geographic coordinates within a GIS environment by collecting matching tie-points on both the older photos as well as more recent digital air photos, which were already georeferenced. While the scales of the photos differed greatly, the pixel size, or "spatial resolution" was standardized for all the photos in a time series in order to standardize the level of detail at which the interpreter would outline the canopy. A minimum bounding study area was set in a neighbourhood that included as large an area as possible but at the same time as many historical photos as possible. Within these reduced study areas, tree cover was manually interpreted and digitized in a GIS environment. The methodology resulted in maps of canopy cover that can be interpreted visually. The resultant GIS layers provided a per cent canopy coverage within each study area over time, by taking the aerial estimates of digitized canopy cover as a percentage of the total area of the neighbourhood being analyzed.

This methodology was adopted from landscape ecology and represents a departure from the dot-grid methodology used for most urban forest analysis. Dot grid, used for political ecology by Nik Heynen and by urban forester David Nowak, has developed from the pre-digital era, when a clear dot grid was placed over an aerial photograph, and the number of dots lying on forest canopy were counted. (See Figure 6.1B.) The results provide an average measure of canopy cover for an entire city but do not provide precise measures of individual neighbourhoods and streets unless intensive sampling is done in small areas.

Our methodology relied on the painstaking manual digitization of the canopy cover of each and every tree, but the results are dramatic. Patterns of canopy cover change can be read visually and can be read at the street-by-street, and even tree-by-tree level. While point based sampling might be faster to carry out, it does not allow for close analysis of changing spatial

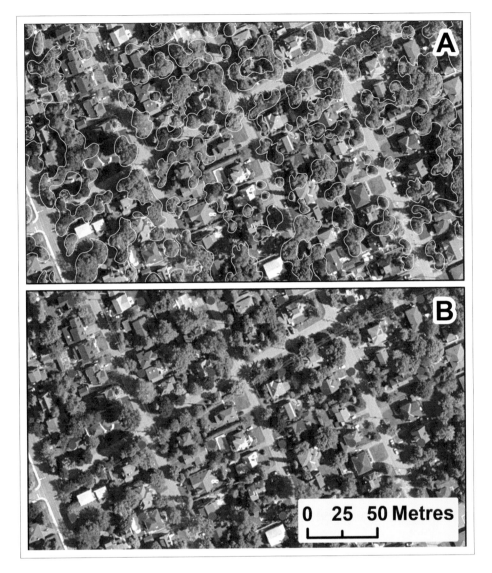

Fig. 6.1. Commonly used canopy cover analysis methods including (A) manual digitizing of canopy cover and (B) point-based sampling.

patterns. (Figure 6.1 demonstrates the difference between the two most common methods.) A second benefit is that individuals with no expertise in geospatial analysis can easily grasp the meaning of the maps. This is important for communicating results with other historians and sharing them with community groups.

We encountered interpretation difficulties that can be classified into three main categories: 1) data availability, 2) data quality, and 3) photo attributes. The availability of historical aerial photographs limits the spatial coverage and depth of analysis that can be performed. In the early years, the flight paths along which photos were taken were sporadic both in terms of spatial coverage and temporal coverage, limiting the neighbourhoods available for analysis, as well as the number of photos available through time for the different neighbourhoods. In many cases, we were able to go back to 1946 photos (and occasionally as far back as 1928); however, regular time intervals were not always

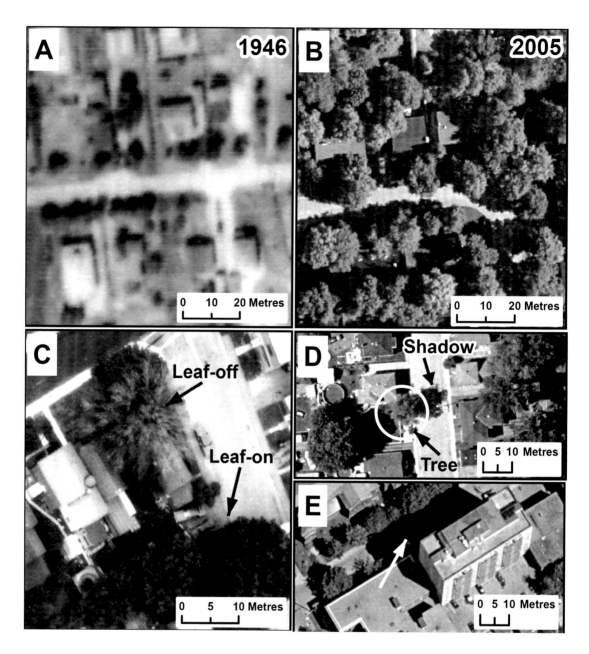

Fig. 6.2. Close-up examples of air photos illustrating some of the issues encountered when digitizing and analyzing. (A) Poor-quality photo (taken in 1946) as a result of technology as well as aircraft height compared with a 2005 photo (B) taken at a lower height using new technology. (C) A comparison between leaf-on and leaf-off conditions demonstrating the difficulty in delineating the crowns of the trees. (D) A large tree-shadow as a result of the low sun angle at the time the photo was taken, demonstrating interpretation difficulties. (E) Shadows created by an apartment building, potentially hiding trees lining the street behind.

Fig. 6.3. Guiges and King Edward Avenue, 1938. Street trees, many of them silver maples, shaded Lowertown in the 1930s. This photograph is from a series of streetscapes used by landscape architect Jacques Gréber that emphasize the beauty of the trees. (Guiges and King Edward [west], Library and Archives Canada/ Department of Public Works fonds/ e010869284.)

possible. In one case, although we had photos from 1946, 1952, and 1956, there were no photos available again until 1985. Such gaps in time have a significant impact on results, especially in such a dynamic environment. Spatial gaps were also problematic. The Canadian government began to gather data for urban "census tracts" in the postwar period, but our study areas rarely matched these tracts, and so the correlations with social indices are not as exact as we might have wished.

Even when photos were available over a specific area, we encountered problems with different acquisition scales (i.e., the aircraft was at a different height during the acquisition, or a different camera lens was used), which resulted in differing levels of image resolution and clarity. (See Figs. 6.2A and 6.2B) While older photos were interpretable by an expert, the results were not as accurate.

The attributes of the photos related to the seasonality and time of day of the acquisitions also caused variability. We sometimes were forced to rely upon photos that were taken in the spring when not all of the tree leaves had opened up and in the fall when some of the leaves had already fallen off. Figure 6.2C demonstrates this issue, with crown delineation hindered when the leaves were not present. Further, as with any work using air photos in an urban environment, shadows are often present. Figure 6.2D provides an example of a large tree shadow cast across the road. Interpretation of these shadows was particularly difficult in older air photos that were only available in greyscale (as opposed to true colour, which is available in more recent photos). Figure 6.2E demonstrates the fact that buildings cast shadows, preventing interpreters from seeing and therefore mapping trees lining the street. Issues of shadows are magnified by photos taken with a low sun angle as well as the presence of taller buildings. Despite these difficulties, we concluded that, while the statistics calculated were by no means exact, the relative measures calculated tell interesting and useful stories about changes over time and provide additional information unobtainable without such techniques.

Analysis of Ottawa's oldest neighbourhood, Lowertown, reveals the impact of natural and social forces on tree canopy cover. Lowertown was built upon a cedar swamp, and it was initially occupied by the Irish and French Canadian labourers employed in building the Rideau Canal. In the 1940s, when our analysis starts, it was a cohesive francophone working class neighbourhood, distinguished by a landscaped federal avenue, King Edward Avenue, leading from the Governor General's residence, and the bustling commerce of the Byward Market. Street-level photographs taken in the 1930s show a dense canopy cover with tree trunks crowding narrow residential streets.[16] (See Fig. 6.3.) Today there are a few remaining massive street trees, some new street tree plantings, and large spreading trees to the rear of the houses. GIS analysis of one small section of Lowertown shows that the canopy cover was 32 per cent in 1946, dropped to 22 per cent a decade later, and further to 10 per cent in 1966 and 1976. It has only risen to 15 per cent in recent years. (Figure 6.4 shows a time series of photos from 1946 to 2005, and Fig. 6.5 shows mapped canopy cover changes from 1946 to 2005).

The study area for Lowertown was small, but this example shows the importance of occasionally working at an even lower scale, on a street-by-street basis. King Edward Avenue is visible on the right in Figs. 6.4 and 6.5: it was landscaped by the Ottawa Improvement Commission in the early twentieth century with a wide central boulevard of American elm trees. The trees grew magnificently (they were planted over a bywash from the Rideau Canal) and

by the 1940s provided deep shade. This avenue was selected by urban planner Jacques Gréber to illustrate the beauties of Ottawa trees; a photograph of King Edward Avenue is featured in his *Plan for the National Capital*. The boulevard of trees was removed in 1965 when Dutch elm disease (DED) struck, to be replaced with additional lanes of traffic, as the avenue became the main traffic artery leading to a new interprovincial bridge.[17] The tree-cutting coincided with a devastating urban renewal project that destroyed the fabric of the neighbourhood, and the avenue has attained iconic status in the memory of Lowertown residents as a symbol of all that they lost.[18] (See Figure 6.6.)

The loss of tree cover on this one avenue skews the results for the entire neighbourhood. If we take this avenue out of the analysis, the tree cover in this working class neighbourhood was only 22 per cent in 1946, declined to 9 per cent in 1956, and then hovered in the low teens until it reached 18 per cent in 2005. Analysis of a second study area in Lowertown showed a similar pattern: 16 per cent cover in 1946; 6 per cent in 1954; 11 per cent in 1976; 8 per cent in 1983, and 15 per cent in 2005. This is the same pattern noted in other neighbourhoods in the inner city. The adjacent neighbourhood of New Edinburgh dropped from 20 per cent in 1928 to 19 per cent in 1956. New Edinburgh, however, was not impacted by urban renewal. (They successfully resisted plans to run the interprovincial artery adjacent to their neighbourhood.) As New Edinburgh gentrified, the canopy cover recovered to 23 per cent in 1966, and 26 per cent in 2005. In a third inner-city neighbourhood, Golden Triangle, the canopy cover dropped from 27 per cent in 1950 to 20 per cent in 1966 and 18 per cent in 1990 and

Fig. 6.4. Lowertown. Close analysis of canopy cover in this inner-city neighbourhood demonstrates that the urban forest is an inequitably distributed environmental benefit. Not only is canopy cover currently lower than that of wealthy neighbourhoods, like Alta Vista (see Fig. 6.7), but the densest cover in Lowertown was for many years along one federal boulevard, King Edward Avenue, on the right. In 1946, overall canopy cover was 32 per cent, but if King Edward Avenue is removed from the analysis, the cover on the adjacent residential streets was only 22 per cent. In the 1960s, the trees on this avenue were removed because of Dutch elm disease, and the avenue is now a six-lane interprovincial highway. For street-level photographs of King Edward Avenue, and a sense of neighbourhood anger at the loss of this boulevard, see the King Edward Task Force website, available at http://www.kingedwardavenue.com. (Aerial photographs courtesy of the National Air Photo Library [Series A10371,2; 1946] and City of Ottawa [2005].)

2005 Canopy Extent
1946 Canopy Extent

0 25 50 Metres

Fig. 6.5. Changes in canopy cover for Lowertown from 1946 to 1950.

has recently risen to 22 per cent in 2005. In all three neighbourhoods, we observe dramatic declines in the middle of the twentieth century and a varying degree of recovery subsequently. These figures, combined with street-level photographs, support the conclusion drawn from textual evidence that the canopy cover was relatively dense in the 1920s and thin in the 1960s.

We might attribute the decline to the management policies of the City of Ottawa. American elm was a favourite tree in Ottawa, as in most cities in eastern North America, and was so thickly planted that it became a problem in the 1920s. Municipal records show that the City engaged in a radical program of tree-trimming and removal to control fast growing "nuisance" trees in the 1930s and 1940s: over 4,000

Fig. 6.6. King Edward Avenue, ca. 1960. Lowertown was subjected to a massive urban renewal project in the 1960s. Photographs justified renewal by documenting poverty in the neighbourhood; they reveal the presence of mature trees planted by an earlier generation. (291–293 King Edward Avenue, ca. 1960. City of Ottawa Archives/2009.0413.1/ CA15960.)

trees were removed between 1921 and 1945.[19] Dutch elm disease brought an unexpected denouement. The disease hit Ottawa in the 1950s, after moving from the United States through Toronto and Montreal, and its impact was only felt in the 1960s and 1970s as the elm trees disappeared from the streetscape. But municipal foresters were cognizant of the impact in other cities and reviewed their street tree policy in 1956. They increased tree planting dramatically, planting a total of 2,600 street trees in 1961 (a significant number in a city with a total of 55,000 shade trees, especially compared to 24 trees planted in 1944) and called for street tree planting and improved management and conservation of existing trees in a 1962 report.[20] Municipal authorities noted the dearth of trees in inner-city neighbourhoods and blamed the residents "in the central wards where too many people are inclined to believe that shade trees are part of the past, or that they are the sole responsibility of the Department."[21]

ALTA VISTA, 1946–2005

Analysis of the Alta Vista suburb, in Fig. 6.7, shows the impact of suburban development on forest cover.[22] Alta Vista was an upper-middle-class suburb, developed in the 1950s to the southeast of Ottawa. We initially selected two areas for mapping: 1) an agricultural area, which had only hedgerows and a few individual trees prior to development; and 2) a forested area. Mapping both areas together, we found that canopy cover nearly halved in ten years of construction, dropping from 57 per cent to 29 per cent between 1946 and 1956. It recovered slowly to 34 per cent over the next thirty years and then twenty years later to 48 per cent. (The numbers are high throughout because the mapping area includes a woodlot that was held in reserve for future highway development.) The lesson here is that canopy cover recovers after suburban development but does so very slowly.

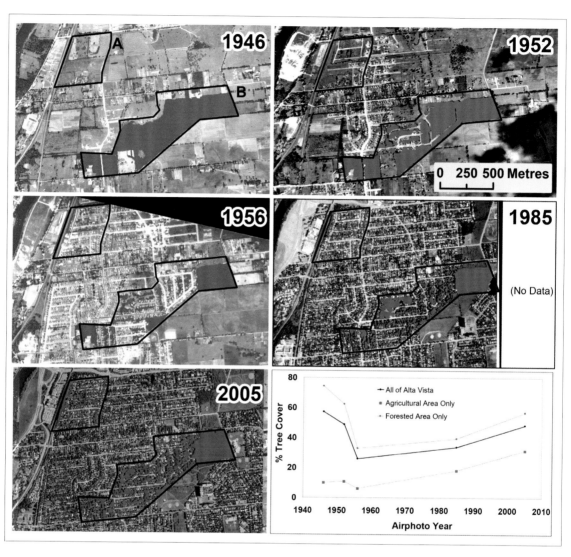

Fig. 6.7. Alta Vista. This prosperous postwar suburb is known for its large lots and mature trees. Two areas were mapped in order to compare the impact of suburban development on open fields and on a dense woodlot. The canopy cover in Area A (see 1946 photo), which began as an agricultural area, was only 10 per cent before construction in 1946. It dropped further to 6 per cent in 1956 with the construction of roads and houses. By 2005, canopy cover had rebounded to 31 per cent. The canopy cover in Area B (see 1946 photo), the original woodlot, was 96 per cent in 1946. It dropped precipitously with construction to 18 per cent in 1956. By 2005, this area had returned to half of its original canopy cover, at 46 per cent. This new forest, however, includes a large proportion of ash trees, now threatened with emerald ash borer. For a complete series of canopy cover mappings of Alta Vista, see http://www.carleton.ca/~jdean/urbanforest.htm. (Aerial photographs courtesy of the National Air Photo Library [Series A 10347, 26; 1946] and City of Ottawa [2005].)

We then analyzed the agricultural area alone (see the region labeled 'A' in 1946 image, Fig. 6.7). Canopy cover here was very low initially, at 10 per cent, and it fell further to 6 per cent before rebounding as new homeowners planted shade trees. It surpassed the original measure in 1985 when it reached 18 per cent and by 2005 had tripled to 32 per cent. It is not perhaps surprising but still useful to know that this kind of low-density suburban development offers a better forest cover than agricultural areas.

Finally, we identified a small densely forested area with 97 per cent canopy cover in 1946 (see the region labeled 'B' in 1946 image, Fig. 6.7). The decline of canopy cover in this area with suburbanization was dramatic, dropping to 18 per cent in 1956, and the recovery fifty years later had only reached about half of the original canopy cover at 46 per cent. The lesson here? The canopy cover in even the greenest suburb is a poor replacement for a forest.

The analysis shows that Alta Vista was the beneficiary of renewed interest in trees in the 1950s and 1960s. Lots in suburban Alta Vista were spacious, especially compared with Lowertown, and thickly planted with large trees. This intent does not show up in the GIS analysis until the trees matured, thirty years after the planting. In Alta Vista, however, the apparent health of the 2005 urban forest is belied by the fact that it is made up of a large number of ash trees. The emerald ash borer (EAB) was first observed in Michigan in 2002 and was confirmed in the Ottawa area in July 2008. This wood-boring beetle has decimated urban forests in southwestern Ontario and parts of the United States and is expected to have a devastating impact in Ottawa. The City of Ottawa has identified Alta Vista as a hot spot

for EAB. Ironically, the ash was often selected as a substitute for the America elm; although the ash did not have the elegant vase shape of the American elm, it was a tall tree with dense foliage and offered many of the benefits of elm. Some streets in Alta Vista were planted with 60 per cent ash trees. Tree removals and replacement began in 2011, and a limited program of injections with insecticides began in 2012, but it is expected that the canopy cover in this neighbourhood will be dramatically reduced.

ENVIRONMENTAL JUSTICE

Although natural forces like DED and EAB dramatically impact urban forest canopy cover, social forces are also significant. HGIS allows us to incorporate an environmental justice analysis, using census returns to draw some correlations between canopy cover and the income levels of residents.[23] Our findings are only suggestive as the census areas are not entirely congruent with our mapping areas, but they do consistently suggest that canopy cover follows wealth. Lowertown was a heavily (72 per cent) francophone neighbourhood in 1941, with only 18 per cent of the residents owning their homes, and relatively low median male income of $1,692, 83 per cent of the city average. By 1961 the median male income was $3,219, a relative decline to 69 per cent of the city average. In Alta Vista, by comparison, the median male income was $5,678 in 1961.[24] Although we have not yet correlated income to canopy cover in the remaining urban neighbourhoods, it appears that the recovery of the canopy is related to the gentrification of New Edinburgh and the Golden Triangle. Urban renewal reports of inner-city neighbourhoods provide further statistics: substandard buildings are mapped, and charts tabulate the number of residents and the state of the plumbing. It would be possible to correlate these statistics with canopy cover if two inner-city neighbourhoods were compared. Alta Vista, however, was not subjected to intrusive urban renewal surveys, so we were unable to develop further correlations.

There was, however, a significant lag to the correlation with income: prosperous Alta Vista residents had to wait for their trees to grow in 1961, while Lowertown residents benefited for years from turn-of-the-century tree planting efforts. The evidence for time lag in tree canopy cover is significant and shows the importance of a long-term historical analysis of environmental inequities. An analysis of vegetation cover in Baltimore in 1960 and 2000 led the authors to argue that "the landscapes we see today are therefore legacies of past consumption patterns." They explain: "[Our] findings suggest that herbaceous or grassy areas, typically lawns, are good reflections of contemporary lifestyle characteristics of residents while neighborhoods with heavy tree canopies have largely inherited the preferred landscapes of past residents and communities. Biological growth time scales of trees and woody vegetation means that such vegetation may outlast the original inhabitants who designed, purchased, and planted them."[25]

Close analysis also shows a second pattern. In the residential streets of Lowertown, the recent recovery in canopy cover is more apparent than real. The large street trees that dominated the canopy in the 1940s were not replaced. The new street trees are smaller species, such as Japanese lilac and choke cherry, which will never attain the height and breadth of the large silver maples and other forest trees that lined the streets (and crowded the front yards) in the 1940s. Much of the existing canopy cover is provided by large Manitoba maples that have grown to the rear of the buildings. Manitoba maple or box elder are weed trees that self-seed in neglected yards and along fence lines. These trees are considered a liability. Although their extensive canopy provides ecosystem services, they also create a disproportionate amount of disservices such as limbs liable to crack during windstorms, extensive sucker growth, foundation damage from invasive roots and prolific seed production. GIS analysis of canopy cover alone misses this kind of qualitative change.

LIMITATIONS OF HGIS

If used in isolation, the statistics provided by canopy mapping can be misleading. Mapping, for example, measures only breadth and does not take into account the depth of the canopy. A squat Japanese lilac might be the geospatial equivalent to a tall columnar oak. Mapping does not distinguish between species of trees or calculate the biodiversity of the forest. The predominance of ash in the Alta Vista neighbourhood is not apparent on the canopy maps. Nor does the method distinguish between trees on public and private property; something that is significant in the weighing of social benefits and costs of urban trees. In Lowertown, the numbers do not reveal that massive silver maple street trees, managed by the municipality, were replaced in part by Manitoba maples that self-seeded along fence lines and were a private responsibility.

The full potential of the method emerges when it is combined with other sources. Urban foresters use site observations to support their statistics; historians cannot do this, but because trees live for decades, we can with some caution read backwards from existing trees, and we can refer to street-level photographs and textual sources to augment the statistical measures.[26]

Finally, statistics do not tell stories. The numbers do not provide the interwoven narratives of urban renewal and dispossession in Lowertown, the play of federal politics that lies behind the landscaping of King Edward Avenue, and the aspirations that led middle-class residents to drive to the leafy havens on the outskirts of the city in Alta Vista. Statistics do, however, play an important role in environmental history by documenting the material presence of trees. The maps produced in this geospatial analysis provide compelling visual evidence of the dynamic patterns of canopy growth and decline. Geospatial mapping ensures that the stories we tell about the city consider the agency of the natural world.

NOTES

1   For the shifts in attitude, see Joanna Dean, "'Said Tree is a Veritable Nuisance': Ottawa's Street Trees, 1869–1939," *Urban History Review / Revue d'histoire urbaine* 34, no. 1 (2005): 46–57.

2   For a discussion of methodology, see J. T. Walton, D. J. Nowak, and E. J. Greenfield, "Assessing Urban Forest Canopy Cover," *Arboriculture and Urban Forestry* 34, no. 6 (2008): 334–40. For an early historical application of this methodology, see D. J. Nowak, "Historical Vegetation Change in Oakland and its Implications for Urban Forest Management," *Journal of Arboriculture* 19 (1993): 313–19. For politically engaged analysis of historical urban forest mapping, see Nick Heynen's work, especially "The Scalar Production of Injustice within the Urban Forest," *Antipode: A Journal of Radical Geography* 35, no. 5 (2003): 980–98; and "Green Urban Political Ecologies: Toward a Better Understanding of Inner City Environmental Change," *Environment and Planning A* 38, no. 3 (2006): 499–516. For a well designed historical analysis of canopy cover in Twin Cities, Minneapolis, see Adam Berland, "Long-term Urbanization Effects on Tree Canopy Cover along an Urban-Rural Gradient," *Urban Ecosystems* 15, no. 1 (2012). For Canada, see Hiên Pham's analysis of canopy cover in Montreal, T-T.H. Pham, P. Apparicio, A-M. Séguin, M. Gagnon, "Mapping the Greenscape and Environmental Equity in Montreal: An Application of Remote Sensing and GIS," in by S. Caquard, L. Vaughan, and W. Cartwright, eds., *Springer Lecture Notes in Geoinformation and Cartography – Mapping Environmental Issues in the City.* Arts and Cartography Cross Perspectives (2011): 30–48.

3   Nik Heynen, "Scalar Production of Injustice" and "Green Urban Political Ecologies."

4   See, for example, the contents of the journals *Urban Forestry and Urban Greening, Arboriculture and Urban Greening, Landscape and Urban Planning*, and *Urban Ecology.*

5   Sudipto Roy, Jason Byrne, and Catherine Pickering, "A Systematic Quantitative Review of Urban Tree Benefits, Costs and Assessment Methods across Cities in Different Climatic Zones," *Urban Forestry and Urban Greening* 11, no. 4 (2012): 351–63.

6   Escobedo et al. define 'urban forest' as: "the sum of all urban trees, shrubs, lawns, and pervious soils located in highly altered and extremely complex ecosystems where humans are the main drivers of their types, amounts, and distribution." F. J. Escobedo, T. Kroeger, and J. E. Wagner, "Urban Forests and Pollution Mitigation: Analyzing Ecosystem Services and Disservices," *Environmental Pollution* 159 (2011): 2078–87.

7   See C. C. St. Clair, M. Tremblay, F. Gainer, M. Clark, M. Murray, and A. Cembrowski, "Urban Biodiversity: Why It Matters and How to Protect It," discussion paper provided for the City of Edmonton (May, 2010) available at http://www.edmonton.ca/city_government/documents/Discussion_Paper_8_Biodiversity.pdf). D. J. Nowak, D. E. Crane, and J. C. Stevens, "Air Pollution Removal by Urban Trees and Shrubs in the United States," *Urban Forestry and Urban Greening* 4 (2006): 115–23. For a review of the literature, see Roy et al., "Systematic Quantitative Review."

8   For health benefits, see J. Maas, R. A. Verheij, S. de Vries, P. Spreeuwenberg, F. G. Schellevis, and P. P. Groenewegen, "Morbidity Is Related to a Green Living Environment," *Journal of Epidemiology and Community Health* 63, no. 12 (2009): 967–73. The research on hospital recovery is the oft-cited S. Ulrich, "View through a Window May Influence Recovery from Surgery," *Science* 224 (1984): 420–21. For a skeptical review of the claims for the health benefits of greenspace, see A.C.K. Lee and R. Maheswaran, "The Health Benefits of Urban Green Spaces: A Review of the Evidence," *Journal of Public Health* 33, no. 2 (2011): 212–22.

9   For disservices, see Francisco Escobedo and Jennifer Seitz, "The Costs of Managing an Urban Forest," University of Florida, IFAS Extension, FOR 217 (2009) available at http://edis.ifas.ufl.edu/pdffiles/FR/FR27900.pdf. For cultural differences in tree appreciation, see E.D.G. Fraser, and W. A. Kenney, "Cultural Background and Landscape History as Factors Affecting Perceptions of the Urban Forest," *Journal of Arboriculture* 26, no. 2 (2000): 106–13, and H. Schroeder, J. Flannigan, and R. Coles, "Residents Attitudes toward Street Trees in the UK and US Communities," *Arboriculture and Urban Forestry* 32 (2006): 236–46.

10 R. F. Young, "Managing Municipal Greenspace for Ecosystem Services," *Urban Forestry and Urban Greening* 9 (2010): 313–21.

11 The survey of the literature on urban trees by Roy et al. ("Systematic Quantitative Review.") found that 91.5% of the papers assessed used quantitative methods.

12 Thi-Thanh-Hiên Pham, Philippe Apparicio, Anne-Marie Séguin, Shawn Landry, and Martin Gagnon, "Spatial Distribution of Vegetation in Montreal: An Uneven Distribution or Environmental Inequity?" *Landscape and Urban Planning* 107, no. 3 (2012): 214–24. Thi-Thanh-Hiên Pham, Philippe Apparicio, Anne-Marie Séguin, Shawn Landry, and Martin Gagnon, "Predictors of the Distribution of Street and Backyard Vegetation in Montreal, Canada," *Urban Forestry & Urban Greening* 12, no. 1 (2012): 18–27.

13 See the description of the trees at "Champlain Oaks" at http://champlainoaks.posterous.com. This tree became the focus of a urban forest history exhibit at the Bytown Museum, from January 2012 to September 2012, "Six Moments in the History of an Urban Forest." Tree dating by Michael Pisaric.

14 H. A. Catton, S. St George, and W. R. Humphrey, "An Evaluation of Bur Oak (Quercus macrocarpa) Decline in the Urban Forest of Winnipeg, Manitoba, Canada," *Arboriculture and Urban Forestry* 33, no. 1 (2007): 22–30.

15 For the delayed impact of drought on Norway maples, see J. D. Apple and P. D. Manion, "Increment Core Analysis of Declining Norway Maples, Acer platanoides," *Urban Ecology* 3–4 (1986): 309–21.

16 See, for example, Dalhousie and Cathcart Streets, 1938, Library and Archives Canada/Department of Public Works, fonds/197-140 CP, 1069-147.

17 For a description of the devastating impact of Dutch elm disease, see Thomas J. Campanella, *Republic of Shade: New England and the American Elm* (New Haven, CT: Yale University Press, 2003).

18 See, for example, D. L. Morny, *Farewell My Bluebell: A Vignette of Lowertown* (Ottawa: 1998), which opens with a description of King Edward Avenue. See the King Edward Task Force website at http://www.kingedwardavenue.com/ for Gréber's and other historic photographs of the iconic avenue.

19 For a full discussion of this campaign, see Dean, "Said Tree is a Veritable Nuisance." For annual numbers, see Julian Smith and Associates, Sandy Hill West Heritage Conservation District Study (Ottawa: September 1993), B10–B12. Joann Latremouille, landscape architect and historian for this report, brought these figures to my attention.

20   J. Alph Dulude, "Shade Trees and Parklands: Planting, Cultivation and Preservation," September 1962, 2, part 3, Specific Surveys and Policies, City of Ottawa Departmental Reports. Submitted to the City of Ottawa Recreation and Parks Committee by the Commissioner of Recreation and Parks, p. 9.

21   Ibid.

22   See Joanna Dean, "Mapping Alta Vista," at http://www.carleton.ca/~jdean/urbanforest.htm.

23   For detailed results, see Joanna Dean, "The Social Production of a Canadian Urban Forest," in Richard Rodger and Genevieve Massard-Guilbaud, eds., *Environmental and Social Justice in the City: Historical Perspectives* (Isle of Harris, UK: White Horse Press, 2011) .

24   Women's incomes correlate less clearly to status or canopy cover because middle class women often did not work outside the home.

25   C. G. Boone, M. L. Cadenasso, J. M. Grove, K. Schwartz, and G. L. Buckley, "Landscape, Vegetation Characteristics and Group Identity in an Urban and Suburban Watershed: Why the 60s Matter," *Urban Ecosystems* 13 (2010): 255; T. R. Tooke, B. Klinkenberg, and N. C. Coops, "A Geographical Approach to Identifying Vegetation-Related Environmental Equity in Canadian Cities," *Environment and Planning B: Planning and Design* 37 (2010): 1040–1056; Lee and Maheswaran, "Health Benefits of Urban Green Spaces"; and N. Heynen, H. A. Perkins, and P. Roy, "The Political Ecology of Uneven Urban Green Space: The Impact of Political Economy on Race and Ethnicity in Producing; Environmental Inequality in Milwaukee," *Urban Affairs Review* 42, no. 1 (2006): 3–25.

26   The method also provides little sense of the large proportion of the tree that exists underground and out of sight. Urban foresters are increasingly aware of the significance of the root system, and the impact of water table changes and soil conditions on tree health. Mapping changes to the built environment, such as changes to the area covered by hard surfaces, might provide some indication.

7

"I do not know the boundaries of this land, but I know the land which I worked": Historical GIS and Mohawk Land Practices

Daniel Rueck

"Who is the rightful owner of a hole in the ground?" That was the question that an 1895 inquiry in Kahnawá:ke, a Mohawk community near Montreal, sought to answer. In 1894, a fertile hill was excavated by the Canadian Pacific Railway (CPR) as a borrow pit to provide gravel and earth for the building of the rail approach to the new bridge across the St. Lawrence. The CPR did not ask for permission; in fact, it did not even ask whose land this was – it simply gave $100 to the Department of Indian Affairs (DIA) and asked it to pass the money along to the rightful owner. Three men came forward to claim the money and an inquiry was called to discover who was the rightful owner. As the inquiry progressed, it became clear that all three men had a valid claim to this lot, labelled Lot 205 on the DIA map, and that it would not be possible to choose one over the others. Why was it so difficult for the department to know the identity of the owner in light of the fact that it had carried out an extensive land survey only a few years before and was in possession of a detailed cadastral map? This chapter sheds light on this situation by exploring what happened when traditional Kahnawá:ke land practitioners were faced with DIA efforts to eradicate their way of relating to the land. It focuses specifically on the most comprehensive land survey in the history of Kahnawá:ke, known as the Wal-bank Survey, and discusses the implications of the survey for Mohawks and their land. The latter half of the chapter moves from the past to the present, as I share some of my experiences in conducting this research. The chapter ends with a discussion of the possibilities and limitations of learning about Indigenous land practices using GIS tools. I argue that, although the western cartographical tradition has been hostile toward Indigenous ecological knowledge and practice, historical GIS (used judiciously) offers a way to turn historical maps against themselves. Histories and knowledges that were undermined by government surveys and maps can be brought to light by the very maps and data created to destroy them.

Kahnawá:ke dates from the 1660s when several hundred Indigenous people settled across the river from Montreal, then a tiny colonial outpost. The community was initially multi-ethnic, including people from more than twenty nations, but within a few decades it took on a primarily Mohawk character. The French were thankful for this powerful and friendly military presence near their poorly defended towns. Kahnawá:ke Mohawks (hereafter Kahnawakehró:non) decided to settle there for a number of reasons: their 50,000-acre seigneury promised them perpetual revenue in the form of rents from farmers; Jesuit missionaries offered spiritual services and a dry community; and the geographic location of the village allowed for a continuation of their vocation as traders. Kahnawá:ke was the largest Indian village in Canada until the latter decades of the nineteenth century, and, at around 10,000 people today, it continues to rank among the most populous Indigenous communities in Canada. Located only ten kilometres from

downtown Montreal (see Fig. 7.1), the 12,000-acre reserve is today surrounded by suburbs and traversed by highly intrusive transportation and hydroelectric infrastructure. The effects of Montreal industrialization were felt as early as the 1820s when stone quarried in Kahnawá:ke was used to construct the Lachine Canal, and in the 1850s when Kahnawá:ke became the terminus for one of the first railway lines to connect Lake Champlain with the St. Lawrence River.[1] Nevertheless, Kahnawá:ke remained relatively independent and self-governing until the 1880s, when Mohawks experienced intensified and interrelated incursions into their lives and lands. This kind of uninvited development on Mohawk territory continued throughout the twentieth century with the construction of high-voltage power lines, highways, bridges, and finally the St. Lawrence Seaway, which cut through the village in the 1950s and is widely seen today as the ultimate environmental and cultural tragedy in the history of the community.[2]

My research for this chapter focuses on the last two decades of the nineteenth century when the magnitude of these incursions, along with heavy-handed DIA interference in Kahnawá:ke affairs, were felt for the first time. One particular DIA project, the Walbank Survey, attempted to define property boundaries and standardize ownership practices along Euro-Canadian capitalist lines in the 1880s. The Mohawk land-management regime, which the Walbank Survey sought to replace, was a customary system that aimed to maximize access of all community members to land and resources while limiting the potential for individual commercial profit from the same. Over the course of the nineteenth century, these customs had been frequently challenged by reformist

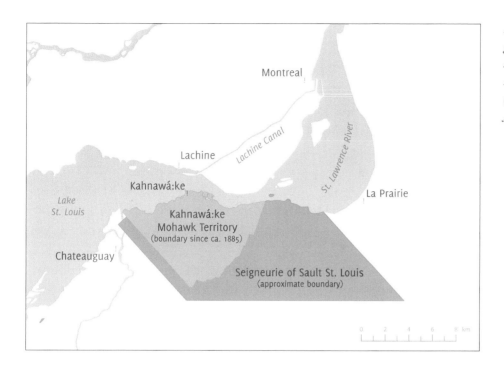

Fig. 7.1. The geographical situation of Kahnawá:ke. (Map by Louis-Jean Faucher.)

Mohawk landowners who wanted a land tenure regime in line with Canadian norms and the DIA which supported those landowners. The Walbank Survey was the culmination of these efforts.

THE WALBANK SURVEY

The DIA initiated the Walbank Survey for a number of reasons. First, DIA officials, like most of their non-Indigenous contemporaries, believed that Mohawks made poor use of their land.[3] They believed that this was due, not only to a supposed Mohawk lack of interest in agriculture, but, more importantly, to the belief that Mohawk landowners lacked security of ownership. Furthermore, the DIA felt it lacked the information it needed to act as an effective arbiter in land disputes between Mohawks.[4]

Creating accurate cadastral maps and a standardized system of titles would encourage owners to 'improve' their land and would give the government the information it needed to effectively intervene. The department also had a mandate to 'enfranchise' Indigenous people, which meant transforming their legal status from Indians into non-Indians. According to contemporary legislation, one of the requirements for enfranchisement was for a male Indian to own and farm land. By giving a piece of land to each Mohawk "head of household," DIA officials reasoned, the entire community could be enfranchised, that is to say, effectively eliminated. Finally, department officials believed that the survey would facilitate the expected expropriations that would accompany the 1885–87 construction of the St. Lawrence Bridge for the Canadian Pacific Railway, one end of which would rest on Mohawk territory.

Although the DIA did not express it in these terms, the Walbank Survey was also an effort to erase a way of living on the land. The existing property regime was the result of Mohawks adapting their ancient land practices to new realities, including a permanently located village, shrinking land-base, and the industrialization of neighbouring Montreal. To outsiders, it appeared that Mohawk land ownership was no different from standard individual free-hold tenure, except that Mohawks lacked a standardized system of land titles and appeared not to respect others' property. But the Kahnawá:ke system of land ownership had its own logic and was in many respects similar to the practices in other Haudenosaunee (Iroquois) communities at the time. Kahnawakehró:non considered their entire territory to be owned collectively, but small pieces could be claimed by individuals as long as they were cultivated. Land left uncultivated became available to others. An individual could not claim more land than he or she could work. Standing trees could not be owned by individuals, the only exception being maple trees actively tapped for sugar. Most other trees were available to all community members who wished to cut them and use the wood for their own purposes. Such wood was not to be taken out of Kahnawá:ke or sold.[5] The consistent articulation of these principles by Mohawk leaders throughout the nineteenth century shows that Kahnawá:ke land practices were not the result of lawlessness, as advocates for the survey insisted. Instead, the Mohawk understanding of their relationship to land was based on the conception of the territory as a commons that limited the possibilities for land-related commercial activities while offering community members free wood and small plots of land suitable for small-scale farming.[6]

In the second half of the nineteenth century, Kahnawakehró:non became increasingly concerned about the steady encroachment of non-Mohawk farmers along the boundary of their territory. Many of these farmers had been in the habit of periodically moving boundary markers, and this had not gone unnoticed. Throughout the 1870s, Kahnawá:ke chiefs called on the DIA to conduct a boundary survey that would restore lost territory and ensure that such encroachment would not happen again. DIA responses to these requests made it clear to the chiefs that the priority of the department was not the boundary but the interior of the territory. The department wanted to subdivide the reserve, but the chiefs always refused proposals of that nature. In 1874, for example, Kahnawá:ke chiefs submitted a powerfully worded petition stating that it was their duty to protect and represent those who would suffer most from such a subdivision. The lots produced by such a subdivision, they argued, would be small and of uneven quality, and, with the loss of the common wood and land resources, the community would no longer be viable.[7] DIA plans for subdivision were thus put on the back burner until the 1880s.

The DIA finally responded to Mohawk calls for a boundary survey in the summer of 1880, when it gave the contract to Provincial Land Surveyor, William McLea Walbank. The son of a lawyer and Conservative member of the Newfoundland House of Assembly, Walbank (1856–1909) studied architecture and civil engineering at Queen's University in Ireland before earning a degree in civil and mechanical engineering at McGill University in 1877. He spent his career in Montreal working as an architect, engineer, and surveyor and was later involved in efforts to develop the

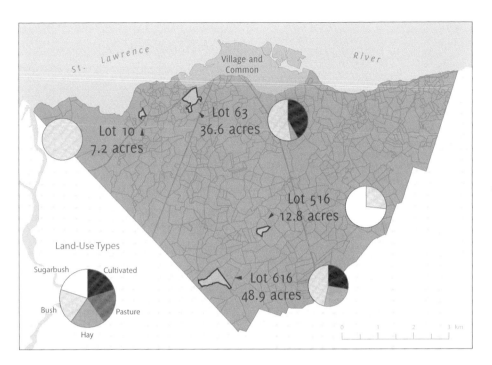

Fig. 7.2. Map of Kahnawá:ke showing the lots owned by Skatsentie (a.k.a. Joseph Williams), ca. 1885, and Walbank's land-use classification for each. (Map by Louis-Jean Faucher.)

hydroelectric potential of the Lachine Rapids. He was twenty-four years old at the start of the boundary survey.[8]

Walbank and his staff completed the survey of the reserve boundary by the end of 1880.[9] Walbank concluded his report on the project by suggesting that subdividing the entire reserve would be the logical next step and that he was well-situated to carry it out himself. Such a survey, he warned, should be planned and executed very carefully because if Mohawks were granted the same land rights as non-Natives, the whole reserve would be in the hands of Whites within five years. Instead, Walbank suggested "it might be beneficial to give rights to sell or exchange their lands with each other (with certain restrictions) and so by degrees educate them into the manners and customs of the more civilized people."[10] In February 1882, a DIA clerk travelled to Kahnawá:ke to announce that the department intended to act on Walbank's recommendations. The clerk reported to the department that the chiefs had manifested their contentment with the plan and that they had expressed a hope that the survey would be carried out quickly.[11] There is no reason to believe that most chiefs were actually happy about the announcement, given that the DIA regularly misrepresented the views of Indigenous leaders and that Kahnawá:ke leaders had consistently opposed subdivision plans in previous years. Even if some chiefs decided to cooperate once it became clear that they could not stop it, such a decision should not be seen to indicate voluntary agreement when the implied threat of state violence lurked in the background.[12]

The one leader who appears to have had a genuine enthusiasm for the project was Chief Skatsentie (Joseph Williams) (1846–1885). He was a young, wealthy trader whose father had done considerable business in Germany selling

Indian curiosities. An anonymous columnist for *The Catholic World* reported in 1883 that the thirty-seven-year-old Skatsentie lived in a luxurious house and was sanguine as to the survey "working well and benefiting his 'braves.'"[13] In a letter to the DIA in 1882, Skatsentie appeared to downplay Kahnawakehró:non opposition to the subdivision, saying that they simply wanted more information "as to the character of such a subdivision."[14] According to Walbank data, he owned at least four lots in Kahnawá:ke, totalling 103.65 acres, as well as two barns. Walbank valued these properties at $1,127.[15] Figure 7.2 shows the location and land uses of each of Skatsentie's lots. Lot 516 included a substantial sugarbush. Both Lot 63 and Lot 616 included cultivated and haying lands. All four lots contained 'bush,' which was valued primarily for the provisioning of firewood.

Walbank, along with three teams each consisting of a surveyor and two Mohawk assistants, began work on the subdivision survey in June 1882.[16] He expected to complete the project the following year. However, Walbank's ignorance of Kahnawá:ke land tenure, the paucity of maps and deeds, and the lack of cooperation from mistrustful community members meant that every part of the project took longer than anticipated. After finally completing a survey of the locations and boundaries of existing lots in 1884 (Fig. 7.5), he began the process of valuing them.[17] The DIA advised him to value lots not as if they were owned by Whites but based on a land market where Indians could only buy from each other. This had the effect of making Kahnawá:ke valuations much lower than valuations for similar lots on non-Indian land.[18]

Before designing the new property grid, Walbank had to know how many lots would be needed so that he could assign one to each "head of household." To this end, Walbank set up a tribunal process in which anyone who claimed to be a head of household could present him/herself to a tribunal, which would be made up of the council of chiefs, the agent, and Walbank himself. Claimants appeared and were asked a series of questions, which were recorded on standardized forms. The questions were designed to gather three types of information: 1) facts about each claimant for the creation of a membership list and list of electors; 2) information that could be used to exclude people from membership; and 3) information about lots and improvements owned by members. Aside from standard questions about names, birth dates, and birth places, Mohawks were asked many questions that reflected DIA concerns about race, monogamy, and absences from Canada.[19] The tribunal operated from February until June 1885. The information for each claimant and lot was recorded in five large volumes of record books, which, along with a map depicting land uses, existing lots, and projected lots, form the basis for my GIS analysis (Fig. 7.4).

Many questions can be raised about the accuracy of the information Walbank gathered in this way. He admitted to filling in answers to question number ten ("Do you hold any land on the Reserve; and how did you acquire such land?") for the claimants because, in his words, "any information I might get from the individual Indians would be very unreliable and inaccurate."[20] Answers to other questions were written down as standardized English phrases and never in the hand of the claimant. Considering Walbank's young age, inexperience, and lack of empathy for those he judged uncivilized, it is likely that many Kahnawakehró:non claims were not fairly represented in his workbooks. On the other hand, there is no reason to think he deliberately falsified information, and the involvement of the four chiefs on the tribunal served as a kind of counterpoint. They may not have had the full support of the community, but they were still subject to the kind of accountability that came with living among their constituents.

After the tribunal heard from all claimants, the DIA (in consultation with the Department of Justice) reviewed all the contested claims. These were claims for which at least one chief had contested the person's right to membership. Of the 610 total claimants (513 men and 97 women), 175 (27 per cent) were contested. Each of the four chiefs on the tribunal had the opportunity to either approve or contest each claim. Chief Skatsentie, mentioned earlier as an enthusiastic promoter, died of unknown causes in May 1885 and did not play a role in the tribunal decisions. In 122 of the 175 contested claims, all chiefs agreed to reject the claim of the individual; 53 disputed claims, however, were not unanimous. It is nowhere made clear exactly what the criteria were or who defined them, but it appears that the chiefs and the department were not working with the same understanding, nor were the chiefs always in agreement. Historical anthropologist Gerald Reid has analyzed the process in detail to better understand political rifts within the community and has offered a number of insights. Claimants were contested for a number of reasons: being underage, being non-widowed women, having been born elsewhere, having parents who were born elsewhere, having been absent for a long time, having been born out of wedlock, and being 'white' or 'half-breed.' The final decisions were a result of the back-and-forth between chiefs and DIA, but the chiefs had little power over the final results. In most cases, the DIA applied *Indian Act* membership rules. The chiefs had the most say when it came to deciding who was excluded from membership on the basis of racial criteria. Since there was no complete membership list, the department usually took the chiefs' word on who was sufficiently Mohawk to belong and who would be evicted. The DIA intervened to contradict the chiefs only in the case of the Delorimier family, who had long been considered interlopers by a majority but whose Indian status had been confirmed by an 1850 court. The department also forced the inclusion of Mohawk women who had married non-Indians before the 1869 law that stipulated that such women lost their status. Reid argues that the chiefs' motivation was primarily to limit the number of band members so that each would receive an adequate share of the small territory.[21] In addition to the contested claimants, there were also approximately 130 cases of disputed ownership that were brought before the tribunal.[22] The tribunal was an important event in the history of inclusion and exclusion at Kahnawá:ke, but it did not resolve the matter. The question of membership remains extremely contentious to this day.[23]

Walbank's tribunal made the subdivision survey very real and personal to Kahnawakehró:non, and by 1885 many openly opposed it. The survey, however, coincided with events that circumscribed Mohawks' ability to resist it. The Canadian public became openly hostile toward Amerindians as Saskatchewan Métis and their allies set up an independent provisional government in the spring of 1885. Ottawa crushed Indigenous forces and executed the leaders of the movement. Following this incident, there was little sympathy for

Indigenous causes in Canada, and Mohawks took this into consideration as they registered their complaints that summer. In July 1885, for example, some fifty Kahnawakehró:non sent the DIA a carefully worded petition expressing concern over the long duration and high cost of the subdivision survey and requesting an investigation into the matter. "It is with anxiety," they wrote, "that we look for the completion of said survey: we are inexpert in the nature of the work, but assuredly one acting faithfully should have finished it by this time, comparing to the small size of the Seigniory."[24] When the department questioned Walbank, he dismissed those behind the petition as nothing but a few alcoholic agitators, a common way of discrediting those who opposed DIA plans. "The complaint does not come from the respectable part of the tribe," he bristled, "but from some whom I have prosecuted for bringing intoxicants on the Reserve, and is composed of some fifty or sixty of the most troublesome men of the tribe, and who take no interest in any matters except opposing all progress."[25]

With the work of the tribunal mostly complete in the fall of 1885, Walbank began the work of creating the new property grid. As long as the claims were in limbo he could not begin laying out lots, so he started by marking out new roads that would serve as the basis for the grid. He urged the DIA to quickly process the disputed claims, but decisions were not forthcoming until the following summer.[26] In the meantime, Walbank drew a map of the new subdivision and invited successful claimants to choose their new lots. He also invited owners of existing lots to review the valuations Walbank had given their land and improvements.[27] These notices, written in both Mohawk and

English, sparked another round of protests, this one mostly from large landowners.

After two clandestine meetings in early June 1886, a group of Kahnawakehró:non sent the DIA a petition protesting the low valuations and the high cost of the survey.[28] These were concerns of the privileged minority who often supported DIA initiatives, and perhaps this is why the department decided to crack down. On June 25, Walbank and Agent Brosseau called a general meeting to denounce the petitioners and to announce a ban on unauthorized public meetings. Some landowners hired arbitrators to attempt re-valuations,[29] but the DIA rejected such actions. Deputy Superintendent of Indian Affairs Lawrence Vankoughnet stated that "the valuation would be on the basis of values of such property in Caughnawaga, as between Indian and Indian and not as between White people," and that there would be no exception to that rule.[30]

One might have expected poor and landless Mohawks to support a project in which they were set to receive thirty acres. There is, however, little evidence to suggest this was the case. While the overt opposition came mostly from large landowners, there were a number of small landowners who also protested the subdivision. A good example is Ohionkoton (Angus Jacob), who returned his notice with the defiant message in the Mohawk language: "Now you gentlemen: I answer. I like the way that I have. I do not sell my land."[31] Ohionkoton had also failed to appear at the tribunal interviews of the previous year. Walbank listed him as the owner of a 1.03-acre lot of cultivated land worth $13.[32] Landless and land-poor Kahnawakehró:non were aware that the new property arrangement included the criminalization of wood-cutting on others' lots and would deprive them of an essential, hitherto free, heating and cooking fuel.

In the fall of 1886, Walbank began the actual subdivision of the land, aiming to make his map a reality. He planned to create 387 lots of about thirty acres each, most of which had already been assigned owners (Projected Lots in Fig. 7.5).[33] The most difficult part of the process would be to reassign land that was occupied and improved, so he started by subdividing the "Grand Park," a 506-acre swampy area on the western side, known today as the "Big Fence."[34] But how would land be transferred from old to new owners? Since the department was legally obligated to compensate owners for improvements, owners of the old lots had to be paid for lost buildings, cleared land, fences, and orchards before new lots could be taken up by new owners. The problem was that this would be very expensive, and, although Walbank's valuations were much lower than they would have been off reserve, the total needed to compensate all landowners was still staggering. At a time when the DIA was facing serious questions about the $15,000 already spent on the survey, Walbank asked the DIA for a $50,000 temporary fund. Owners of new lots would be instantly indebted to the department for any improvements found on the new lot and would be asked to pay down this debt in instalments. If they could not make their payments, Walbank suggested the DIA reserve for itself the right to lease out that lot until the debt was paid.[35] The largest landowners would lose more improvements than they would gain and would thus end up with less land and more money. Without such a fund, Walbank wrote, he was at an impasse because people were not willing to abandon their current holdings without compensation. The DIA told him that no such money would be made available but that he should continue surveying.[36]

Walbank was probably right in saying there was no other practical way to complete the process, and the department's unwillingness to support him suggests that either officials failed to see his logic or they already saw the subdivision as doomed. With the prime minister facing questions on the cost and duration of the survey in the House of Commons in 1887-89, the survey was taking on the appearance of a political boondoggle.[37] In 1890, an opposition Member of Parliament asked the government why this incomplete survey cost $1.80 per acre when the cost of the Dominion Lands Survey on the Prairies was 4 cents per acre.[38] In the face of this kind of questioning, it is not hard to understand why the DIA had to stop spending money on the survey, even though it was not complete.

Figure 7.3 visually represents the way almost every proposed new lot took in lands from more than one previously existing lot. One can only imagine the tragic comedy that would take place on the ground if such a redistribution were to take place. Existing roads and paths would cease to be useful, barns would be separated from fields, and the ecological, geographical, and cultural logics that had determined the original layout of the lots would become subservient to the bureaucratic logic of the rectangle and the grid. People would lose land, buildings, and improvements, and other people would gain those same things. The idea was to give everyone an equal portion of land, but the lots were not equal in quality. Aside from geographical differences, there were also great differences in how different lots had been used over the long occupation of the territory.

Fig. 7.3. Detail of a part of the territory of Kahnawá:ke, ca. 1885. Dark boundary lines represent existing lot lines drawn by Walbank. Rectangular light lines represent the projected lots. (Map by Louis-Jean Faucher.)

St. Lawrence River

It was inevitable that some would lose a great deal in the exchange while others would gain thirty acres for which they would be indebted. Walbank's scheme would transform the land-rich into money-rich (although landowners felt their improvements had been seriously under-valued), and would indebt the poor for land they did not request.

In a situation where legal forms of pro-test had been taken from them, Mohawks increasingly turned to other means. Walbank informed the DIA in September 1887 that one of his surveyors had been impeded by a num-ber of Mohawks who had "offered obstruction to the running of the new lines of Lots [and] threatened personal violence," as well as re-moving pickets and destroying marks.[39] Wal-bank specifically accused three men of these actions. The first, Thiretha (Peter Diome), was listed as owning four lots totalling 194 acres of all land use categories (cultivated, bush, hay,

pasture, sugarbush), valued at $1,473. When Thiretha had earlier failed to appear at the tri-bunal, Walbank noted that "this man resides here upon his land which is very extensive; he refuses to attend here to make his statement." The second man was Kataratiron (Joseph Jacob), born 1842, who was listed as owning an 80-acre lot, of which a significant portion was cultivated, valued at $1,804.[40] The third man was a certain Doctor Jacobs, whose identity cannot be verified from the Walbank data. It is worth noting that two of these men were land-rich while the third was a medical doctor. For these men, the issue was apparently not access to firewood but the imminent loss of property without adequate compensation.[41]

In May 1887 Walbank staked out sixteen lots, lined up a new owner for each, and asked the DIA for money to compensate them for lots they would be giving up. The department refused to make money available.[42] In July,

Walbank tried again, this time proposing to transfer only one title. In this way, Walbank intended at least to set a precedent, but the department refused to grant the claimant a title before he had been compensated for lost lots.[43] In other words, the department would not pay to compensate owners, nor would it allow new owners to take up lots. This left Walbank with no way forward. He stopped working on the subdivision fieldwork in December of 1887 and said he was "extremely glad to be finished with it." He went on to say "it is one of the most difficult and unsatisfactory surveys one could possibly have."[44] He finished his paperwork in spring of 1888, filed a lengthy report with Prime Minister Macdonald, also the head of Indian Affairs, and soon distanced himself from the project.[45]

The redistribution never took place – not even for one person – and the standardized thirty-acre lots never became a reality. Nevertheless, Walbank's map and data were used as the basis for most real estate transactions from then on. The lot lines were not rectangular but at least the DIA had something to work with. Technical problems were later found with Walbank's work, but the department used it as if it had been perfectly executed. The DIA drained the Kahnawá:ke band account to pay Walbank the $22,000 for the survey,[46] and when the account was empty, the DIA found money through a feat of creative accounting involving loans from Temiskaming and Sarnia Chippewa band accounts.[47]

The enclosure of the Mohawk commons, like the enclosures in the British Isles one hundred years earlier, had occurred in a piecemeal fashion. Throughout the nineteenth century, individual Mohawks claimed exclusive ownership to lots in ways that were at odds with customary rules, but those who did this were in the minority and often poorly regarded. The Walbank Survey greatly accelerated this process. In the context of a customary land-ownership regime in which boundary lines were not fixed, Walbank's lines represent a snapshot in time. Mohawks continued to dispute the legitimacy of Walbank's lines for years and continued to use the land in ways they formerly had, but the lines were there to stay. Mohawk land ownership customs had prevailed as long as they did in part because outsiders simply did not have the cultural knowledge and data to understand and adjudicate conflicts. Walbank gave the DIA a powerful tool with which to take possession of the land and govern it according to its own rules.

MAPPING THE WALBANK SURVEY WITH HGIS

In addition to giving the DIA a powerful tool for surveillance and control, Walbank also provided this historian with a rich data set with which to better my understanding of the history of Kahnawá:ke. When I started my doctoral research, I knew very little about GIS, but when I discovered the Walbank map at Library and Archives Canada, I sought ways to organize and interpret the information it contained. My friend, Louis-Jean Faucher, a geographer and a GIS instructor for ESRI, the leading producer of GIS software, offered to help. Together, we georeferenced the digital image of the map (Fig. 7.4) using government-produced topographic vector layers, historical maps, and

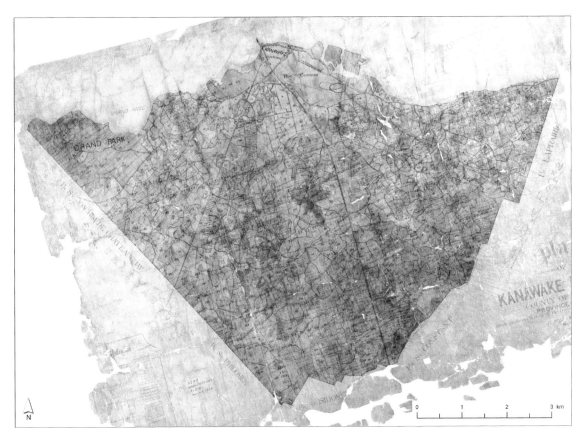

Fig. 7.4. Map of Kahnawá:ke produced by Walbank, 1885–89, after geo-referencing. (Source: Plan of Kanawake reserve, county of Laprairie, Province of Quebec, made under authority of the Indian Act 1880 47 Vic. C ....by W. McLea Walbank, C.E., P.L.S., Montreal, Sep. 1885 and 29th Aug. 1889, Library and Archives Canada.)

current satellite photographs. I spent a number of months entering the Walbank data into a spreadsheet and using ArcGIS to tease out the different elements of this extremely complex map. Faucher then used the plotted lines and spreadsheet data to create the maps that are included in this chapter.

The Walbank map is a composite of three main layers: existing lot boundaries, projected lot boundaries, and land-use categories. I plotted all existing lot boundaries as irregular polygons and attached the associated lot number to each (Existing Lots in Fig. 7.5). This was painstaking work as it was often difficult to

know the difference between a road, a land-use line, and a property boundary. In many cases, the only way to make these distinctions was to compare the map with reference book data and twentieth-century maps. The map created from this data shows a cadastre of irregularly shaped and sized lots. Plotting the boundaries of projected lots (rectangular polygons) was much less time-consuming and considerably less difficult. The resulting map is a striking contrast to the map of the original lots, and the two placed side-to-side give a clear picture of what was at stake (Fig. 7.5).

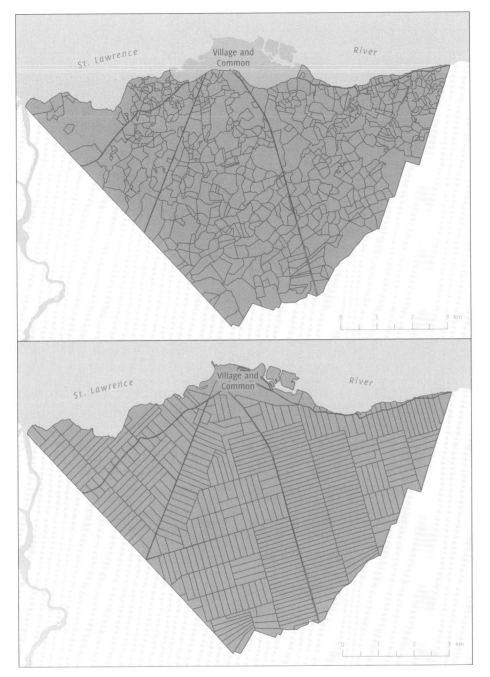

Fig. 7.5. Maps of Kahnawá:ke showing existing and projected lots, ca. 1885. (Maps by Louis-Jean Faucher.)

The map divides land-use into seven categories: 1) bush and hay; 2) bush; 3) bush and swamp; 4) cultivated; 5) pasture; 6) beaver hay; and 7) sugarbush. Unfortunately these seven categories differ from the five categories used in the reference books (cultivated, pasture, hay, bush, sugarbush), making direct comparisons difficult. Land-use data is coded on the map with a combination of colours and patterns, but these are extremely difficult to discern because

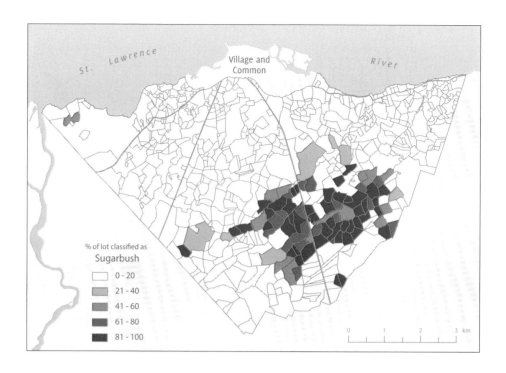

Fig. 7.6. Map of Kahnawá:ke showing percentage of lots classified as sugarbush, ca. 1885. (Map by Louis-Jean Faucher.)

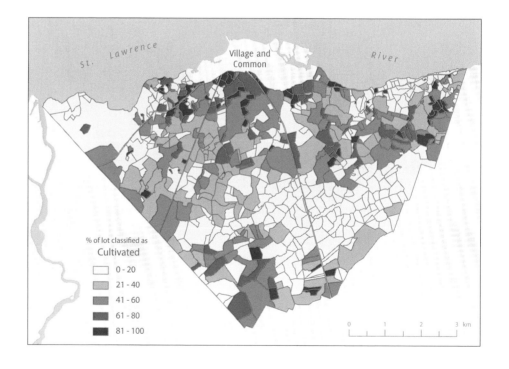

Fig. 7.7. Map of Kahnawá:ke showing percentage of lots classified as cultivated, ca. 1885. (Map by Louis-Jean Faucher.)

the colours have faded and the map is ripped and water-damaged. Reference book data was often useful for interpreting the map, but there are many inconsistencies and conflicts between the map and the reference books. These discrepancies may have resulted from the collection of data at different times and from the changing land uses from year to year. Given these problems I was able to visualize land uses only based on percentages for each lot, which reveals certain patterns over the entire territory. For example, Fig. 7.6 shows that sugarbushes were concentrated toward the southeast and Fig. 7.7 shows cultivated land clustered near the village, along the western border, and on the northeastern side of the reserve.

The completed project to date represents only a small part of what can be done to further develop a historical GIS for Kahnawá:ke. There is potential to incorporate much more, and many other types of information to the existing GIS, including historical photographs, stories, videos, and aerial photographs. Conscious of my place as a non-Native researcher telling the story of an Indigenous community, I have maintained regular contact with Mohawk scholars and other community members while conducting this research. Their comments have been essential in assisting me to interpret archival texts in ways that truthfully and respectfully reveal the actions of their ancestors. It remains to be decided exactly how this material should be made available to the community, but I believe it is important that current and future Mohawk cartographers and historians, professional and amateur, should have easy access to copies of digital images of historical maps I obtained from various archives, as well as the Walbank dataset and vector layers. With the memories of the Oka Crisis still fresh, and

with the knowledge of ongoing CSIS (Canadian Security Intelligence Agency) monitoring of Haudenosaunee communities, many Kahnawakehró:non would be uncomfortable with an outsider making detailed maps of the territory that could have military or police applications.[48] This historical mapping project, however, has not to my knowledge been perceived as controversial in Kahnawá:ke.

HGIS AND THE HISTORY OF INDIGENOUS LAND PRACTICES

Flawed though it is, the Walbank data is a unique source in that it provides a highly detailed cartographic snapshot of the nineteenth-century land practices of an Indigenous community, one that is unparalleled in Canada. In the big picture, however, the emergence of GIS technology has raised a number of questions about its usefulness for Indigenous peoples. There are potential pitfalls in representing historical Mohawk land practices using the texts of a non-Mohawk surveyor who was decidedly ignorant of, and antipathetic toward, Mohawk ways. Surveyor texts are widely recognized to be some of the most effective tools of colonization.[49] Historian Raymond Craib identifies surveyors as one of the important "faces" of the state for people who had previously kept the state at arm's length:

> Surveyors were neither passive extensions of objective instruments nor an homogeneous and transparent group of lackeys in the service of the state or

landlords.... They often appeared in rural areas as intermediaries between an abstract state (and its policies) and local populations who were affected by those policies. People experience "the state" as they experience "the market" or "capitalism," not as a broad abstraction but as a series of manifestations with a very human face: judges, notary publics, police squads, tax collectors. And surveyors.[50]

Laws concerning land may have existed, but it took a surveyor to make the law reality. In the case of Kahnawá:ke, Walbank's job was to make the DIA vision a reality, at the expense of Mohawk visions. A number of scholars have also pointed out the problems inherent in using western cartographical tools to represent Indigenous territory. Robert Rundstrom, an early critic of GIS, protests that "from an Indigenous point of view, history is suffused with domination and disenfranchisement at the hands – and maps – of the inscribers."[51] "Inscribing cultures" are ones that place great emphasis and value on the objects they produce – objects such as maps – and de-emphasize the process of making those objects. Indigenous peoples, to whom Rundstrom refers as "incorporating cultures," have tended to emphasize oral communication and performance-based ways of expressing territoriality instead of valuing the object of the map.[52] Rundstrom argues that for non-Indigenous societies, "storage is crucial, and leads to stasis and fixity," whereas Indigenous peoples have valued nuance, ambiguity, and flexibility. Writes Rundstrom, "the history of cartography is replete with examples of people from inscribing cultures appropriating geographical information from those of

incorporating cultures, and ultimately using it to disenfranchise, if not completely obliterate them." For Rundstrom, GIS is part of the same process of "Western technoscience" that is disenfranchising Indigenous peoples.[53] It would seem to follow that incorporating state-produced survey data into an HGIS could contribute to the hegemony of the status quo, to the detriment of Indigenous peoples.

It is important to note, however, that Rundstrom is writing about GIS used by government agencies to represent Indigenous lands in the present, not in the past. In fact, it appears that historians have been largely absent from this debate, which has raged for two decades. One scholar who has made a strong case for the use of GIS by historically disenfranchised communities is environmental sociologist Nancy Peluso. She argues that local people and activists in Indonesia have been able to use GIS technology to successfully press their land and resource claims. The goal of these efforts, she writes, is for local people to appropriate technologies previously used to consolidate state power and to use them for their own benefit. By creating their own maps, these people have been able to undermine the authority of government-produced maps. Indigenous "counter-maps" she argues, "greatly increase the power of people living in a mapped area to control representations of themselves and their claims to resources. Counter-maps thus have the potential for challenging the omissions of human settlements from forest maps, for contesting the homogenization of space on political, zoning, or property maps, for altering the categories of land and forest management, and for expressing social relationships in space rather than depicting abstract space in itself."[54] Thus, the problem for Indigenous peoples is not necessarily the technology of mapping itself but who controls the technology and in whose interests. When GIS is used in the interests of, and in consultation with, Indigenous peoples, it can serve to counter colonialist and statist narratives.

Many Indigenous peoples in Canada and elsewhere are already using GIS and mapping as tools of empowerment.[55] They are using cartography to press their claims, imagine their own spaces, and manage their territories. Indigenous communities across Canada have employed land-use and occupancy mapping to assert their territorial and resource rights since at least the 1970s,[56] and have made use of GIS software for these and other reasons since the software became available.[57] Some Inuit communities, for example, now use GIS to ensure that the traditional ecological knowledge of their elders is available to younger hunters who are increasingly dependent on the Global Positioning System (GPS) for navigation.[58] It is now common practice for Indigenous communities to create their own maps or to have maps made according to their own needs. Terry Tobias, an expert on Indigenous use and occupancy mapping, emphasizes that Indigenous communities do not have the luxury of choosing *not* to engage with the Western cartographical tradition in this way. He argues that, if presented in a way that non-Indigenous people can understand it, "land use and occupancy information warrants respect, even a level of reverence." Tobias sees Western science as a potentially "powerful tool in the hands of First Nation governments."[59] Although creating a GIS requires a certain expertise, the technology can be seen as a step toward the democratization of mapmaking. Peluso makes the comparison between GIS software and the printing press. As

the latter technological breakthrough brought the possibility of writing for publication to a large number of people, so also maps are no longer being made only by state bureaucracies and corporations.[60]

Another way to counter the colonialist effect of state-produced historical maps is to place them in context. Creating historical maps of Kahnawá:ke land practices using Walbank data and situating these maps in the context of Mohawk narratives of land and territory is a step in turning the Walbank Survey against itself. Walbank translated Mohawk land practices into data and maps, into the cartographical language he understood. In plotting and entering his data into a GIS, the purpose is not to simply make his work more accessible and malleable but to make his work understood in the context of Canadian and Mohawk history. Taken alone, the Walbank data entered into a GIS would only serve to further the purposes of Walbank and the DIA. Reinterpreted and placed in historical and cultural context, however, it can take on new meanings. In discussing the difference between data (information) and data placed in context (knowledge), Robin Boast et al. argue that we must "avoid information systems becoming knowledge deserts."[61] In the sense that maps cannot be neutral (i.e., in that they are inherently ideological), they are no different than other kinds of text. Just as historians must place archival documents in context in order to paint a richer and truer picture of the past, the same must also be done with maps and related data.[62]

Boast et al. warn against the potential maps have for "freezing" the dynamic social processes of customary law. The Walbank map is a case in point. Creating maps of Indigenous territories, in much the same way as codifying customary law or writing down oral traditions, can take away the flexibility inherent in traditional practices and distort their nature. Boast et al. argue, however, that people's actions are often completely at odds with the dictates of maps and that maps need to be constantly updated to take these actions into account.[63] Rundstrom's argument that "GIS technology, when applied cross-culturally, is essentially a tool for epistemological assimilation, and as such, is the newest link in a long chain of attempts by Western societies to subsume or destroy Indigenous cultures" is, in my reading, overstated.[64] While I agree that GIS has potential for harm, Rundstrom fails to recognize its great potential for good. As a point of comparison, the proliferation of reading and writing among Indigenous peoples greatly changed their lives, but few would today advocate for illiteracy as a way forward. A GIS is unlike a map in a way that Rundstrom does not acknowledge. While a map is printed on paper, a GIS can constantly evolve: new data can be added, and the process need never end. Information can also be intentionally left out of a GIS if silences and blank spaces are preferable. I believe that the great strength of HGIS for Indigenous history is its flexibility in this regard. Many projects across a number of disciplines have recognized the potential of GIS for visually representing Indigenous categories, stories, and place-names.[65] Historians have thus far not played an important role in these efforts, but there is no reason why these spatial representations of Indigenous spaces cannot be expanded to include historical information as well.

One final problem scholars have raised in conjunction with mapping Indigenous lands concerns the incompatibility of geographical categories. Categories used by nation states are often at odds with categories employed by local people. Peluso gives the example of Indonesian forestry maps, which represent six forest land-use categories based on climate, topography, and soil. In deciding on these categories, government cartographers did not consider local peoples' land uses nor the composition of existing plant species. They characterized shifting cultivation as a "non-permanent" land-use, whereas local people did not see their actions in those terms.[66] Of course, differences in geographical categories are also related to a host of other factors, including language and environment. Geographer Nadine Schuurman illustrates this point by describing the problems inherent in creating a standardized vegetation classification system for the European Union.[67] Rundstrom suggests, however, that Indigenous categories may be so different from those of the dominant society that representing them on maps may be impossible. Indigenous people "often exhibit a trust in, or, if motivated, a *quest* for ambiguity in the meaning of naturally occurring features." Rundstrom argues that the complexity and difference of Indigenous geographical knowledge, tied as it is to kinship, spirituality, and experience, makes GIS mapping of such knowledge a practical impossibility.[68] Many Indigenous people understand the world in terms of ubiquitous relations between humans and nonhuman beings (for example, hunters' conceptions of prey as kin), and Rundstrom contends that GIS cannot but treat plants, animals, and land-forms as "manipulable objects under varying degrees of human control." "At present," Rundstrom wrote in 1996, "GIS does not capture relatedness, but constructs it."[69]

It is widely acknowledged, however, is that maps also fall short in portraying the complexity, spirituality, and experientiality of people who belong to the dominant society. Maps do not and cannot show everything. They are by definition a simplified representation of reality. But, contrary to the relatively static maps of the past, GIS technology allows for the incorporation of more and more information of all kinds, virtually without limit. With each additional layer of text, narrative, and data, more opportunities present themselves for more accurate and creative portrayals of reality. Another response to Rundstrom is that both GIS software and GIS practitioners are becoming more sophisticated in their ability to deal with the challenges of representing different ways of perceiving the world. Technical progress has been made in GIS computing methods to allow for the inclusion of context-related categories and ontologies, geographical concepts that are relative to places and societies.[70] "It is increasingly accepted," wrote Duerden and Kuhn in 1996, "that the integrative abilities of GIS can effectively replicate the eclectic way in which First Nations describe their world."[71] Boast et al. agree that GIS tools are now flexible enough to allow for "the cultural diversity of knowledge resources" while still "incorporating sufficient systematic information to enable effective retrieval."[72] None of this is intended to minimize the challenges inherent in cartographically representing Indigenous land practices, but I believe that GIS technology gives scholars and Indigenous communities a way to start doing so. I am excited to see the many ways historians will narrate environmental histories of Indigenous peoples using this powerful tool, and the fruitful inter-disciplinary and inter-community relationships that will result.

CONCLUSION

Despite all the potential of HGIS, it also has its limits. A case in point is the 1895 dispute I described in my introduction wherein three men all claimed to have been owners of a particular hill, labelled Lot 205 on Walbank's map. The relationship between each of these men and the land in question cannot be accurately mapped. The only data available are from the proceedings of the 1895 inquiry and the Walbank Survey. Walbank categorized three quarters of the 9.49-acre lot as haying land and the rest as bush. He valued the lot at $73.[73] According to Walbank, the owner was Saionesakeren (Peter Montour) who died soon after the survey, and whose apparent successor was Jacques Lachandière.[74] But two other men came forward to claim the compensation money, and the department considered their claims legitimate enough to conduct a hearing and to call witnesses. It emerged from the testimonies that each man used the hill, or part of it, in different ways and at different times. Neither claimants nor witnesses were sure about the location of Walbank's lot boundaries, but they did remember what they had done in that general area. The hill had been used for cutting hay, gathering wood, and planting different crops. One witness, fifty-six-year-old Satekarenhas (Matthias Hill), who had planted peas on the hill a number of times, was asked about the boundaries of the lot. He responded, "I do not know the boundaries of this land, but I know the land which I worked."[75] Satekarenhas knew the land intimately, but Walbank's lot boundaries meant nothing to him.

Fig. 7.8. Lot 205 in summer of 2011. Once a fertile hill, it was turned into a borrow pit, and later filled in. Today it is a maintenance area for the Kanawaki Golf Course. (Photo by Daniel Rueck.)

In his report to the DIA, Indian Agent Alexander Brosseau claimed that no ownership dispute existed for the lot previous to the CPR expropriation. He wrote, "qu'il n'y a en aucunes disputes pour ce terrain avant que la compagnie du C.P.R. ait en l'intention de prendre du terrain à cet endroit, que les propriétaires inconnus jusqu'alors ont commencés a faire leurs réclamations." This statement paints a remarkably peaceful picture of pre-Walbank land practices that contradicts the official rhetoric from the early 1880s asserting that a subdivision survey was necessary to resolve land conflicts. Brosseau also reported that none of the three claimants worked the entire lot simultaneously but that they all used it in a number of ways at different times.[76] In the end, the DIA was forced to conclude that the witnesses on all sides were credible and split the compensation money three ways. In a subsequent internal memo, the Secretary of Indian Affairs admitted that the Walbank data had not been accurate for this lot, but the department dared not deviate from it because "otherwise the expense involved in Mr. Wallbank's [*sic*] survey … would be of little value."[77] This incident brings to light, not only a shortcoming in Walbank's data, but also the limits of GIS for mapping Mohawk land practices. We know that a number of people used this lot in different ways, but the archives do not contain the kind of information needed to construct a GIS for this lot. Although Walbank got a few things wrong, his data do give us an idea of what he saw at a particular moment in time. While his rectangular grid never materialized, the existing lots he mapped eventually became a cadastral reality, on the ground as well as in people's minds. This is testament to the power of land surveys and maps to transform landscapes and mindscapes. It remains to be seen how effective HGIS will be as a tool of decolonization, but I am optimistic as to its potential. Mr. Walbank would have been surprised to learn that one day his survey would be employed to explain and illustrate the Indigenous practices he was seeking to stamp out.

NOTES

1 Daniel Rueck, "When Bridges Become Barriers: Montreal and Kahnawake Mohawk Territory," in *Metropolitan Natures: Urban Environmental Histories of Montreal*, ed. Stéphane Castonguay and Michèle Dagenais, pp. 228–44 (Pittsburgh: University of Pittsburgh Press, 2011).

2 See Stephanie Phillips, "The Kahnawake Mohawks and the St. Lawrence Seaway" (Master's thesis, McGill University, 2000); *Kahnawà:ke Revisited: The St. Lawrence Seaway* (film), dir. Kakwiranó:ron Cook (Kakari:io Pictures, 2009), http://www.kahnawake.com/community/revisited.asp.

3 This was (and still is) a trope commonly applied to indigenous people. For examples, see James C. Scott, *Seeing Like a State: How Certain Schemes to Improve the Human Condition Have Failed* (New Haven, CT: Yale University Press, 1998), 273. For more on the trope of the indolent Indian, see John S. Lutz, *Makúk: A New History of Aboriginal–White Relations* (Vancouver: UBC Press, 2008), 147. For a study of how the DIA made farming on the Prairies next to impossible for indigenous people, see Sarah Carter, *Lost Harvests: Prairie Indian Reserve Farmers and Government Policy* (Montreal: McGill-Queen's University Press, 1990). Peasants and small-scale farmers have often faced the same kind of criticism. For examples, see J. M. Neeson, *Commoners: Common Right, Enclosure and Social Change in England, 1700–1820* (Cambridge: Cambridge University Press, 1993).

4 For more on modern states' need to "see" land, see Craib's excellent study on the role of surveying and cartography in Mexican nation-building: Raymond B. Craib, *Cartographic Mexico: A History of State Fixations and Fugitive Landscapes* (Durham, NC: Duke University Press, 2004), 91.

5 I have distilled these principles from the Twenty-One Laws passed by Kahnawá:ke chiefs in 1801. These principles were reiterated repeatedly by members of all factions throughout the nineteenth century. "Règlements établis par les chefs du Sault Saint-Louis," Feb. 26, 1801, RG10, vol. 10, p. 9446–9454, reel C-11000, LAC. It should also be noted that beginning around 1800, there existed a small number of Mohawks who disagreed with these principles or tried to re-interpret them to their advantage.

6 For more on Mohawk land practices, see Joseph François Lafitau, *Moeurs des sauvages ameriquains comparées aux moeurs des premiers temps*, 4 vols. (Paris: Chez Saugrain l'aîné et al., 1724); Harmen Meyndertsz van den Bogaert, *A Journey into Mohawk and Oneida Country, 1634–1635: The Journal of Marmen Meyndertsz van den Bogaert*, ed. Charles T. Gehring and William A. Starna, trans. Charles T. Gehring and William A. Starna (Syracuse, NY: Syracuse University Press, 1988); Lewis H. Morgan, *League of the Ho-de'-no-sau-nee or Iroquois* (North Dighton, MA: JG Press, 1995 [1851]), 306–17; Matthew Dennis, *Cultivating a Landscape of Peace: Iroquois–European Encounters in Seventeenth-Century America* (Ithaca, NY: Cornell University Press, 1993).

7 Chiefs Jarvis A. Dione, Jos. K. Delisle, Jos. T. Skey, and Thos. Asennase to E. H. Meredith, May 14, 1875, RG10, vol. 1917, file 2764, LAC.

8 "Walbank, Matthew William," *Encyclopedia of Newfoundland and Labrador*, vol. 5, p. 496; "William McLea Walbank" Canadian Architecture Collection, McGill University, accession 20, http://cac.mcgill.ca, accessed Feb 16, 2011; "William McLea Walbank," CA601, S139, ANQ, http://pistard.banq.qc.ca, accessed Feb 16, 2011; Larry

S. McNally, "Pringle, Thomas," *Dictionary of Canadian Biography Online*, www.biographi.ca, accessed Feb 16, 2011.

9 The history the boundary itself is a fascinating and contentious issue, but it is not the subject of this chapter.

10 Correspondence between Walbank and the Department of Indian Affairs, 1880–1881, RG10, vol. 2109, file 20,131, LAC.

11 J. V. de Boucherville was clerk in charge of land sales. De Boucherville to Dept of Indian Affairs, March 11, 1882, RG10, vol. 7749, file 27005–1, LAC.

12 Robert M. Cover, "Violence and the Word," *Yale Law Journal* 95, no. 8 (1986): 1607. Nicholas Blomley, "Law, Property, and the Geography of Violence: The Frontier, the Survey, and the Grid," *Annals of the Association of American Geographers* 93, no. 1 (2003): 130.

13 n.a., "At Caughnawaga, P. Q.," *Catholic World* 37, no. 221 (1883): 612–13.

14 Joseph Williams to De Boucherville, February 28, 1882, RG10, vol. 7749, file 27005–1, LAC.

15 Caughnawaga Reference Books, 1885, RG10-B-8-aj, vols. 8968–8972, LAC. Skatsenhati (claimant 407) was born 1846, died 15 May 1885, married Anen Katsitsarokwas in 1869, and owned four lots totaling 103.65 acres: 27.83 acres cultivated ($529), 14.5 acres hay ($149), 51.54 bush ($159), 10.32 sugarbush ($290). Total value assessed: $1,127.

16 Walbank to Vankoughnet, Sept. 27, 1882, RG10, vol. 7749, file 27005–1, LAC.

17 W. A. Austin Report, Feb 25, 1887, vol. 7749, file 27005–1, LAC.

18 Vankoughnet to Walbank, July 28, 1884, RG10, vol. 7749, file 27005–1, LAC.

19 Caughnawaga Reference Books, 1885, RG10-B-8-aj, vols. 8968–8972, LAC.

20 Walbank to Vankoughnet, Feb 28, 1885, RG10, vol. 7749, file 27005–1, LAC.

21 Gerald Reid, *Kahnawà:ke: Factionalism, Traditionalism, and Nationalism in a Mohawk Community* (Lincoln: University of Nebraska Press, 2004), 40–45.

22 W. A. Austin report to Deputy Minister of Indian Affairs, Feb. 25, 1887, RG10, vol. 7749, file 27005–1, LAC.

23 See the 2008 NFB film "Club Native" directed by Tracey Deer, and numerous newspaper reports on attempts to evict non-Natives in January and February 2010.

24 Petition from about fifty Kahnawakehró:non (in general council) to Vankoughnet, July 7, 1885, RG10, vol. 7749, file 27005–1, LAC.

25 Walbank to Vankoughnet, July 27, 1885, RG10, vol. 7749, file 27005–1, LAC.

26 Walbank to Vankoughnet, Oct. 19, 1885, RG10, vol. 7749, file 27005–1, LAC.

27 LAC, RG10, vol. 2181, file 36,622-3A.

28 Petition from several Kahnawakehró:non to Vankoughnet, June 15, 1886, RG10, vol. 7749, file 27005–1, LAC.

29 Walbank to Vankoughnet, [July 26], 1886, RG10, vol. 7749, file 27005–1, LAC.

30 Vankoughnet to Walbank, July 30, 1886, RG10, vol. 7749, file 27005–1, LAC.

31 Caughnawaga Reference Books, 1885, RG10-B-8-aj, vols. 8968–8972, LAC. Translated by a contemporary interpreter, probably Owakenhen (Peter Stacey).

32 Caughnawaga Reference Books, 1885, RG10-B-8-aj, vols. 8968–8972, LAC. Ohionkoton (claimant 582) owned 1.03 acres of cultivated land, valued at $13.

33 Not included in the area of these lots were 550 acres for the village and common (a community managed pasture), 30 acres for quarrying, and 60 acres for roads.

34 Walbank to Vankoughnet, Sept. 10, 1886, RG10, vol. 7749, file 27005–1, LAC.

35 Walbank to Vankoughnet, Mar. 29 and May 18, 1887, RG10, vol. 7749, file 27005–1, LAC.

36 Vankoughnet to Walbank, May 21, 1887, RG10, vol. 7749, file 27005–1, LAC.

37 Canada, House of Commons, *Debates*, June 15, 1887; March 28, 1888; May 21, 1888; March 7 & 8, 1889.

38   Canada, House of Commons, *Debates*, Mar. 18, 1890, p. 2158.

39   Walbank to Vankoughnet, Sept. 13, 1887, RG10, vol. 7749, file 27005–1, LAC.

40   Caughnawaga Reference Books, 1885, RG10-B-8-aj, vols. 8968–8972, LAC.

41   Walbank to Vankoughnet, Sept. 13, 1887, RG10, vol. 7749, file 27005–1, LAC.

42   Walbank to Vankoughnet, May 18, 1887; Vankoughnet to Walbank, May 21, 1887, RG10, vol. 7749, file 27005–1, LAC.

43   Walbank to DIA, July 4, 1887; R. Sinclair to Walbank, July 13, 1887, RG10, vol. 7749, file 27005–1, LAC.

44   Vankoughnet to Walbank, Dec 31, 1887; Walbank to Vankoughnet, Jan 2, 1888, RG10, vol. 7749, file 27005–1, LAC.

45   Walbank report to John A. Macdonald, June 14, 1888, RG10, vol. 7749, file 27005–1, LAC.

46   Internal memo, Feb 5, 1890, RG10, vol. 7749, file 27005–1, LAC.

47   R. Sinclair memo to Deputy Minister, Jan 30, 1891, RG10, vol. 7749, file 27005–1, LAC.

48   For more on the Oka Crisis, see Geoffrey York, *People of the Pines: The Warriors and the Legacy of Oka* (Toronto: Little, Brown, 1991); Brenda Katlatont Gabriel-Doxtater and Arlette Kawanatatie Van den Hende, *At the Woods' Edge: An Anthology of the History of the People of Kanehsatà:ke* (Kanesatake, QC: Kanesatake Education Center, 1995).

49   Giselle Byrnes, *Boundary Markers: Land Surveying and the Colonisation of New Zealand* (Wellington, NZ: Bridget Williams Books, 2001), 10.

50   Craib, *Cartographic Mexico*, 10.

51   Robert A. Rundstrom, "Mapping, Postmodernism, Indigenous People and the Changing Direction of North American Cartography," *Cartographica* 28, no. 2 (1991): 4.

52   For more on inscribing and incorporating cultural practices, see Paul Connerton, *How Societies Remember* (Cambridge: Cambridge University Press, 1989).

53   Robert A. Rundstrom, "GIS, Indigenous Peoples, and Epistimological Diversity," *Cartography and Geographic Information Systems* 22, no. 1 (1995): 50–51.

54   Nancy Lee Peluso, "Whose Woods Are These? Counter-Mapping Forest Territories in Kalimantan, Indonesia," *Antipode* 27, no. 4 (1995): 386–87.

55   For examples, see: Rundstrom, "Mapping, Postmodernism, Indigenous People and the Changing Direction of North American Cartography": 8; Frank Duerden and Richard G. Kuhn, "The Application of Geographic Information Systems by First Nations and Government in Northern Canada," *Cartographica* 33, no. 2 (1996): 49–62; Kenneth G. Brealey, "First (National) Space: (Ab)original (Re)mappings of British Columbia" (PhD thesis, University of British Columbia, 2002).

56   For an excellent early example see, Hugh Brody, *Maps and Dreams: Indians and the British Columbia Frontier* (Vancouver: Douglas & McIntyre, 1981).

57   Duerden and Kuhn, "Application of Geographic Information Systems.

58   Claudio Aporta and Eric Higgs, "Satellite Culture: Global Positioning Systems, Inuit Wayfinding, and the Need for a New Account of Technology," *Current Anthropology* 46, no. 5 (2005): 729–53.

59   Terry N. Tobias, *Chief Kerry's Moose: A Guidebook to Land Use and Occupancy Mapping, Research Design and Data Collection* (Vancouver: Union of BC Indian Chiefs and Ecotrust Canada, 2000), 20–22.

60   Peluso, "Whose Woods Are These?" 387.

61   Robin Boast, Michael Bravo, and Ramesh Srinivasan, "Return to Babel: Emergent Diversity, Digital Resources, and Local Knowledge," *The Information Society* 23 (2007): 400–401.

62   Nadine Schuurman, "Formalization Matters: Critical GIS and Ontology Research," *Annals of the Association of American Geographers* 96, no. 4 (2006): 731.

63   Boast et al., "Return to Babel": 400–401.

64   Rundstrom, "GIS, Indigenous Peoples, and Epistimological Diversity": 45.

65   For an example, see Christopher Wellen, "Ontologies of Cree Hydrography: Formalization and Realization" (Master's thesis, McGill University, 2008).

66 Peluso, "Whose Woods Are These?": 389–90.

67 Schuurman, "Formalization Matters," 734.

68 Rundstrom, "GIS, Indigenous Peoples, and Epistimological Diversity": 50–51.

69 Ibid., 46–47.

70 See Pragya Agarwal, "Operationalising 'Sense of Place' as a Cognitive Operator for Semantics in Place-Based Ontologies," in *COSIT 2005, LNCS 3693*, ed. A. G. Cohn and D. M. Mark (2005); Nadine Schuurman, "Formalization Matters," 726–39.

71 Duerden and Kuhn, "Application of Geographic Information Systems": 49–62.

72 Boast et al., "Return to Babel" 397.

73 Caughnawaga Reference Books, 1885, RG10-B-8-aj, vols. 8968–8972, LAC.

74 Probably Tanekorens (James Lachiere) in the Walbank Survey.

75 Matthias Hill sworn statement, Feb. 15, 1895, RG10, vol. 2774, file 155,133, LAC.

76 Alexander Brosseau letter to DIA, Nov. 3, 1894, RG10, vol. 2774, file 155,133, LAC.

77 J. D. McLean memo to Deputy Minister, Apr. 9, 1895, RG10, vol. 2774, file 155,133, LAC.

8

Rebuilding a Neighbourhood of Montreal

François Dufaux and Sherry Olson

INTRODUCTION

From experience with the latest flu virus, most of us are well aware that people co-evolve with minute organisms, and aware, too, that people are constrained or impelled by the actions of bigger entities – the nation, the giant corporation, or the oscillation of the jet stream. In the biological and earth sciences today, as well as the social sciences, scholarly excitement centres on issues of scale and relatedness, on how relatedness at one scale affects relatedness at another, how size and shape affect the ability of the virus or the whale to move, to signal and respond. The sciences depend on instrumentation and imagery that make it possible to detect, measure, magnify or reduce, and to move between large and small. The story we tell here, of the rebuilding of a burnt-out neighbourhood of Montreal in the 1850s, points to the value of HGIS as a research instrument – for the curious citizen, the adventurous amateur, or the advanced scholar.

For nineteenth-century Montreal, an instrument was already available to us, with city blocks, lots, and buildings of 1825, 1846, 1881, and 1901, and we have taken advantage of it to infer change and to explore the responses of property owners to a single dramatic event, the Great Fire of 1852.

Fig. 8.1.
The zooming
opportunities of
GIS: territorial,
urban, building,
and dwelling
scales. (Sources:
MAP project,
François Dufaux,
BAnQ, Atlas
of Montreal,
Goad, 1881;
expropriation
de la rue Notre-
Dame Est.)

1825-1900 / urban expansion

1825 / neighbourhood land subdivision

1846 / neighbourhood built form

1852-1892 / building scale

Owners' responses highlight the meaning of design as a part of social history. As shown in other papers in this collection, GIS offers a platform to gather information about spaces and the people who occupy them. Although GIS performs its role in management and display of our data, we direction our attention in this chapter to the "zoom" as a feature for analyzing nested spaces. The historical sequence reveals the attention property owners were giving to both global and local considerations. How did citywide traffic and investment opportunities affect the owner's decision to rebuild? How did the investor juggle commercial space or dwelling space? a choice of materials? or a particular layout of rooms? From a social perspective, the zoom allows us to observe the person or family group who occupy a room, the room as part of a dwelling, the several dwellings stacked in a building, the buildings aligned in terraces and streets, and the streets arranged as a network with distinctive topology. The street we feature here, once known as St. Mary Street and today as Notre-Dame East, was the backbone of an 1830s neighbourhood destroyed by fire in 1852, rebuilt, destroyed forty years later to widen the street, and again rebuilt. As we shall see, large features, slow to change, are "environments" for the smaller ones, and small adjustments of boundaries and design contribute to the viability of a neighbourhood and success of the city as a whole. To consider changes at every scale, the "zoom" of the HGIS makes possible an interactive approach to urban form.

For research on nineteenth-century Montreal, the convenient instrument is the HGIS "MAP, *Montréal, l'avenir du passé*," a collective project in which dozens of people have invested their skills, each with a different purpose in mind.[1] Over the last ten years, applications have included studies of infant and child mortality, social mobility from one generation to the next, the catchment of a school in relation to parents' objectives, and the gendering of investments in the business centre.[2] Such studies draw on the massive databases integrated into MAP. The spatial links place the household on the cadastral lot – 7,000 lots in the 1850s, 30,000 in 1900 – with sufficient precision to measure buildings as small as the privy or the stable.

Since the neighbourhood of St. Mary Street was destroyed a third time in the 1970s, nothing remains of the great wooden city that existed before the fire – plank houses *pièce-sur-pièce*, shingled roofs, wooden stables, and boardwalks. Our analysis depends on archival documents, pieced together by insertion in the GIS. The habitat, as it was newly rebuilt in the 1850s, was displayed in elegant architectural renderings made when it was torn down in 1891. The earlier habitat – before the fire – we recovered from three sources: a cruder map of 1846, phantom walls and foundations that survived as vestiges on the 1891 drawings, and verbal specifications attached to the original construction contracts.[3] In addition to the conventions of a "geographic" display, the architectural questions invited application of SketchUp, a design-based software, and analytic techniques known as "space syntax."[4] With the enlarged tool kit, we take advantage of diverse and scattered sources – engravings, photographs, old maps, legal contracts, and newspaper clippings – to interpret the production of urban space.[5] In the process we uncover the logic of design – or re-design – at the several scales of the streetscape, the block, the lot, and the building. Those choices are related. We discover also social relatedness that is associated

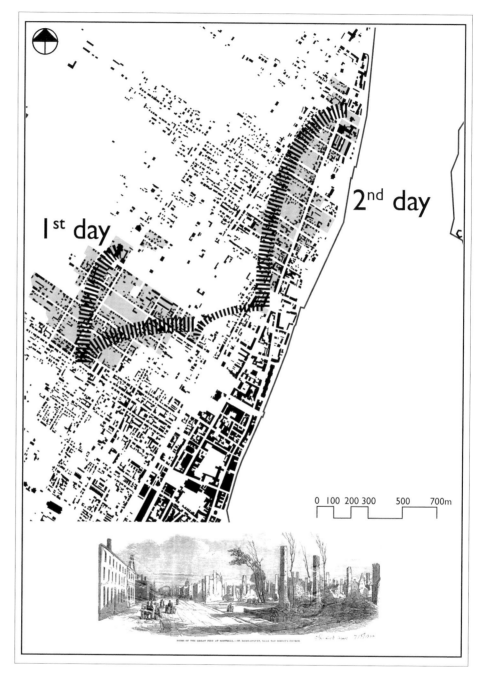

Fig. 8.2. The extent of the 1852 fire. (Sources: Cane map of 1846, corresponding MAP layers; and newspaper accounts; Ruines du grand incendie à Montréal, rue Saint-Denis, près de Bishop's Church, 1852, Image M7411.1.1, Courtesy of McCord Museum.)

with the evolution of the built environment: the everyday circulation of the family in a dwelling, the business transacted in the street, and what it meant to be "neighbours."

The history of a city is a succession of imagined futures. On Thursday, 8 July 1852, a spark interrupted all the various futures imagined by the people who lived in St. Mary Street, worked there, owned, rented, or financed the properties. All day the fire ran along the east side of St. Lawrence. On Friday morning, re-kindled, it swept east with a fury along St. Mary,[6] destroying one fifth of all the houses in the city and leaving 12,000 people homeless. Figure 8.2, reconstructed from newspaper accounts, indicates the unusual meteorological conditions that nursed the furious spread of the fire. On Saturday afternoon, a town meeting, after venting recriminations and accusations of negligence, struck a committee for relief, and a week later the committee identified three classes of sufferers – substantial property owners who would have to look after themselves, small property owners who would need some assistance (presumably loans) in recovering their footing, and tenant families to whom the community would, through the clergy, furnish bread and soup and blankets and tents – under no circumstances cash![7] The emergency made clear the interdependence of all elements of the community, but most of the people were left to their own devices. In the face of tragedy, how did they re-build their assets? The investments of families in St. Mary Street give some idea of their strategies. In owners' efforts at reconstruction, we shall see the importance of collaborations of family networks and negotiations between neighbouring property owners.

A PARTICULAR KIND OF STREET

St. Mary Street was laid out as a country road leading to the town centre, the fortified Montreal of the French regime. The Adams map of 1825 (Fig. 8.3) shows it as the route of south-shore farmers from the ferry across the St. Lawrence to the central market, and, since it linked installations of the British army (the barracks east of the docks, and the cannon on St. Helen's Island), it was the only road authorized for conveyance of gunpowder. By the 1830s, it was "the Great Main Street" of a fast-growing suburb, and on the Cane map of 1846 its increased density is apparent in the building footprints that form a near-continuous façade.

In a "typo-morphology" of streets, St. Mary was a "matrix road," backbone of development. At intervals across it, "planned building routes" were opened to promote subdivision of land into urban lots; and eventually, connecting streets were created for movement between subdivisions. That sequence is widely observed in cities, and when the urban fabric has become densely structured to a point of congestion, a "breakthrough street" is cut through the existing network.[8] In this case, Craig, the breakthrough street, was conceived as a firebreak 80 feet wide and developed 1853–1856, challenging the advantage of St. Mary Street for access to the centre. St. Mary was still perceived as an axis of centrality, and some of the most ambitious rebuilding projects asserted a vision of the street as an extension of the historic town centre.

At the scale of the city, centrality affects the economic value of a piece of land. This is

Fig. 8.3. St. Mary Street and Ward, 1825, 1846. (Sources: Adams map of 1825, Cane map of 1846, corresponding MAP layers, BAnQ, McGill University Libraries, Special Collections.)

1825

1846

0 50 100 150 200m

Fig. 8.4. After the fire of 1852, creation of the Craig Street "firebreak" created competition to the merchants in Notre Dame Street properties. (Sources: Cane map of 1846, Sicotte map of 1874, corresponding MAP layers, BAnQ.)

Craig streets plots

0 50 100   200   300m

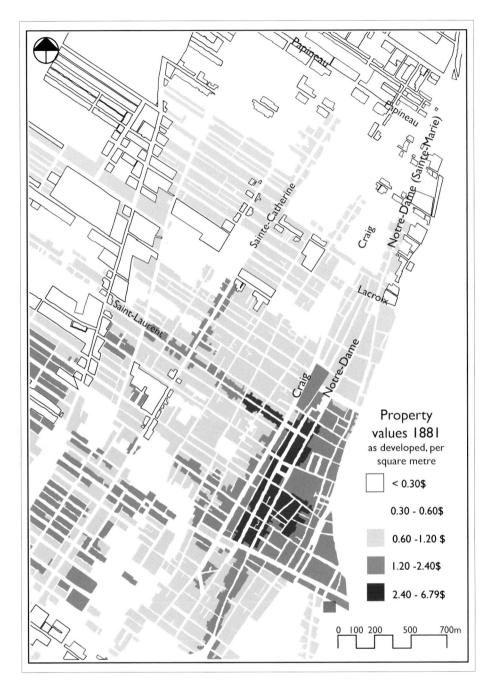

Fig. 8.5. Property values per square metre as developed in 1881. (Sources: Goad Atlas of 1881 and MAP databases from taxroll of property, Montreal, June 1880.)

Property values 1881
as developed, per square metre

< 0.30$

0.30 - 0.60$

0.60 - 1.20 $

1.20 - 2.40$

2.40 - 6.79$

0   100 200   500   700m

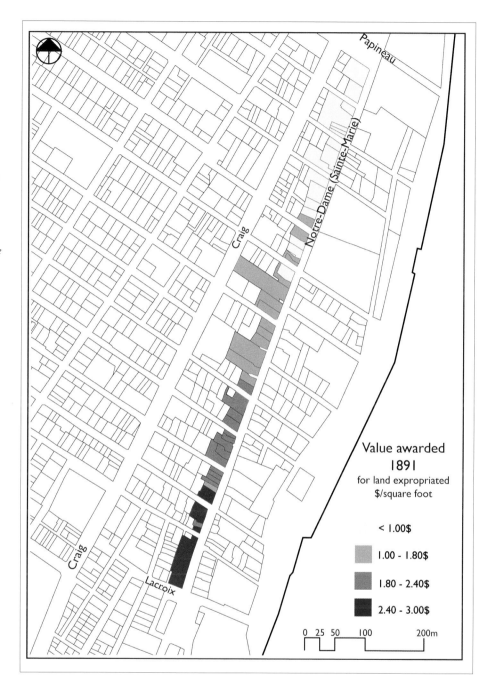

*Fig. 8.6.
Awards for land
expropriated in
1891 per square
foot. Owners
were compensated
also for the
estimated value
of brick, pipe, and
other materials.
(Sources: BAnQ,
Expropriation de
la rue Notre Dame
Est.)*

Value awarded
1891
for land expropriated
$/square foot

< 1.00$

1.00 - 1.80$

1.80 - 2.40$

2.40 - 3.00$

0 25 50   100   200m

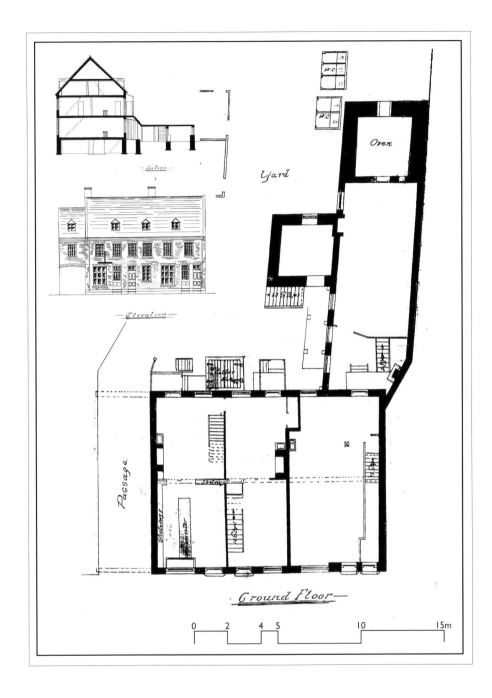

Fig. 8.7. William Clarke's house was promptly rebuilt. (Source: BAnQ, Expropriation de la rue Notre Dame Est, TP11,S2, SS2, SSS42, D184-1891-30.)

apparent in Fig. 8.5 for 1881. The cross streets close to the centre underwent earlier and more intense development; by the 1850s, they were longer and better connected to the fast-developing northern section of town, so that along St. Mary the compensation awarded for land in 1891 gives a good indication of the value of centrality (Fig. 8.6). From northeast to southwest, that is, from fringe to the centre of town, it ranged from one to three dollars per square foot; and the Desautels family, for example, holders of a dozen properties in the vicinity, promptly rebuilt their more valuable corner lots, while their mid-block lots were rebuilt more slowly and with smaller structures. Some of the burnt-out lots farthest from the centre remained undeveloped for more than a decade, listed on the tax roll at zero value, with "ruins of a stone house."

Consistent with its "main street" functions, St. Mary Street was built up in the 1830s, and rebuilt in the 1850s, for a great diversity of activities – shops, storage cellars, in the rear a smithy or oven, above the shops comfortable residences for the grain merchant, hardware dealer, or pharmacist, three- or four-room apartments for families, a room behind the shop for the marginal shoemaker or trunkmaker, and little attic rooms for apprentices, shop clerks, and servants. Clark's bakery is a good example (Fig. 8.7) with its grocery shop in front and bakehouse in the rear. On numerous corners were eating places, drinking places, and boarding houses. Judges of the 1840s repeatedly fined François Chef for leaving his horse and cart standing in the street at midnight, or allowing his tenants to disturb the peace. We'll see more of Chef and Clark and their neighbours, but the diversity of their interests and activities stems from the advantageous situation of the street in the overall network.

The variety of structures and layouts ensured persistence of that social diversity, with a wider range of rents than on most streets, and a wider range of ages, family sizes, and origins. With St. Lawrence "Main," whose diversity is still famous and functional today, and several other strips, the axial streets or "main drags" of the town were distinctive habitats of diversity. Amounting to one street segment in ten, they functioned to knit together people, activities, and sites to capture those distinctive urban advantages of agglomeration.[9]

LAYOUT OF A NEIGHBOURHOOD

The lots along the original matrix road were laid out in the French-régime tradition as "long lots" stretching back from the waterfront, by an easy-to-implement survey (chain measure from a baseline). The layout gave each farmer access to road and river, with roughly equal amounts of bottomland or woodlot. The long lots can be seen on the oldest maps. Subdivided over the years to accommodate heirs, the boundary lines did not meet St. Mary Street at a square corner, but at an angle of about 70 degrees, producing city lots with peculiar diamond shapes. In the 1830s, to exploit the frontage on "the Great Main Street of the Faubourg," owners built diamond-shaped houses, and half a century later, when city council ordered the buildings on the north side torn down for the widening, they compensated several of the evicted tenants for their rugs – cut on the bias to fit those angled rooms, rugs they could not use in the space they would rent next.[10]

Those property lines proved valuable in our construction of GIS layers for 1880. To warp and anchor historic maps for overlay on the modern engineering map of the city, we could not rely on the corners of landmark buildings as control points because so few buildings had survived. But the property lines matched. Thus the most permanent and dependable features were purely imaginary lines, imposed in the seventeenth-century survey and jealously guarded as markers of ownership. After providing soup and bread for fire victims, the next most pressing demand was restoration of those property lines. On the 26th, two weeks after the fire, baker William Clarke demanded that the city surveyor forthwith stake out the alignment of St. Mary Street so that he and his neighbours could survey and rebuild their fences in this vast "field of ruins."

RECONSTRUCTION AT THE SCALE OF A LOT

The destruction of buildings on neighbouring lots created unexpected opportunities to square up the diamond-shaped lots and consolidate the buildings. Customary law framed the mutual accommodation of owners in sharing the costs of their *mitoyenneté*: building and maintaining the fence along the line between them, a party wall or *pignon*, sometimes jointly owned chimneys. The adjustments, by an "an exchange of triangles," had a decided payoff in terms of the solidity, maintenance, and convenience of the buildings. Louis Chef (brother of François) and his neighbour Trefflé Goyet

agreed to trade triangles, and, after several weeks of dispute and the arbitration of skilled masons, Goyet accepted Chef's plan to raise the party wall to a greater height than his own. At least a dozen pairs of neighbours negotiated such adjustments of boundaries and party walls (Fig. 8.8).

The layout of streets and lanes gave each property a front and a back. All that diversity of activities was hidden behind a continuous line of façades designed, little by little, to convey urbanity, a certain civility, respectability, and order.[11] For greater elegance, neighbouring owners also seized opportunities to cooperate on the design of their façades. Nextdoor neighbours Terroux and Sénécal signed an agreement to match the height of the stone course, the style of stonework (boucharde, demi-boucharde, piqué), the same seven-foot setback from the street, the colour of the paint, and the design of the dormer windows.[12]

Ordinarily, an owner developed or re-developed within the rigid lines of authority over a single lot and maintained or renovated the rooms within the envelope of the existing walls. But when the capital invested in roof and walls was destroyed, the range of choices was suddenly larger. The owner could consider putting up a taller building, or building deeper into the lot, to profit from greater demand for rental space. And when destruction extended over an entire neighbourhood, as it did in 1852, neighbouring owners were faced with a narrowing of their financial options and a widening of their design choices, subject, of course, to negotiation.

Reconstruction on such a scale was handicapped by shortages of materials and labour. Within a month after the fire, local lumber dealers were ordering cedar beams by the

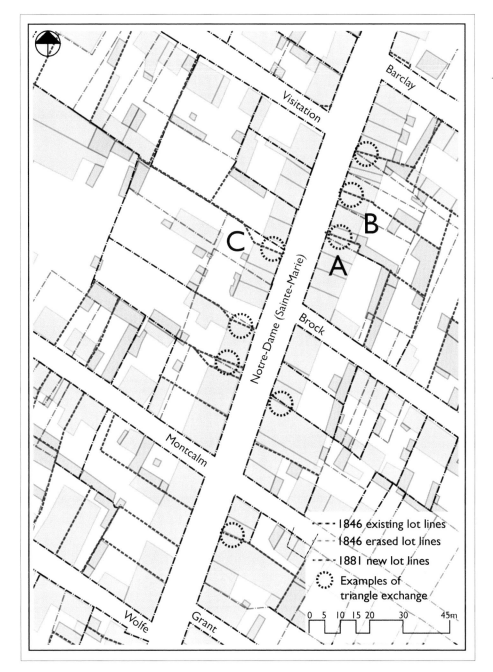

Fig. 8.8. Pairs of neighbouring properties involved in exchanges of triangles. Near Brock Street are the properties of Louis Chef (A), his neighbour Trefflé Goyette (B), and his brother François Chef (C). (Sources: Map layers of lots are based on the Cane Map of 1846 and the Goad Atlas of 1881.)

1846 existing lot lines
1846 erased lot lines
1881 new lot lines
Examples of triangle exchange

0 5 10 15 20   30   45m

Fig. 8.9. Plan and section of J. Terroux house. The presence of stone on the rear wall versus brick on the street façade suggests the reuse of ruins in reconstruction after the 1852 fire. (Source: BAnQ, Expropriation de la rue Notre Dame Est, TP11,S2,SS2,SSS42, D184-1891-10.)

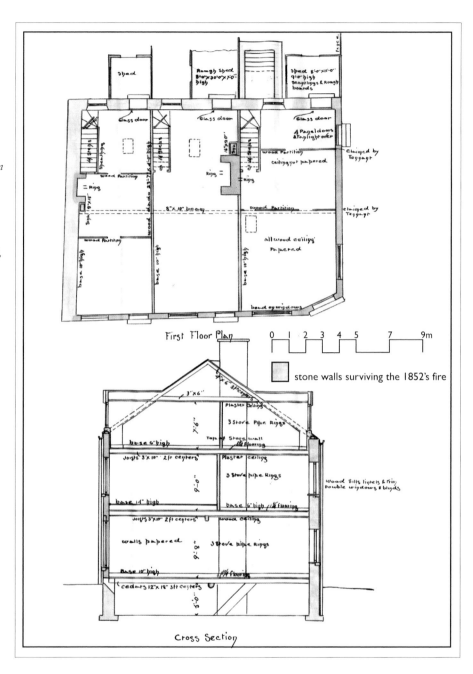

hundreds "to be delivered in the spring," in lengths of 40, 36, or 32 feet. The contracts mentioned here, rather early on the scene, advanced larger-than-usual sums to the builders, and some of the owners arranged to supply the stone or lime. Charles Terroux, to rebuild a property at the corner of Montcalm, arranged for re-use of the tumbled stone and brick from another property he owned in the next block (see Fig. 8.9). Masons and carpenters promptly moved into Montreal from surrounding villages, along with others who could deploy a horse and cart for hauling stone and gravel.

A more powerful constraint was the scarcity of capital, since there were alternative opportunities for investors in the mid-1850s: development of the Grand Trunk Railway, industrial expansion at hydraulic sites along the Lachine Canal, and larger bank buildings, stores, and warehouses in the commercial centre. Despite pressure to rebuild and restore income as fast as possible, it was four years before property values in St. Mary Street reached the level before the fire, and a decade before all the lots were rebuilt. The time it took to rebuild was a function of the condition of the building, the value of the location, and the assets the owner could mobilize. Clarke was one of the few who rebuilt that very summer. Fire is capricious, and his house seems to have suffered less damage than most; by January 1853 he had made repairs enough to rent part of the building and operate his bakery. Contracts like his indicate the high priority for repairing buildings damaged but not utterly destroyed, and re-erecting smaller ones on side streets that could be roofed before winter. (In the Montreal climate, most construction was scheduled April through October, and it was essential to roof the building by 1 November.) The layout

of Clarke's building (Fig. 8.7) reveals the original arrangement of connecting rooms – *en enfilade* – with numerous doors to the street, yard, and back lane.

A large property owner scattered his assets in a neighbourhood. Their concentration made it easier to collect the rents, oversee repairs, and negotiate annual leases, but too tight a concentration incurred a risk of severe losses in even a small fire.[13] This is the kind of information available from the MAP databases for property taxrolls of 1848, 1880, and 1903. An owner of multiple properties could sell one lot to raise money for reconstruction on another. Jason Sims, a lumber merchant, prospered from the situation but was nevertheless short of cash and sold his interest in the lumberyard to his partner. The wealthiest owner in the Faubourg was the Molson family; their wealth was older, more exclusive; their investments extended to the centre of town, rooted in high-level political connections, and they controlled long lines of credit in rural trades in liquor and groceries. The fire stopped short of the substantial brewery John Molson owned at the intersection of Papineau Square but destroyed extensive properties his brother William owned in St. Mary Street. John purchased a competitor's ruined foundry, and William took advantage of the situation to acquire additional land; he rebuilt a church he had endowed, but he did not give priority to (re)development of his lots in St. Mary Street. The fact that William Molson waited so long to redevelop and never fully rebuilt all his lots is probably a result of his greater investment opportunities. As an industrialist (distillery, gas works, and brewery), he operated on a city-wide scale, while Clarke, Chef, and Desautels, confined to a neighbourhood, needed to restore their rental assets as quickly as possible. One of the surprises of our exercise was the evidence that smaller investors played such important roles in the reconstruction.

The largest of the local insurance companies had promptly collapsed under the weight of demands. To supply additional capital, an unprecedented arrangement was devised. A firm incorporated as the Trust and Loan Company of Upper Canada borrowed money from British lenders associated with Baring's and made loans to Montreal property owners. The company was a client of Prime Minister John A. MacDonald and his law partner, and the banking arrangements in Britain were negotiated by Francis Hincks, the high-ranking inspector general of Canada. The municipal corporation (that is, the mayor and city council) certified each borrower and guaranteed the loans, with a further guarantee from the legislature of the United Province of Canada.[14]

In mid-March 1853, François Chef and his wife Catherine Roussin were awarded four of those guaranteed loans (1,850 pounds) to rebuild a block of four houses on St. Mary. On the 4th of May, their contractor started excavation, and by September the construction site extended the full breadth of the lot so that their neighbour William Clarke complained they were blocking the cartway they were legally bound to share. By January François Chef's property was producing income. Little "clumps" of loans suggest that neighbouring owners were discussing the options and operating in close touch. On the other side of St. Mary Street, Chef's brother Louis also began building in May: he signed with a mason on the 17th, took out four loans on the 18th, and in September obtained a fifth loan (400 pounds), adding up to the largest sum on record.[15]

DESIGN OF A BUILDING

The interdependence of choices produced readaptations affecting at once the scale of the block, the lot, the building, and its internal arrangements.[16] The size of the building footprint is constrained by dimensions of the lot and its frontage on the street, and this necessarily frames the internal layout and the shapes and dimensions of the rooms. Conversely, an owner's notions of comfort, convenience, and market demand called for certain sizes of rooms, their position at front or rear, their contact with outdoor light and ventilation, and relative access or privacy. The structural choices affected also the choice of materials and techniques: wall composition, and exterior cladding, structure and covering of the roof – at costs based on availability of materials and labour. The new regulations for fire-resistant brick or stone cladding added to the cost of "replacement" and invited close attention to the market.

Let us consider first the specifications for a very modest building, strictly residential, in a smaller cross street. Just a week after the fire, Marguerite Forté signed a contract with builder Joseph Turcot. The attached specifications outline a wood structure, the walls to be clad with brick, the roof with sheet iron, to satisfy the new fire regulations. The agreement described a good weatherproof house. It is supported on brick posts and hemlock beams without expensive deep foundations. From the details and measurements in the "specs," we can draw the plan and apply a SketchUp model to see how it must have looked (Figs. 8.10 and 8.11). The house was small (16 feet by 20 feet), divided in the traditional way into three rooms, one at the

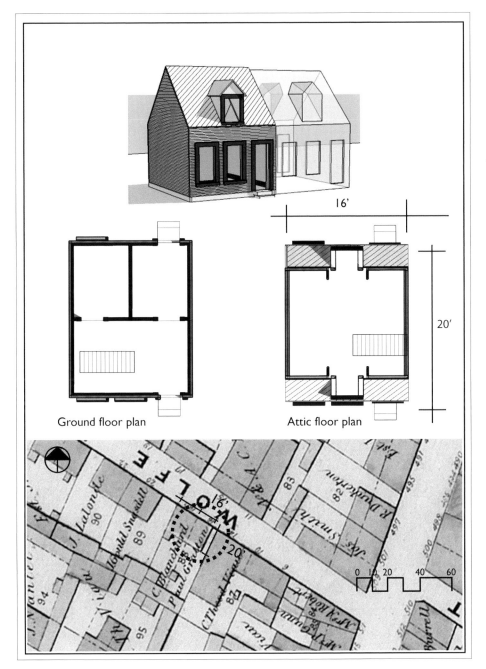

Fig. 8.10. House of Marguerite Forté: a reconstruction based on specifications. (Sources: BAnQ, Act of Simard 14 July 1852; Goad Atlas of 1881.)

Ground floor plan

Attic floor plan

front and two small ones at the back. Details of window frames and baseboards underlined the goal for a fine windproof construction, and the dormer windows indicate that the attic will be inhabited, suggesting about 600 square feet of interior space in all. Further details, such as the stone landing at front and back entrances, baseboards, and double windows, make it clear that this little house is no temporary shack but a building intended to last. Marguerite Forté was at once answering an immediate need and investing in the future.[17]

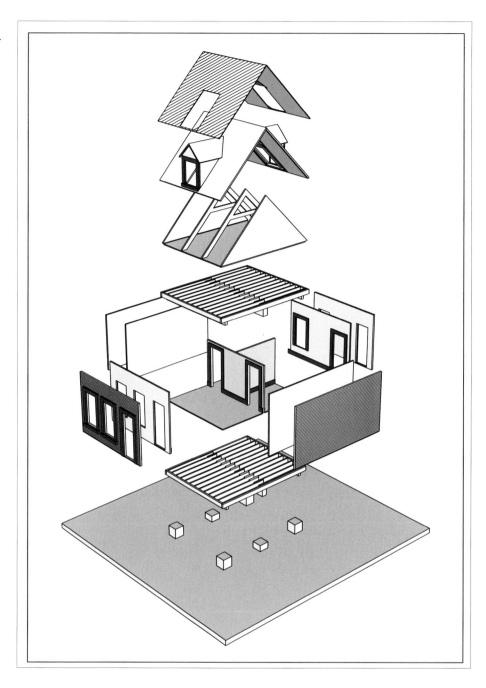

Fig. 8.11. House of Marguerite Forte: model based on the specifications. (Source: BAnQ, Act of Simard 14 July 1852.)

What the Chef brothers built was more substantial and illustrates the magnitude of the risks and the role of personal ambitions. François Chef, on his lot facing Brock Street (48 feet front), replaced several small wooden houses with a single building, thereby increasing total floor area to twice what it was before, and ensuring a greater stream of rents. His building (Fig. 8.12, lower drawing) met the new rules for fire-resistant materials but employed a traditional design: cut stone lintels, sash windows, a third-storey attic with dormer windows, shops with separate entry, and windows a little larger. Across the street, his younger half-brother Louis Chef put up a much larger building, 80 feet front by 36 feet deep, three storeys, with four shop fronts, four dwellings above, storage cellars, and stylish ornament: an elegant rounded corner, moulded capitals, pilasters *bouchardés*, architrave, bevelled consoles, and cut stone fireplaces. The two brothers shared some features: ground floor commercial, and two floors of dwellings above, much the same layout, with masonry walls and metal roof. But François chose a cautious design – sturdy, traditional, with stone façade and smaller shop windows – while Louis adopted an ostentatious brick décor with larger show windows, entablatures and arches, like what could be seen at the very centre of town.

The examples illustrate two important principles of real estate development. First is the pertinence of the saying, "Time is money." Because the builder's contract and reconstruction loan specified conditions of future payments, an owner needed time to pay off the lot to a former owner (often over five or ten years), to repay the construction loan, and collect rents from tenants in quarterly or – more often – monthly instalments, often higher in summer than winter. Each party to a contract – purchase, lease, or loan – was concerned with protection for the capital owing. People like Mrs. Forté and the Chef brothers were juggling short- and long-term considerations: the arrival of winter, the horizon of the loan, the vacancy rate, the cost of insurance, the risk of recession, the risk of dying, and the expectation of inheriting.

Second, design is a circular process. It implies compromise and adjustment among four objectives: (1) the composition (apparent in plan and elevation), (2) the configuration of rooms, their number and connections, (3) techniques of construction, and (4) the financial and legal constraints on feasibility. The designer may address one objective after the other, moving from the larger urban scale down to the look and the details, considering first, perhaps, the budget, then moving to the program, the composition, and the construction, but re-visiting each objective to achieve a synthesis in the final plan. The zoom displays the encounter of the several concerns. What tenants would they hope to attract to this particular site? Who would extend credit for the project? An owner's social network may provide assets to support its borrowing power or impede it. The architectural choices tell us about both the means and the ambitions of a family. The process is essential to understand the thinking of Louis and François Chef and Mrs. Forté and to appreciate their strategies of design and the social position to which each of them aspired.

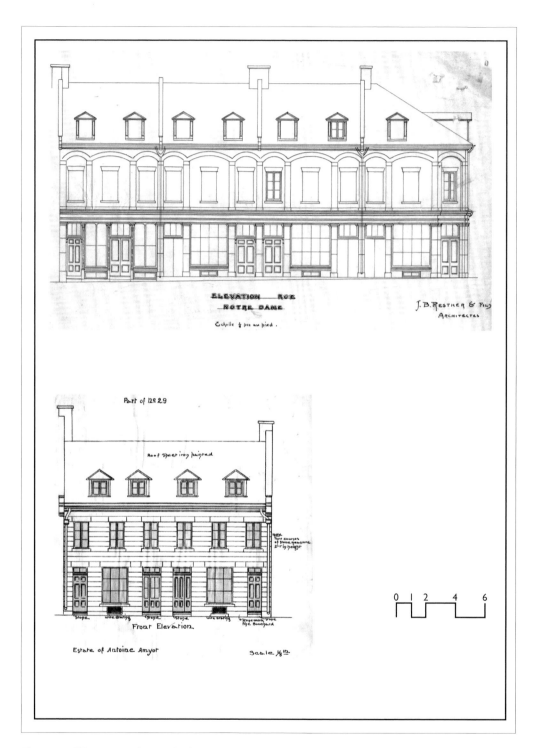

Fig. 8.12. Elevations of houses built in 1853 by the brothers Louis Chef (above) and François Chef (below). (Sources: BAnQ, L.Chef dit Vadeboncoeur, vs – La Cité de Montréal, TP11,S2,SS2,SSS42,D375; Expropriation de la rue Notre Dame Est, 1891, TP11, S2, SS2, SSS42, D184-1891-39.)

There was in St. Mary Street a network of Irish Catholic families, a network of Protestant families, and several extensive networks of French Canadians.[18] The Chef dit Vadeboncoeur family offers a good example of manoeuvres observed in all the others to achieve a degree of upward mobility. Because real estate in a fast-growing city tends, over the long run, to appreciate, a hold over a scrap of land made possible "success stories." Fewer than 15 per cent of householders inside the city limits owned a property, fewer still owned more than one lot. Like François Chef (father of François and Louis), many of the people who subdivided these blocks in the 1790s were independent craftsmen – master baker, master mason, or master carter. Little by little, they became "rentiers," deriving a significant share of their income from their property, and several of their sons-in-law and grandsons became notaries and doctors.

A legal mechanism critical to their strategy for holding onto property was the *substitution*, by which a couple made a legacy or a gift of property to their *great-grandchildren*, "born or to be born." The children and grandchildren were mere "substitutes," caretakers or managers, able to enjoy the use and income from the property, and obliged to maintain its value for their offspring, the ultimate owners. In 1840, that was the strategy of François Chef (the father) and his second wife Monique Brousseau, who wrote wills that gave their son Louis the use and enjoyment of properties they had acquired in the 1830s. The will put ownership in the hands of Louis' children and placed Louis himself under an obligation to conserve the value of the property.[19] This legal arrangement

influenced the design of a building. It usually favoured a sound construction that was better served by simpler and solid details, in order to last longer and minimize maintenance, rather than an impressive stylistic statement. It favoured also a flexible interior layout to insure long-term adaptations for a wide range of tenants, households, and street-level shops.

Their trusteeship involved oversight by the court-appointed curator and a family council weighted with uncles and great-uncles. Rooted in the customary law of France, and carried into the Civil Code of 1867, the mechanism was employed as well by the Irish Catholic and Protestant families of St. Mary Street.[20] It was often accompanied by clauses in the will that insisted that the legacy was intended to feed and support these children and should therefore be regarded as untouchable by creditors.

When the City cut that unusually wide swath for Craig Street, it trimmed some of the properties "*grevés*" under the substitution mechanism. Under court supervision, these moneys, too, were applied to rebuild on the remaining lots, to restore the incomes of aging parents and the value of the assets destined for the ultimate owners, the great-grandchildren still minors or not yet born.[21]

The elder François had set up both sons, François and Louis, in the same way, with small properties under a substitution, but at the time of the fire the two sons were not in the same financial position, nor did they have the same aspirations. François by this time owned fourteen properties, most of them with small wood buildings; he was able to sell or borrow on the properties that did not burn; he invested with prudence and ultimately left a valuable estate to his three children. While the masons were still at work, his wife, Catherine Roussin, died, having re-affirmed on her deathbed that she wanted her property divided equally among their children. Each would have a usufructuary enjoyment and an obligation to transmit it to their children "insaisissable" by any creditors. Her husband carried out the rebuilding and the division.

Louis, on the other hand, had just emerged from a bankruptcy proceeding. In a rash of business failures in 1848, he had lost the contents of the grocery he had operated for ten years. He chose to live in the handsome new building and operate his shop there, but he soon found himself unable to make the payments on the five loans he had taken out, in trouble again over the line of credit to the grocery. The creditors sued, branding him an "Absconded Debtor"; the trust company turned the new building over to the City (guarantor of the five loans), and the City eventually sold it at auction. So far as we know, this is the only loan that was not repaid. Louis' wife Sophie Guilbault retained at least the modest assets she brought into the marriage, and their son Louis Napoléon was emancipated at fifteen to try his skill in the grocery trade independent of his father. Thus, despite their parents' strategy of equal endowments, one son was able to provide houses for all of his grandchildren, the other none.[22]

THE AFTERMATH

The reconstruction of the 1850s is impressive; at the end of four years the taxable value of the whole array of properties on St. Mary Street was restored. In 1891 that success nourished an illusion on the part of city council and engineers that the same kind of uses and buildings could again be profitably rebuilt. To cope with heavier traffic on St. Mary Street, they wanted to widen it by thirty feet. This was an understandable response to the growth of the city, the extension of the overall street network, accompanied by heavier paving, street-based utilities like larger water mains, gas mains, trunk sewer, and in 1892 introduction of the electric tram. (The first line ran along St. Mary Street.) This was just one of a large number of widenings carried out.[23] The authorities imagined that the foreshortened lots would be of greater value and yield greater taxes than ever, and they tried to ensure this by requiring three-storey construction in brick or stone with shop front windows.

But those expectations were not realistic. As the city grew, St. Mary Street did not acquire the higher-order centrality of St. Lawrence Main. Railways had made the ferry obsolete. Craig Street (the new firebreak street) was diverting traffic from St. Mary. On the waterfront lots, industries were expanding, notably Molson's brewery and the Canadian Rubber factory; and they did not provide customers for "main street" shops.[24] The expropriation of 1891 produced a flurry of selling and long delays in reconstruction. Some of the owners were affected by another expropriation in 1894 to widen and "tunnel" Brock Street (one of the cross streets), another project that further disrupted tenancy, traffic, neighbouring relations, and everyday life,[25] and in the 1910s the provincial legislature backed massive acquisitions by the Canadian Pacific Railway for a marshalling yard and a grand passenger station. As more distant suburbs developed, the street carried more and more truck and auto traffic. It has since been re-imagined as an expressway, and in the 1970s the fronting properties were, for a third time, destroyed.

From the St. Mary Street case, one would have to ask: How common were these contingencies – fire and municipal expropriation? Every square metre of the city is owned by somebody, and the way the land is developed depends on who owns it, the capital the owner can deploy, the advantages of access to other parts of town, the kinds of services the city provides, and the obligations and taxes the owner pays. The municipal administration chartered in 1840 was structured as a corporation of property owners. The property tax was its principal source of revenue, and to this day amounts to two thirds of the revenues of the City of Montreal. Implicit in that structure is a "growth machine" dynamic that in 1890 was already in contradiction with family interests.[26] The episodes we've described challenged family strategies and smashed expectations of a great many people, but the neighbourhood-wide impact also synchronized the options, creating opportunities to rearrange spaces and reconstruct the city.

CONCLUSION

We can draw some tentative conclusions of three types. First, with respect to HGIS, we have pointed out the value of the "zoom." It's now a familiar technique: your query to find a restaurant, a zoom to the address, and another click to zoom out to see how it's situated in relation to the route you might take. The ability to move in close may not be so obvious to the historian who is considering whether to invest in mastering the techniques of GIS or

to an academic team engaged in designing the scope of a new venture into HGIS. For MAP, we adopted a rather high level of precision – individual addresses, accurate street widths and buildings as small as the privy – in order to exploit particular sources for a particular city (notably the Goad Atlas of 1881). That choice, as it turns out, has allowed the integration of a large array of data and collaborations toward a great variety of scholarly objectives. Given a city so rich in sources, we wanted to open up the many possibilities. Some of the original maps are fragile; they appear at different scales, and they are housed in different archives. We wanted to allow for a wide range of queries, to lead to deeper inquiry, essentially a pedagogy of research into the urban habitat.

Second, from the standpoint of the history of architecture, the HGIS has suggested some possibilities we are excited about. One is the retrieval of those "ghosts" of buildings such that we can recover the dimensions of the all-wood habitat that has disappeared from Montreal. For most of those little wooden houses that constituted the vernacular architecture of the first half of the nineteenth century, no formal plans were preserved, but the GIS sharpens clues that lie hidden in the foundations and cellars of subsequent buildings whose plans have survived. From the construction contracts and specifications, we can recover room sizes, layouts, materials and techniques of building, and some of the reasoning behind the design, economic objectives, or even whims of the owner. For Mrs. O'Brien, the builder promised the best brick Mr. Adams makes, laid in such a way as to produce "un joli devant de maison."

Inquiry invites theory, and the GIS serves well the space syntax theory that interprets the habitat as a set of nested spaces in which we move and circulate. That literature suggests useful schematics and numeric codes for the way spaces are connected and nested, pointing to the connectivity of Mrs. Forté's kitchen, or the strategic introduction of a hallway when a house is made into a boarding house.

A third set of findings are practices of interest to social historians as well as historians of architecture. The GIS makes it possible to watch the city developing over time and to move back and forth between the urban plan as a whole and a particular neighbourhood or property. This calls for a theory of design, inviting us to situate each singular building project in its social context as a response to opportunities and constraints, macro and micro. This is a promising development if we want to discern the significance of the myriad minute changes implemented by individuals and families in their real estate commitments and strategies.

GIS offers a platform sensitive to choices of scale and design. It allows the user to discern the effects of legal and economic constraints and to interpret the response of a social network to spatial opportunities.

By exploiting the tools of the geographer and the urban planner, and by sharing lines of inquiry from both social history and the history of architecture, we have uncovered collaborations among neighbouring property owners and family strategies for protecting assets. The small players of St. Mary Street introduce some nuance to the grand narrative of Montreal fortunes. The expropriation of 1891 shows they had little influence on municipal affairs, but they were not without resources – a scrap of real estate, some craftsmanship, some tools, skills in networking and negotiation, family loyalties, and a planning horizon of four generations.

NOTES

1   For a broader view of the MAP project, see Robert C. H. Sweeny and Sherry Olson, "MAP: Montréal l'avenir du passé, Sharing Geodatabases Yesterday, Today and Tomorrow," *Geomatica* 57, no. 2 (2003): 145–54; http://mun.ca/mapm/. A dozen Montreal scholars and a dozen graduate students initiated the project in 2000. A grant from Geoide, one of Canada's National Centres of Excellence, supported for one year a full-time technician, Rosa Orlandini. We are grateful for the continuous participation and insights of Robert C. H. Sweeny and Jason Gilliland; for initial insights of David Hanna, Jean-Claude Robert, Kevin Henry, and Rosalyn Trigger; for sources and technical support from Ville de Montréal (Service des archives, Service de géomatique); McGill Libraries (Special Collections, Digital Collections, and Electronic Resources); archivists of the Bibliothèque et Archives nationales du

Québec (BAnQ), and grants from the Social Science and Humanities Research Council of Canada to Olson, Gilliland, and Danielle Gauvreau. Sammy Yehias, student in Architecture, Université Laval, created the SketchUp models.

2   Sherry Olson and Patricia A. Thornton, *Peopling the North American City: Montreal, 1840–1900* (Montreal and Kingston: McGill-Queen's University Press, 2011); Danielle Gauvreau and Sherry Olson, "Mobilité sociale dans une ville industrielle nord-américaine: Montréal, 1880–1900," *Annales de Démographie Historique* no. 1 (2008): 89–114; Roderick MacLeod and Mary Anne Poutanen, "Proper Objects of This Institution: Working Families, Children and the British and Canadian School in Nineteenth Century Montreal," *Historical Studies in Education* 20, no. 2 (2008): 22–54; Robert C. H. Sweeny, "Risky Spaces: The Montreal Fire Insurance Company, 1817–20," In

The Territories of Business, edited by C. Bellavance and P. Lanthier (Sainte-Foy: Les Presses de l'Université Laval, 2004), 9–23.

3   The key cartographic sources are J. Adams, Map of the City and Suburbs of Montreal, 1825; James Cane, Topographical and Pictorial Map of the City of Montreal, 1:5100 (Montreal: Robert W.S. Mackay, 1846); Frederick W. Blaiklock, E. H. Charles Lionais, and Louis-Wilfrid Sicotte, Cadastral Plans, City of Montreal (Montréal, E.H.C. Lionais, 1880); C. E. Goad, *Atlas of the City of Montreal showing all Buildings and Names of Owners* (C.E. Goad Ltd., 1881). Of particular importance for situating addresses, owners, residents, and events on the lots are municipal taxrolls of property and rental valuations of occupants (MAP digital databases for 1848 and 1880 and 1903); census records (MAP digital databases for 1842, 1881, and 1901); and Lovell's Montreal Directory (accessible at http://bibnum2.banq.qc.ca/bna/lovell/index.html). For this stretch of St. Mary Street, we collected additional data from municipal taxrolls for 1852 (before the fire), each year 1853–1857, and at five-year intervals 1861–1906; and from microfilms Census of Canada runs for 1861, 1871, 1891, and 1911.

4   At various stages we have employed GIS softwares ArcView 3x and ArcGIS 10; architects' softwares AutoCad and Google SketchUp 8 Pro (the freeware version will serve); and the relational database FoxPro 2.6. For the last, any database format such as Paradox or MS-Access can be substituted, and databases can be created or independently accessed in Excel.

5   For economic and social theorization of the production of urban space, see Henri Lefebvre, *Espace et politique, Le droit à la ville II* (Paris: Anthropos, 1972); Pierre Bourdieu, *Esquisse d'une théorie de la pratique* (Paris: Droz, 1972); Alan Pred, "Place as Historically Contingent Process: Structuration and the Time-Geography of Becoming Places," *Annals of the Association of American Geographers* 74, no. 2 (1984): 279–97; Jeremy Boulton, *Neighbourhood and Society: A London Suburb in the Seventeenth Century* (Cambridge: Cambridge University Press, 1987).

6   "English extracts, conflagration at Montreal," *Sydney Morning Herald* (NSW), 4 November 1852, 6, as reprinted from the *Illustrated News* (London) of 17 August. See also *La Minerve*, 13 and 15 July 1852; *The Montreal Witness*, 12 July 1852.

7   *Proceedings of the General Relief Committee*. The new reservoir had been emptied for repairs; property owners and city officials had ignored repeated earlier regulations that called for brick or stone cladding and fire-resistant roofing in "main streets" as well as the central core now known as "Old Montreal." At the end of twelve months, the committee made an accounting of the 26,000 pounds they had spent on relief. Further details of citizen responses and references on the fire are reported in François Dufaux and Sherry Olson, "Reconstruire Montréal, rebâtir sa fortune," *Revue de Bibliothèque et Archives nationales du Québec* 1 (2009): 44–57.

8   We refer to two types of analysis: A "typo-morphological analysis" sorts streets in terms of the phase of development in the historical process of urban expansion and transformation, as in G. Caniggia and G. L. Maffei, *Interpreting Basic Building: Architectural Composition and Building Typology* (Firenze: Aliena Editrice, 2001). A "space syntax" analysis provides a mathematical index of the relative accessibility of a street segment as part of the street network. For the logic and methods of space syntax, see Bill Hillier, *Space Is the Machine: A Configurational Theory of Architecture* (New York: Cambridge University Press, 1996); Julienne Hanson, *Decoding Homes and Houses* (Cambridge: Cambridge University Press, 1998); Laura Vaughan, "The Spatial Syntax of Urban Segregation," *Progress in Planning* 67, no. 3 (2007): 205–94.

9   J. Peponis, E. Hadjinikolaou, et al., "The Spatial Core of Urban Culture," *Ekistics* 334/335 (1989): 43–55, comparing various Greek cities, argue that 10 per cent of streets make up the core of a city, with higher integration values according to space syntax measures. On segregation in Montreal, see Jason A. Gilliland and Sherry Olson, 2010. Residential segregation in the industrializing city: a closer look. *Urban Geography* 31, no. 1 (2010): 29–58.

10 BAnQ, Expropriation de la rue Notre Dame Est. We found details for 59 properties (all on the northwest side), building plans and elevations for 39 of them. The properties discussed here, at the

corner of Brock street (now Beaudry), come from files 29 (François Chef) and 325 (Louis Chef).

11 All of the notarized acts refer to BAnQ, Fonds Cour supérieure, District judiciaire de Montréal, Greffes de notaires, identified by the name of the notary and date of the act.

12 BAnQ, Act of Dagen, 19 August 1852.

13 On the risks associated with fire insurance see Robert C.H. Sweeny, "Risky Spaces: The Montreal Fire Insurance Company, 1817–20," In *The Territories of Business*, edited by C. Bellavance and P. Lanthier (Sainte-Foy: Les Presses de l'Université Laval, 2004), 9–23.

14 The arrangements were presumably lucrative for all parties and gave the company a toehold for loans in rural areas of Lower Canada (Quebec) over the next half-century. The loans for reconstruction are recorded in the repertory of notary T.-B. Doucet, 10 January to 10 June 1853.

15 The Canadian pound was discounted at about 20 per cent below the British pound sterling; and in 1854 a new currency was introduced, the Canadian dollar, at $4 to the Canadian pound.

16 These are the kinds of questions asked in urban morphology. See Jason Gilliland, "Redimensioning the Urban Vascular System: Street Widening Operations in Montreal, 1850–1918," in *Transformations of Urban Form*, edited by G. Corona and G. L. Maffei (Firenze: Alinea Editrice, 1999), FK2.7–11; G. Cataldi, "From Muratori to Caniggia: The Origins and Development of the Italian School of Design Typology," *Urban Morphology* 7, no. 1 (2003): 19–34; François Dufaux, "A New World from Two Old Ones: The Evolution of Montreal's Tenements, 1850–1892," *Urban Morphology* 4, no. 1 (2000): 9–19. GIS has made a valuable addition to the arsenal of computer aided design (CAD) that architects use to handle the mass of engineering data, spatial dimensions of the many components of a project, and the temporal sequencing of the construction process.

17 BAnQ, Act of Simard, 14 July 1852. The specification for *pruche* usually refers to hemlock, *Tsuga canadensis*, relatively rot-resistant. Pine was used for floors and ceilings as well as beams. The minimal foundations reserve the option for moving a building, as was rather common

for wood structures, e.g., Act of Simard, 23 September 1853.

18 The Irish network included the Cuddy and Morley families and another lineage of O'Briens. The Protestant network was centred on the Lambs and Shortleys, intermarried, bakers like Clarke and the French Canadian O'Brien. Other French Canadian networks are those of the Martel and Dumaine, the Dufresne, the Desautels and Goyet. The modest percentage of owner households (ca. 15%) can be appraised from the taxrolls of 1848, 1880, and 1903. See Robert C. H. Sweeny, "Property and Gender: Lessons from a 19th-Century Town," *Journal of Canadian Studies* (London) 22 (2006/7): 9–34.; Stephen Hertzog and Robert D. Lewis, "A City of Tenants: Homeownership and Social Class in Montreal, 1847–1881," *The Canadian Geographer* 304 (1986): 316–23.

19 François Chef had also an elder son, Jean-Baptiste, also a carter, married about 1819, and Jean-Baptiste had seven children. The management of the properties can be traced in BAnQ, Tutelles (acts of guardianship) between 1837 and 1894.

20 BAnQ, Acts of Labadie 14 May 1840, 26 June 1845, and 14 December 1853; and Act of Belle, 28 September 1852. Other St. Mary Street owners who used the substitution mechanism were J. H. Richelieu (Act of C. F. Papineau 23 December 1852); Pierre Desautels (Act of N. B. Doucet 23 March 1817); François Désautels (Act of J. Belle 11 November 1851); Marguerite Désautels (Act of Hétu 26 February 1880); John Hogarth (Act of Busby 23 February 1847; Tutelles, 8 June 1853); Charles O'Brien (Act of Baron 4 October 1822); and Joseph Léveillé. Genealogy of the Chef family was compiled using the Drouin index to Catholic marriages of men and women, microfilmed registers of Paroisse Notre-Dame, and indexes compiled by members of the Quebec Family History Society and the Latter-day Saints.

21 See also BAnQ, Expropriation de la rue Craig. Financing objectives are explicit in a sale confirmed by Act of Lapparé, 21 November 1853. The holograph will of Joseph Augustin Labadie, recorded for probate by Act of Gaudry, 1 July 1854, makes explicit his intent to cover legacies and debts by selling all his burnt properties in

the Faubourg, except those at the corner of Saint-Ignace and the corner of Panet, and to use the surplus to rebuild on those two lots.

22 In the first bankruptcy, Louis Chef lost three properties he thought he had secured but held onto three others that were hedged by belonging to his children, two and five years old. He was released by his creditors fourteen months before the fire (BAnQ, Dossiers des faillis). At the time of the widening of Brock Street, he made an unsuccessful attempt to recover the property from the party to whom the City had sold it (BAnQ, Expropriation de la rue Brock, 14 May 1894).

23 An even larger array of widenings was contemplated, as shown by homologation lines visible on the Goad Atlas of 1881. A line established by city surveyor and approved by city council warned owners that further investment within the zone to be "taken" would not be compensated. On the expropriation process, see Jason Gilliland, "The Creative Destruction of Montreal: Street Widenings and Urban (Re) development in the Nineteenth Century," *Urban History Review* 31, no. 1 (2002): 37–51.

24 A suite of space syntax integration measurements reveals the growing importance of St. Catherine Street as the main east-west artery and, conversely, the declining importance of St. Mary.

25 BAnQ, Expropriation de la rue Brock.

26 See Harvey Molotch, "The City as a Growth Machine: Toward a Political Economy of Place," *American Journal of Sociology* 82, no. 2 (1976): 309–32; Robin L. Einhorn, *Property Rules, Political Economy in Chicago, 1833–1872* (Chicago: University of Chicago Press, 2001).

9

Growth and Erosion: A Reflection on Salt Marsh Evolution in the St. Lawrence Estuary Using HGIS

Matthew G. Hatvany

INTRODUCTION

Salt marshes are the focus of attention of scientists and government actors studying the impact of climate change on Quebec's St. Lawrence Estuary.[1] Viewed as some of the world's richest ecosystems, under optimal conditions salt marshes are capable of producing several times more organic material per acre than the best agricultural lands. Historically, they were highly valued by Amerindians and colonial peoples as important hunting and fishing grounds and as fertile sources of natural pasture and fodder. Since the industrial period, however, they were frequently treated with disdain, being diked, drained, or filled for intensive agriculture, road construction, and urban expansion. In post-industrial times, however, appreciation of the marshes has come full circle, and they are now recognized for their "ecosystem services" as producers of organic material that forms the base of the estuarine food chain, filters pollutants from water, sequesters carbon, and buffers coasts against storm events.[2]

In the St. Lawrence Estuary, there are some 44 km² (4,418 ha) of salt marsh. Since the beginning of the post-industrial period (1970–), there has been growing anxiety over the future of the marshes because of a perceived erosion of the coastline. Marsh erosion, coastal managers argue, menaces not only important ecosystem services but also the built environment (houses, roads, railroads, dikes, etc.).[3] According to a survey of some eighty studies published between 1977 and 2004, the magnitude of that erosion is unsurpassed in modern history and driven by human activities.[4] Anthropogenic causes, the survey argues, are responsible for intensifying storm events, accelerating sea-level rise and reducing protective winter ice cover, all of which result in erosion of coastal marshes. Additionally, activities like dam construction on major rivers traps sediments upstream, reducing their downstream availability in the estuary for marsh maintenance and building. As a result, one specialist notes, the sedimentary coastlines of the St. Lawrence Estuary and gulf have been eroding at a rate of 0.5 m to 2 m annually since 1931.[5]

As a citizen profoundly interested in society's relationship with the environment, I am alarmed by the threat that erosion poses to the future of the St. Lawrence salt marshes. However, as a historical geographer of wetlands, I am also interested in their natural and cultural evolution. Regrettably, that concern with looking both backward and forward in time is not widely embraced by most natural scientists. The problem of "presentism" – or only looking forward – stems in large part from the underlying interest of funding agencies and governments in better understanding current and future problems. And yet, the study of the past has made me keenly aware that the current

discourse of marsh erosion is quite novel – a product of the post-industrial era. In fact, it contradicts the historical understanding shared by medieval, colonial, and industrial period observers that salt marshes are constantly growing as a result of geomorphological processes like sedimentation and biological processes like plant succession. In the process of studying the medieval and early modern history of prominent salt marshes like those of the bays of Mont-Saint-Michel and the Aiguillon in France, it has been impossible for me to ignore the fact that the marshes of Western France have been growing at impressive rates. So rapid, in fact, that many marshes have been successively diked over the last millennium – new dikes being regularly constructed in rhythm with the seaward advance of the marshes.[6]

Many of the early colonists settling the banks of the St. Lawrence Estuary came from areas in France where successive marsh diking was known; thus, it is not surprising that in my study of the Kamouraska region of Quebec (Fig. 9.1) – where some of the largest salt marshes in the St. Lawrence Estuary are located – farmers employed many of the same marsh diking techniques as those employed in the Old World.[7] From the 1850s until as late as 1983, lay and scientific observers all noted that the marshes of Kamouraska were growing at annual rates varying between 30 to 75 cm and more. Underscoring this point is the fact that the word "erosion" (*érosion*) is rarely mentioned in the plethora of archival documents on the marshes written before the 1970s.[8]

This conspicuous dichotomy between past and present discourses on salt marsh evolution is a conundrum that coastal researchers have failed to address.[9] Could hundreds of years of historical observation have misrepresented

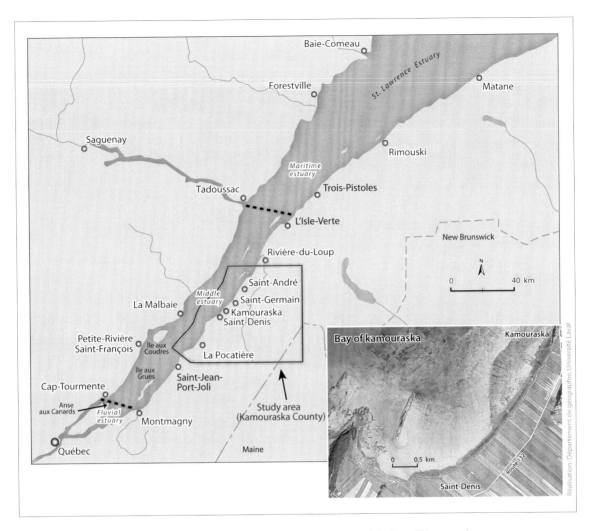

Fig. 9.1. St. Lawrence Estuary and case study area of Kamouraska County and the Bay of Kamouraska.

marsh growth? Could modern science, with highly sophisticated tools for observing and measuring erosion, be distorted? Or, might there be other explanations? Have anthropogenic activities provoked a radical environmental shift in marsh evolution over the last forty years that might explain the apparent contradiction between the historical and current literature? Looking backwards and forwards along the environmental timeline can provide a more nuanced perspective on this question, on the impact of anthropogenic activities, and on the future of the marshes.

One way to address these questions is to turn to Historical Geographical Information Systems, or HGIS. With HGIS, it is possible to trace the physical evolution of one of the largest salt marshes in the St. Lawrence Estuary – the Bay of Kamouraska – using historical maps from the eighteenth and nineteenth centuries, aerial photographs from the twentieth century, and GPS (Global Positioning Systems) coordinates from the twenty-first century. HGIS is an evolving tool and methodology that over the last decade has shown substantial promise in answering pressing environmental questions. Outstanding examples from France and the United States of the use of HGIS in the study of salt marshes include Virginie Bouchard et al.'s work on the evolution of the marshes of the Bay of Mont-Saint-Michel; Fernand Verger's study of the progression of the marshes of the Bay of the Aiguillon; Franchomme and Schmitt's analysis of wetlands in the Nord Pas-de-Calais; and Keryn Bromberg Gedan and Mark Bertness's quantitative study of human impacts on New England salt marshes.[10]

Clearly, the capacity to wed historical records with avant-garde computer-assisted mapping technologies makes HGIS an appealing research method for today's environmentalists. And yet, as innovative as HGIS appears, it has a proven pedigree rooted in nearly a century of historical and geographical thinking about how best to visually represent natural and cultural change of landscapes over time. In the 1940s, historical geographer Andrew Hill Clark encouraged his students to combine historical and spatial data, arguing that the geographer's "muddy boots" must go to the archives in order to master the temporal aspects of landscape evolution.[11] Conversely, using HGIS obliges the temporal specialist to go into the field to observe and confirm spatial patterns revealed in historical maps, aerial photographs and textual descriptions. It is in this sense of a holistic vision of the natural and cultural landscape informed by field observations and archival work that I see the ultimate potential of HGIS as a tool for the study of environmental change.

METHODOLOGY

The salt marsh of the Bay of Kamouraska (Fig. 9.1) is an ideal study site to explore the utilization of HGIS as a tool and method of environmental analysis. Located on the South Shore of the St. Lawrence Estuary, some 150 km east of Quebec City, the marsh is considered "intermediate" in size for the St. Lawrence Estuary, about 6.5 km long and up to several hundred metres wide. Thanks to the presence of an active seigneur who mapped the marshes in the colonial era, the nearby establishment of an agricultural society and an agricultural school in the early industrial period, and significant interest in the marshes of the region by a succession of farmers, government ministries, historians, environmentalists, and geographers in the last century, the Bay of Kamouraska is the most historically documented salt marshes in the St. Lawrence Estuary in regard to both natural and cultural evolution.[12]

While most people intrinsically understand what a salt marsh is, there is ongoing debate over precise definitions and delineations. Marsh characterization is highly dependent upon disciplinary background.[13] In this study, I use a simple description inspired by the interdisciplinary work of Desplanques and Mossman (2004) on the salt marshes of the Bay of Fundy. "Salt marsh" refers to a flat low-lying coastal area of vegetated marine soils that can be a few metres to several kilometres long and wide. It is periodically inundated by tidal salt water so that only plants adapted to saline conditions survive. This definition excludes unvegetated tidal flats below mean sea level.[14] Salt marsh evolution is analyzed through a time-series study of the lateral movements of the lower limit of the marsh using methods inspired by Champagne et al., Bouchard et al., and Verger and based upon data obtained from historical maps, aerial photographs, GPS coordinates, and my own field observations of the Bay of Kamouraska over seventeen years (Fig. 9.2).[15]

Tidal regimes play an important role in the geological and biological delineation of marshes. In the Kamouraska region, tides are macrotidal (several metres), rising and falling twice daily (semi-diurnal) with mean tides of 4.2 m. High water (HW) is 6.2 m, low water (LW) is 0.1 m, and mean sea level (MSL) is 3.16 m.[16] In relation to these measures, the salt marsh is divided into low and high marsh according to the amount of time (daily, monthly, and annually) that the marsh is submerged by the tides. The inferior limit of the low marsh, submerged twice daily by the tides, begins around mean sea level and is associated with the growth of *Spartina alterniflora*, a salt-tolerant plant adapted to daily submersion. The carpet-like growth of *Spartina alterniflora* is easily identified on aerial photographs, making it possible to accurately determine the area covered by the low marsh.[17] The high marsh, situated above the line of average high tides (known as mean high water or MHW), is submerged irregularly by only the highest tides and is dominated in Kamouraska by *Spartina patens*, *Spartina pectinata* and other plants capable of tolerating periodic submersion in salt water. *Spartina patens* and *Spartina pectinata*, while harder to individually identify than *Spartina alterniflora*, can nevertheless be distinguished on aerial photographs and used to accurately determine the area and limits of the high marsh.[18]

Over the last seventeen years of field observations, I have noted three forms of marsh limit, or seaward edge of the marsh, on the Bay of Kamouraska. These limits are here represented in schematic form using Adobe Illustrator (Fig. 9.2). The first form (Fig. 9.2A) is denoted by an abrupt scarp, varying between 10 and 30 cm in height, located at mean high water (MHW). This scarp neatly separates the high marsh from tidal flat and is indicative of a marsh undergoing active erosion. A second form (Fig. 9.2B) features a gradual transition between the low marsh and the tidal flat where isolated colonies of marsh grass are sometimes seen colonizing the tidal flat. This form is indicative of an actively growing marsh. The third form (Fig. 9.2C) combines both a scarp at high water (HW) and a gradual transition between the low marsh and the tidal flat. While this third type of marsh form is not documented in the St. Lawrence literature, since the 1960s Verger has noted in French studies that this form of marsh limit is indicative of simultaneous movements of both erosion and growth.[19]

Historic cadastral plans of the Bay of Kamouraska salt marsh exist in the provincial archives for 1781 and 1826, while a series of aerial photographs of the salt marsh at low tide exist for 1929, 1948, 1974, and 1985 at various provincial and national repositories. After scanning and importing into a GIS program the 1826 cadastral plan and all of the aerial photographs, my research team georeferenced them onto a standard projection to a scale of 1:20,000.[20] Because of problems of distortion with the 1826 map and older photographs, overall precision is estimated to be ± 5 m.[21] Once these spatial data were loaded into a GIS, it was possible to determine the overall area of salt marsh and to identify and trace the

Fig. 9.2. Forms of salt marsh limits on the Bay of Kamouraska. While many texts provide examples of growing (A) and eroding (B) marshes, field observations proved to be vital to identifying the existence of both process on the same marsh (C). These illustrations were drawn from field notes and photographs and then produced using Adobe Illustrator.

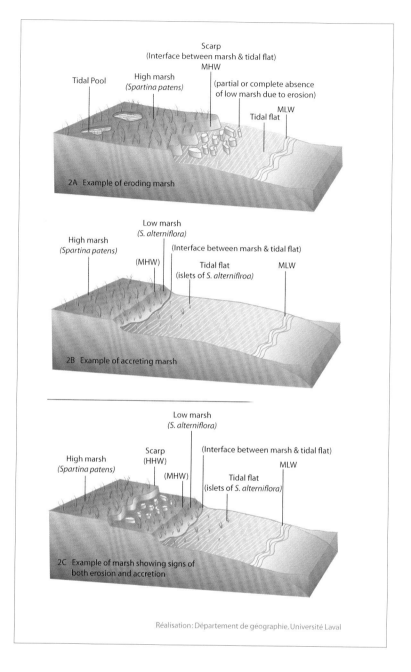

geographic evolution of the seaward edge of the salt marsh from 1929 to 1985.

The next step was to determine the seaward edge of the marsh in the present landscape. To do so, my research team made several field excursions to the Bay of Kamouraska in July 2008. We traced the seaward marsh edge on foot using a hand-held differential GPS (dGPS) that registered geographical coordinates every few metres. The final step in the research procedure was to combine the dGPS coordinates with the spatial data from the aerial photographs. GIS software allowed us to view the spatial data from the aerial photographs and GPS coordinates as a separate layer and then to merge all of the layers into a single image showing the evolution of the salt marsh limit over the period 1929–2008. Then, using the measuring tool in ArcGIS, we were able to note the differences in marsh limit at diverse dates and tabulate those changes in an Excel spreadsheet.

ANTHROPOGENIC CHANGE (1781–2008)

Most study of the evolutionary trends of the salt marshes of the St. Lawrence Estuary is limited in time to the last eighty years, corresponding with the beginnings of aerial photography in the late 1920s. To go further back in time is possible but requires the use of more complicated and significantly less precise methods relying on proxy sources (paleoecologic and stratigraphic analysis of soil cores) combined with costly and imprecise radio-carbon dating techniques. However, in the case of the Bay of Kamouraska, I turned to another source of information – historical maps – that allowed me to extend the spatiotemporal range of study by an additional 140 years to the late eighteenth century. The detailed cadastral plan of part of the Seigneurie de Kamouraska by Jeremiah McCarthy in 1781 (Fig. 9.3A) was ideal, providing significant spatial information on the central and western sections of the Bay of Kamouraska. Clearly visible on the plan is the seaward edge of the salt marsh at mean sea level (MSL), in addition to information allowing me to identify the mean high water level (MHW) and the high water level (HW) on the salt marsh.[22]

A second cadastral plan of the entire Seigneurie de Kamouraska by Joseph Hamel in 1826 (Fig. 9.3B) affords the first complete view of the entire salt marsh, or *prairies de grève*, of the Bay of Kamouraska.[23] This plan likewise provides information making it possible to identify the seaward edge of the marsh at mean sea level (MSL), and the upper limits of the salt marsh at mean high water (MHW) and high water (HW). These data are exceptional for reconstructing the environmental history of the marsh because they can be used to establish a baseline for delimiting the total area of the salt marsh before intensive anthropogenic activities began in the industrial period (after 1850). By using ArcGIS to draw a polygon around the entire marsh as represented in the plan by Hamel, I determined that the total area of the salt marsh in 1826 was 309.8 ha with a length of 6.7 km and mean width of 535 m.

The first aerial photographs of the salt marsh come from a 1929 mosaic archived in the Université Laval Centre GéoStat. Unfortunately, the coverage of this mosaic provides only a partial view of the central and western sections of the marsh.[24] Fig. 9.4A shows the geographical coordinates of the seaward edge of the marsh in 1929, indicated in this HGIS layer by an overlain dashed line. The next series of aerial photographs covering the entire marsh, taken in 1948, was obtained from the National Air Photo Library in Ottawa (Fig. 9.4B). Here again the coordinates of the seaward edge of the marsh were noted in the HGIS layer and indicated by a dashed line.[25]

By placing the 1948 mosaic under magnification (using the zoom function in ArcGIS), I identified a dike (*aboiteau*) that was constructed on the interface between high and low marsh. Using archival documents, I ascertained that

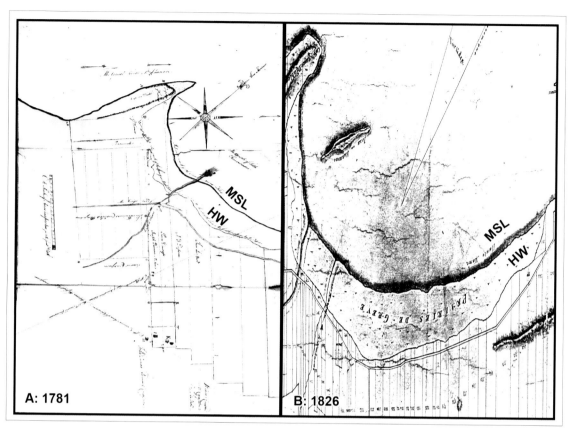

Fig. 9.3. A: the McCarthy plan of the western half of the Bay of Kamouraska in 1781. B: the plan by Hamel of the entire salt marsh in 1826. Note the northern orientation of the McCarthy plan and the southern orientation of Hamel's plan; standardization of map orientation occurred in the late nineteenth century. (Source: J. McCarthy, "Plan and Survey of Cap-au-Diable in Kamouraska," 1781, Bibliothèque et Archives nationales du Québec à Québec, CA301, S45, no. 3; J. Hamel, "Plan de la seigneurie de Kamouraska," 1826, Bibliothèque et Archives nationales du Québec à Québec, P407, famille Taché, no. 2.)

it was constructed in 1937–38 by the provincial government (Departments of Agriculture and Colonization) in collaboration with local farmers in order to augment agricultural production during the Great Depression. In addition, I noted a series of smaller secondary dikes several hundred metres south of the primary dike. Archival information revealed that these secondary dikes were built after 1941, following the breaching of the main dike by a severe storm event. These secondary dikes remained in place for over thirty years until the main dike was rebuilt in 1979–80.[26] With this information, it was possible using GIS to draw a polygon around all of the remaining salt marsh in 1948. In doing so, I determined that diking had decreased the average width of the marsh from 535 m in 1826 to 428.6 m in 1948, reducing overall marsh area by about 18 per cent from 309.8 ha to 254 ha.

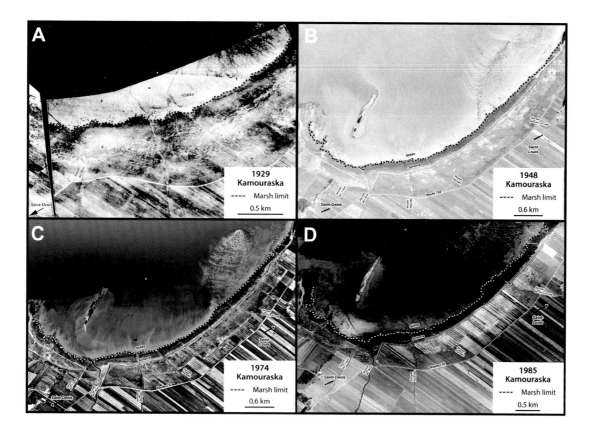

Fig. 9.4. Evolution of the seaward edge of the salt marsh of the Bay of Kamouraska, captured on aerial photographs from 1929, 1948, 1974 and 1985. They are viewed here as separate layers that will be integrated into a single layer GIS (see Fig. 9.5) in order to determine shifts in marsh location over time.

The next two series of aerial photographs analyzed come from the Université Laval Centre GéoStat (1974 series) and from the Quebec Ministry of Natural Resources (1985 series).[27] As with the previous photos, I noted the coordinates of the seaward edge of the salt marsh in the GIS, creating separate layers for 1974 and 1985 (Figs. 9.4C and 9.4D). Placing the 1985 series of photos under magnification revealed the reconstruction of the primary dike in 1979–80 and the demolition of the secondary dikes (*digues dormantes*) after 1980. By drawing a polygon around the total area of marsh in the 1985 GIS layer, I determined the extent of marsh destruction due to reconstruction of the primary dike. By 1985 the average width of the salt marsh had been reduced to 214 m, while the area of the marsh decreased to 115 ha. This equals a loss of nearly 63 per cent of the marsh since 1826 as a result of diking and draining.

Traditionally, when data on environmental change are presented textually, readers are forced to imagine landscape evolution. Mapping data provides the reader with an optical representation that enhances understanding by directly engaging the reader's sense of sight.[28] This is the objective of Fig. 9.5, where I combined the 1948, 1974, and 1985 GIS layers into a HGIS map and then added the dGPS coordinates obtained in 2008. By merging these separate layers into a single image, my goal was to produce a dynamic map illustrating the natural evolution of the salt marsh over the last seventy-seven years. To facilitate interpretation, the data were colour-coded by year and six observation points (A, B, C, D, E, and F) were overlaid on the map in order to measure and tabulate the movements of the seaward edge of the marsh at logically spaced intervals. The resulting map and embedded table (Fig. 9.5) provide a dynamic representation of the marsh, clearly illustrating both erosion and growth events.

Erosion is plainly visible on the map in the western half of the marsh at points A (–162 m), B (–70 m), C (–38 m) and D (–55 m). These findings are supported by the field observations made in 2008, where I noted a clearly visible scarp in the western section of the marsh separating the high marsh from the tidal flat (Fig. 9.6A). Equally visible on the map are areas of growth of the salt marsh in the eastern half of the bay at point E (107 m). Here, I noted in my field observations, both a scarp near the high water line and a gradual transition between the low marsh and the tidal flat at mean sea level (Fig. 9.6B). This combination of forms, as noted earlier, is indicative of both growth and erosion. Finally, the map and table illustrate significant lateral growth of the marsh at point F of more than 200 m. In my field observations for this sector, I noted the absence of any kind of scarp. On the contrary, there was abundant evidence of sedimentation and a gradual transition between the low marsh and tidal flat (Fig. 9.6C).

In regard to natural processes, the data in Fig. 9.5 illustrate that over the last seventy-seven years there has been both lateral growth and erosion of the salt marsh. In the western half of the salt marsh, overall change to the marsh edge is negative, with a loss of some –175,254.68 m². In the eastern half of the salt marsh, on the contrary, change to the marsh edge has been positive, with an advance of some 153,650.68 m². When these two data sets are combined, we note that changes in location of the marsh edge resulting from natural phenomena (erosion and sedimentation) have resulted in a net loss of about 2.1 ha (–21,604 m²).

By referring to the historical plans, it is possible to compare the visual form, location, and curvature of the salt marsh in 1781 and 1826 with that represented in Fig. 9.5. Through such a comparison, one immediately notes that the semi-circular form of the lower marsh edge has remained remarkably stable for nearly two centuries. This is especially poignant when I contrast the evolution of the Bay of Kamouraska salt marsh with that of other well-studied salt marshes in the United States and France. For instance, in the 1970s, Redfield demonstrated that the salt marshes of Cape Cod (Barnstable, Massachusetts) experienced radical changes in form over the last two centuries primarily due to sedimentation and growth resulting from

Year	Point A (metres)		Point B (meters)		Point C (meters)		Point D (meters)		Point E (meters)		Point F (meters)	
	Erosion	Accretion	Erosion	Accretion	Erosion	Accretion	Erosion	Accretion	Erosion	Accretion	Erosion	Accretion
1929-1948	No data	No data		22		19	-76		No data	No data	No data	No data
1948-1974	-81		-7			6		5		29		72
1974-1985	-69		-33		-15			21		40		74
1985-2008	-12		-52		-48		-5			38		59
Subtotal	-162	0	-92	22	-63	25	-81	26	0	107	0	205
Total	-162		-70		-38		-55		107		205	
	Change western area				Change eastern area				Total change in marsh area			
	-175,254.68 m²				153,650.98 m²				-21,604 m²			

Table: Evolution of the marsh - tidal flat interface; Bay of Kamouraska, 1929-2008

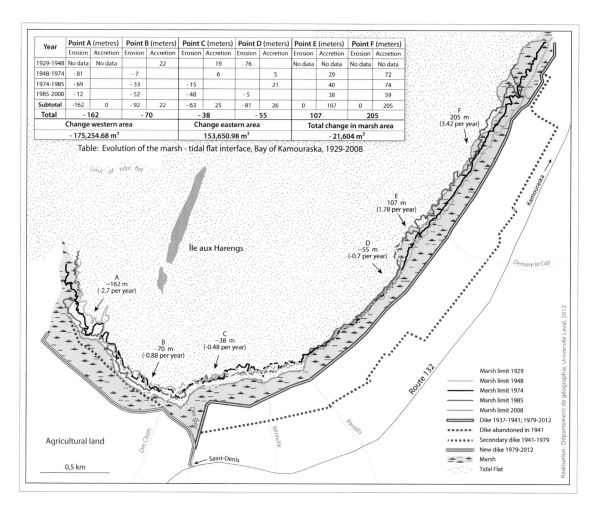

Fig. 9.5. Dynamic HGIS map of the lateral movements of the seaward limit of the Bay of Kamouraska salt marsh, 1929–2008. By combining information on marsh limits at several different moments in time, this map provides a dynamic vision of how the marsh has both grown and eroded over time at different locations on the bay.

Fig. 9.6. Forms of marsh limits photographed during field research in 2008 on the Bay of Kamouraska. Fig. 9.6A shows a scarp between the high marsh and tidal flat, denoting erosion. In 9.6B, there is both a scarp and a gradual transition between marsh and tidal flat, denoting erosion and growth while 9.6C shows a gradual transition between marsh and tidal flat, denoting marsh growth.

natural processes.[29] Similarly, Verger's work on the Genêts marsh on the Bay of Mont-Saint-Michel (Normandy, France) demonstrates that it has eroded and regrown several times during the last century due to natural processes.[30]

OVERALL CHANGE TO THE BAY OF KAMOURASKA SALT MARSH SINCE 1781

The data illustrate that over the last 230 years sedimentation and erosion on the Bay of Kamouraska salt marsh have been in relative equilibrium, the lower marsh limit retaining its basic form since the late eighteenth century. The same argument is not true when it comes to discussing direct human impacts on the marsh landscape. During Amerindian and colonial periods, human impacts on the marsh were negligible. However, with industrialization, anthropogenic activities played a significant role in reducing the area of high marsh as a result of diking and draining. According to the baseline data derived from the 1826 plan, the overall area of salt marsh before intensive agricultural practices began was approximately 309.8 ha. By 1948, the marsh had been reduced to 254 ha by diking and draining. Reconstruction of the primary dike in 1979–80 led to a further reduction of marsh area to 115.3 ha by 1985. Since then, an additional 28.3 ha were lost due to the complete drainage of all former tidal marsh behind the primary dike. Contrasting the data on anthropogenic and natural change of the salt marsh over the last two centuries, it is clear that humans – and specifically industrial society – have had the greatest impact on the evolution of the salt marsh. Over the last 182 years, 0.6 per cent (2 ha) of total marsh area changed as a result of natural processes (sedimentation and erosion). On the other hand, nearly 72 per cent (223 ha) of the marsh was transformed due to diking and draining.

CONCLUSIONS

What are we to make of the findings revealed by the HGIS, especially when juxtaposed with the historical discourse of marsh growth due to natural processes and the current discourse of marsh erosion resulting from climate change? Why does the evolution of the Bay of Kamouraska salt marsh correspond with neither interpretation? The answer is in part methodological. The combining of archival materials (historical texts and maps) with spatial data (aerial photographs and GPS coordinates) within a GIS provides us with the ability to look both forwards and backwards on the environmental timeline. What surely appeared to previous scholars as unidirectional marsh movements when viewed through the lens of a single lifetime or context of society becomes considerably more complex and multidirectional when viewed across several lifetimes and historical contexts.

The answer is also in part epistemological. Stephen J. Gould, in *Time's Arrow, Time's Cycle*, suggests that in dealing with complex questions human nature tends to reduce problems to more simple dichotomies – yes or no, black or white. Yet these dichotomies, he warns, have the unintended result of reducing the richness of reality. [31] Gould's understanding of this problem perfectly fits the historical and current literature on the salt marshes of the St. Lawrence Estuary where marsh evolution has been reduced to simple dichotomies – unidirectional movements of either growth or erosion.

Building on Gould's thesis, the nonconformity of the data generated by this HGIS suggests the vital importance of situating the interpretation of marsh evolution within the time and place out of which it emerges. In the industrial period, one cannot separate the prevailing scientific interpretation of growing marshes from the socioeconomic discourse of land expansion through diking, draining, and mastery of nature. The transformation of salt marshes to benefit human needs that occurred at this time influenced how those who came afterward – post-industrial scholars – interpret marsh evolution. In other words, the post-industrial discourse of generalized marsh erosion as a result of climate change and sea-level rise cannot be easily separated from the discourse of anthropogenic destruction of a fragile environment. In the case of the Kamouraska salt marshes, the negative response to the destruction of 72 per cent of the salt marsh in the industrial period created a discursive echo that today reverberates in the interpretation of the impact of climate change and sea-level rise on marsh evolution. [32]

Objective minds, history tells us, do not exist outside of particular contexts of time and space. Unidirectional understanding of marsh growth and erosion, therefore, must be viewed as mental constructions imposed on data rather than demanded by them. HGIS, by combining the spatial preoccupations of geographers with the temporal preoccupations of historians, makes it possible to study environmental change, not only from the present context in time, but equally from the vantage point of other contexts. In so doing, HGIS brings to the study of the environment, not just the ability to empirically observe change over time, but also variations in the interpretation of those changes according to different contexts of society. In the case of the growing and eroding salt marshes of the St. Lawrence Estuary, combining empirical observation of the physical environment with the epistemological preoccupations of the social sciences is key in explaining why salt marsh evolution in Kamouraska County corresponds with neither the historical nor the current understanding of marsh growth and erosion.

NOTES

1   J.-É. Joubert and É. Bachard, *Un marais en changement, caractérisation du marais salé de la baie de Kamouraska* (Rimouski, QC: Comité ZIP du Sud-de-l'Éstuaire, 2012).

2   M. Hatvany, "Wetlands," *Encyclopedia of American Environmental History* 4 (2011): 1380–83; C. Desplanque and D. Mossman, "Tides and Their Seminal Impact on the Geology, Geography, History, and Socio-Economics of the Bay of Fundy, Eastern Canada," *Atlantic Geology* 40, no. 1 (2004): 1–118.

3   P. Bernatchez and C. Fraser, "Evolution of Coastal Defence Structures and Consequences for Beach Width Trends, Québec, Canada," *Journal of Coastal Research* 28, 6 (2012): 1550–1566.

4   P. Bernatchez and J.-M. M. Dubois, "Bilan des connaissances de la dynamique de l'érosion des côtes du Québec maritime laurentien," *Géographie physique et Quaternaire* 58, no. 1 (2004): 45–71.

5   Comité ZIP de la Rive Nord de l'Estuaire, *Forum citoyen 2007 sur l'érosion des berges et l'occupation du territoire en Côte-Nord* (Baie Comeau, QC: Webcréation, 2007), 11.

6   M. Hatvany, "Wetlands and Reclamation," in *International Encyclopedia of Human Geography*, ed. R. Kitchen and N. Thrift, vol. 12 (Oxford: Elsevier, 2009), 241–46; V. J. Chapman, *Salt Marshes and Salt Deserts of the World* (New York: Interscience); V. Bouchard, F. Digaire, J.-C. Lefeuvre, and L.-M. Guillon, "Progression des marais salés à l'ouest du Mont-Saint-Michel entre 1984 et 1994," *Mappemonde* 4 (1995): 28–33; F. Verger, *Zones humides du littoral français : estuaries, deltas, marais et lagunes* (Paris: Belin).

7   H. Charbonneau and N. Robert, "Origines françaises de la population canadienne, 1608–1759," in *Atlas historique du Canada, I : Des origins à 1800*, dir. R. C. Harris (Montréal : Les Presses de l'Université de Montréal, 1987), plate 45; M. Hatvany, "The Origins of the Acadian *Aboiteau*: An Environmental Historical Geography of the Northeast," *Historical Geography* 30 (2002): 121–37; Y. Suire, *Le Marais poitevin : une écohistoire du XVIᵉ à l'aube du XXᵉ siècle* (La Roche-sur-Yon : Centre vendéen de recherches historiques).

8   J.-D. Schmouth, "Mise en culture des terrains envahis par les eaux salées," École d'agriculture de Sainte-Anne, Sainte-Anne-de-la-Pocatière, unpublished text, 1874, reprinted in *La Gazette des Campagnes*, 15 septembre 1942, 152–54; A. Hamel, "La récupération et la mise en valeur des alluvions maritimes du St-Laurent," *Agriculture* 20, no. 3 (1963): 77–83; A. C. Redfield, "Development of a New England Salt Marsh," *Ecological Monographs* 42, no. 2 (1972): 201–37; G. Gourde, *Les aboiteaux : comté de Kamouraska* (Québec : Ministère de l'agriculture, des pêcheries et de l'alimentation, 1980); J.-B. Sérodes and M. Dubé, "Dynamique sédimentaire d'un estran à spartines (Kamouraska, Québec)," *Le naturaliste canadien* 110 (1983): 11–26; P. Champagne, R. Denis, and C. Lebel, *Établissement de modèle caractérisant l'équilibre dynamique des estrans de la rive sud du moyen estuaire du Saint-Laurent*, Rapport manuscript canadien des sciences halieutiques et aquatiques no. 1711 (Quebec : Ministère des pêches et des océans, 1983); M. Hatvany, *Marshlands: Four Centuries of Environmental Change on the Shores of the St. Lawrence* (Sainte-Foy : Les Presses de l'Université Laval, 2003).

9   See, for instance, Joubert and Bachard, *Un marais en changement*.

10 Bouchard et al., "Progression des marais salés"; Verger, *Zones humides*; M. Fanchomme and G. Schmitt, "Les zones humides dans le Nord vue à travers le cadastre napoleon: les Systèmes d'Informations Géographiques comme outil d'analyse," *Revue du Nord* 36 (2012): 661–80; K. D. Bromberg and M. K. Bertness, "Reconstructing New England Salt Marsh Loss Using Historical Maps," *Estuaries* 26, no. 6 (2005): 823–32.

11 A. H. Clark, "Field Work in Historical Geography," *Professional Geographer* 4 (1946): 13–23.

12 Hatvany, *Marshlands*, 164.

13 Hatvany, "Wetlands"; J.-C. Dionne, "Âge et taux moyen d'accrétion verticale des schorres du Saint-Laurent esturaien, en particulier ceux de Montmagny et de Sainte-Anne-de-Beaupré, Québec," *Géographie physique et Quaternaire* 58, no. 1 (2004): 74–75.

14 Desplanque and Mossman, "Tides and Their Seminal Impact."

15 Champagne et al., *Établissement de modèle caractérisant l'équilibre dynamique des estrans*; Bouchard et al., "Progression des marais salés"; F. Verger, *Marais et wadden du littoral français : etude de géomorphologie* (Bordeaux: Biscaye Frères, 1968); Verger, *Zones humides*.

16 Fisheries and Oceans Canada, "Tides, Currents, and Water Levels (Pointe aux Orignaux)," http://www.waterlevels.gc.ca/; Champagne et al., *Établissement de modèle caractérisant l'équilibre dynamique des estrans*, 7.

17 K. L. McKee and W. H. Patrick, Jr., "The relationship of smooth cordgrass (*Spartina alterniflora*) to tidal datums: A review," *Estuaries* 11, no. 3 (1988): 143–44.

18 M. Hatvany, *Paysages de marais : Quatre siècles de relations entre l'humain et les marais du Kamouraska* (La Pocatière: Société historique de la Côte-du-Sud et Ruralys, 2009), 21.

19 On marsh limits, see Champagne et al., *Établissement de modèle caractérisant l'équilibre dynamique des estrans*, 11; Verger, *Marais et wadden*, 277, 289–93; and Verger, *Zones humides*, 57–58.

20 The preliminary collection and analysis of aerial photographs and GIS mapping of the Bay of Kamouraska was done under my supervision and that of D. Cayer by C. Careau as part of his thesis entitled "Les marais intertidaux du Saint-Laurent : Complexités et dynamiques naturelles et culturelles," MS thesis, Dépt. de géographie, Université Laval, 2010.

21 Québec, Base de données topographiques du Québec (BDTQ) à l'échelle de 1/20 000, Normes de production, version 1.0, Québec, Direction de la cartographie topographique, Ministère des Ressources naturelles et de la Faune, 1999.

22 J. McCarthy, "Plan and Survey of Cap-au-Diable in Kamouraska," 1781, Bibliothèque et Archives nationales du Québec à Québec, CA301, S45, no. 3.

23 J. Hamel, "Plan de la seigneurie de Kamouraska," 1826, Bibliothèque et Archives nationales du Québec à Québec, P407, famille Taché, no. 2.

24 Centre d'information géographique et statistique, Bibliothèque générale de l'Université Laval, Mosaïque aérienne du Québec, 1929, F82, p. 3.

25 National Air Photo Library (Ottawa, Canada), A11660-290, 1948.

26 Hatvany, *Marshlands*, 134–38.

27 Centre d'information géographique et statistique, Bibliothèque générale de l'Université Laval, Q74316-94-95-96 and Q74313-133-134-135 (1974); Ministère des Ressources naturelles, Gouvernement du Québec, Q85913-143 (1985).

28 D.C.D. Pocock, "Sight and Knowledge," *Transactions of the Institute of British Geographers* 6 (1981): 385–93.

29 Redfield, "Development of a New England Salt Marsh," 235.

30 Verger, *Zones humides*, 191.

31 S. J. Gould, *Time's Arrow, Time's Cycle: Myth and Metaphor in the Discovery of Deep Geological Time* (Cambridge, MA: Harvard University Press, 1987), 199–200.

32 This discursive echo is most clearly seen today in Joubert and Bachard, *Un marais en changement* and in Bernatchez and Dubois, "Bilan des connaissances de la dynamique de l'érosion des côtes du Québec maritime laurentien."

10

Top-down History: Delimiting Forests, Farms, and the Census of Agriculture on Prince Edward Island Using Aerial Photography, ca. 1900–2000

Joshua D. MacFadyen and William M. Glen

INTRODUCTION

This chapter examines Prince Edward Island's agro-ecosystems from the top down using remote sensing data from aerial photographs in a Geographical Information System (GIS). The total area of crop land and other "improved" land on farms has been investigated by censuses and other routinely generated sources for almost two centuries. The quantity of improved agricultural land has thus been a common way to measure deforestation.[1] The development of historical forest inventories using GIS and aerial photography, however, shows that the *Censuses of Agriculture* were at best an approximate measure of deforestation until the mid-twentieth century. There was slightly more cleared land and far less forest than what appeared in the census, and the discrepancies were quite large in some areas. In the post-war period, the presence of abandoned farms and a growing number of non-census farms made census estimates even less useful for understanding changing land use patterns on Prince Edward Island. The inventories also offer a new perspective on the process of land reverting to forest in areas of agricultural decline. When examined at the property-level, the data reveal acute inequalities in the processes of farm settlement and abandonment. This methodological examination of new sources calls into question the previously held estimates of agricultural activity, its ecological impact, and the rates of forest regrowth in agro-ecosystems.

Remote sensing sources are only beginning to influence the way historians see the Canadian environment, and Prince Edward Island offers a unique case study due to its rich collection of historical aerial photographs and the provincial government's ability as a small province to create complete land-use and forest inventories. The province's small size meant that it has been entirely photographed in multiple intervals since 1935. Complete inventories were created that now allow historians to examine land-cover and land-use change on all crown and private land, including the Prince Edward Island National Park. These processes were not possible in larger jurisdictions, but the example of property sampling in Prince Edward Island offers a spatially explicit model for studying land-cover and land-use change anywhere in Canada. The necessary sources for this analysis, aerial photographs and cadastral maps, exist in varying degrees of completeness for most populated areas of the country, and they offer students of environmental and social

history a new glimpse of the country from above.

In Prince Edward Island, inventories allow the measurement of forest areas and compositions at a number of points in time and for any defined geographic area. Preliminary analysis of three sample watersheds and a selection of private parcels suggests that the extent and composition of forest cover are indicators of both the limitations and the ecological impacts of intensive farming in the early twentieth century. When forest cover is examined at the property level, representative samples can be created to study the full range of productive capacities of farms. The Canadian *Census of Agriculture* was designed to capture these elements of rural life, and its estimates of "improved land" have been a crucial part of scholarly research in these areas. Exploring history from the top down, from state-produced aerial photos, uncovers another side of rural production and a significant segment of agricultural land use that was overlooked by enumerators.

METHODS AND LITERATURE

Geographers N. Ramankutty and J. A. Foley have argued that new historical data are needed to fully understand the human land-use activities that drive global environmental change. Using a combination of satellite-borne remote sensing data and historical statistics, their research over the last decade has modified the estimates of cropland change globally.[2] Ramankutty, E. Heller, and J. Rhemtulla recently established a critique of the "regrowth" narrative of twentieth-century forest history in the United States.[3] They provide useful new estimates of historical cropland change in the United States, accounting for changes in definitions and political boundaries, but they ignore the problem of land use on the growing margin of private land that did not fit either forest or cropland categories. We address this by questioning the meaning and accuracy of "improved land" in the census and by offering a unique example of forest reversion rates in Prince Edward Island.

Like most colonies based on primary industries, Canada was preoccupied with "improved land" as a measure of agricultural progress in the nineteenth and much of the twentieth centuries. This preoccupation generated historical documents valuable not only for examining land-use activity and society in a period of extensive agricultural settlement but also for modelling the influence this activity had on the environment.[4] In woodland ecosystems, it is generally assumed that land that was not "improved" by settlers or cleared for other forms of development remained, or returned to, some form of forest cover. What the census called "natural forests" were far from undisturbed environments, but nevertheless environmental historians can make some basic assumptions about their composition. Occupied farmland was usually divided into "improved" and "unimproved" categories, and the latter included privately owned woodlots or "natural forest" as well as "marsh or wasteland."[5] Definitions varied slightly according to the governing body performing the censuses or assessments, but "improved land" typically meant cleared land on farms that had been ploughed at one point for the purpose of agriculture.

Great pains were taken by the state to ensure that its lands were carefully surveyed and measured. Yet, historians have identified many limitations and inconsistencies. Variables such as the number of farms are misleading in some censuses and inconsistently defined in others.[6] O.F.G. Sitwell identified a discrepancy in earlier counts of improved acreage in Canada that he argued was caused by confusion about whether some pasture and summer fallow land qualified as improved acreage.[7] In some years, improved land was called "arable land" and appeared to mean land that had the potential for cultivation. In other years, the definition seemed to indicate any land that had been ploughed in the past, even if it was only fit for grazing.[8] Different censuses explained the terms to farmers in varying levels of detail, and Donald Akenson noted that census takers were "more strict in defining 'improved' land than were the assessors in defining 'cultivated' land for tax purposes."[9] Beginning in 1931, census officials listed the acreage and improved acreage of abandoned farms, although it was only summarized for census districts and it is curious how they determined the figure if the previous occupants were absent. As Ruth Sandwell has argued, the tabulations in the early twentieth century Canadian printed censuses accorded with one definition of agricultural activity and ignored others, such as subsistence farming, gardening, and other forms of production usually attributed to women.[10]

To land owners and protected tenant farmers, the work of improving farmland translated to improving one's social and economic prospects,[11] but, to the state, "improved land" had a different kind of importance. James Scott argues that the primary objective of census projects and other forms of cadastral reckonings in the period of high-modernist agriculture was to make local knowledge legible to the state, especially for the purpose of levying taxes. The real incentive behind the scientific system was, according to Scott, "the precondition of a tax regimen that comprehensively links every patch of land with its owner – the taxpayer."[12] But making local knowledge available to the Canadian state was about more than just good fiscal policy. Real property taxes were only a distant secondary tax revenue for the Upper Canadian government, and for new immigrants they paled in comparison to what they would have paid for property in the British Isles.[13] In the twentieth century, the Maritime provinces were the only Canadian jurisdictions to make real property taxes a serious form of revenue.[14] Mapping Canadian land and resources was also about the emergence of the scientific and nationalist state, and Suzanne Zeller has shown how the development of several agencies that employed scientific cartography was an important part of nation-building.[15]

Gathering data on improved acreage increased in importance toward the end of the nineteenth century, and some censuses have been extremely diligent in showing the quality and quantity of improved land. Some, like the 1871 *Census of Prince Edward Island*, categorized all enumerated land as first-, second-, or third-rate farmland. In 1883, Manitoba paid careful attention to the rates of settlement, even differentiating between land that had been cultivated in earlier years and land that was recently broken.[16] Early censuses of prairie agriculture recorded the number of acres broken in the previous crop year as well as the acres seeded in the spring when the census was taken.[17] However, for historians who are interested in how land use influenced the environment, improved land alone does not tell the whole story. Geoff Cunfer's research on

the Great Plains demonstrated the importance of studying land use as a percentage of total area. He showed that farmers actually altered a relatively small portion of the plains and were unable or unwilling to expand beyond certain levels of grassland cultivation.[18] Agricultural censuses are also problematic for studying land-use change over time because the basic units of analysis, the census divisions and subdivisions, changed frequently in areas of rapid settlement or other population growth.

This paper is the first to compare Canadian statistics with land-use and forest inventories over a long period, but it is only one of several recent studies to focus on the accuracy of historical statistics. In little more than the last dozen years, scholars have updated many standard sources for historical statistics. The *Historical Statistics of the United States* corrected errors and brought an entire set of agricultural statistics up to date. Ramankutty and others have established an updated estimate of land use for North American crops and forests and have identified inconsistencies in the data for the United States. M. L. Liu and H. Q. Tian have created new estimates for Chinese land-cover and land-use change using similar spatially explicit models.[19] Many of these scholars have helped isolate and reinterpret obvious errors in data gathered by routinely generated sources.

Remotely sensed data from aerial photographs were added to the cartographer's toolbox in the early twentieth century, and the documents offer a new perspective to the question of land-cover and land-use change. The historical GIS research in this chapter was made possible by a series of photographic images taken from the skies above Prince Edward Island at various points in the twentieth century, beginning with a relatively high-quality and practically complete coverage of the province in 1935 and 1936. This remarkable source, and most of the aerial photographic surveys that followed, have been used for a range of academic and official projects on Prince Edward Island, from land-use studies and partial surveys to soil maps and complete land-use and forest inventories created by the provincial departments of Agriculture and Forestry.[20]

Canadian surveyors had used photography for mapping and surveying on land since the late 1880s.[21] The first aircraft used by foresters were flown in 1915 in Wisconsin, and by 1919 Canadian businesses such as Laurentide Paper in Grand Mère, Quebec, used two planes owned by Department of Marine to spot forest fires and survey forest resources. In the latter year, Canadian pilots and surveyors conducted the world's first extensive aerial survey by photographing over 13,000 images of southern Labrador for a Massachusetts pulp and paper company.[22] Within five years, forest surveys using both oblique and vertical aerial photography had become standard practice in the Department of the Interior, the Dominion Forest Service, and several provincial departments and private companies. In the late 1920s, Canadians experimented with winter air surveys, developed new methods for calculating the volume of forests using shadow lengths and locations, and agreed to perform a national forest survey. The meeting was struck in 1929 by Minister of the Interior Charles Stewart, and the provinces agreed to gather aerial survey data while the Dominion Forest Service would collate and compile the national project. Unfortunately, the Great Depression interfered with the national survey, but Richard Rajala argues that in the 1930s Canada's aerial photographic mapping was unrivalled in

both technique and coverage. Several federal agencies and the Royal Canadian Air Force had by then photographed over 800,000 square kilometres of land and mapped about 320,000 square kilometres of forest from the air.[23] The federal government's photographic surveying was obviously widespread and included lands from western Manitoba to southern New Brunswick.

In the summers of 1935 and 1936, the Geodetic Photographic Detachments of the Royal Canadian Air Force photographed the surface of Prince Edward Island, presumably as part of the Dominion Forest Service's larger surveying projects. Fortunately, the photographs, negatives, and flight reports survived and have been reproduced and indexed by the Forestry Division of the Prince Edward Island Department of Agriculture and Forestry. The original photographs were taken using a Bellanca Pacemaker aircraft, a single-engine, high wing aircraft fitted with floats. The flights were operated out of Shediac (New Brunswick) and Charlottetown, and they photographed 99 per cent of the island's surface. Some areas were missed because the lightweight plane had blown off course.[24] The pilots attempted to maintain an altitude of 10,000 feet for all photographs, and the resulting documents produced a geospatial time slice of nearly the entire province at a scale of approximately 1:14,300.[25]

The 1935/36 imagery was followed in June 1958 by another complete photographic survey of the island. The 1958 photographs were contracted by the RCAF to a local company operating a single-engine 1956 Cessna 180. Flying at 8,200 feet, the pilot and photographer produced a second set of photographs for the entire province over three months and with a scale of approximately 1:15,840. With the acquisition of the 1958 photographs, Prince Edward Island became the first province to have two complete sets of aerial photography. Further sets of aerial photos were also taken in 1968, 1974, 1980, 1990, 2000, and 2010.[26]

The remotely sensed data in these photos were critical for a broad range of surveys and topographical maps created by forestry companies and federal offices such as the Department of National Defense, the Department of the Interior, and the Geological Survey of Canada (GSC).[27] A well-known photo interpretation technique meant that most photographs were taken with enough overlap to allow stereo imagery. Stereoscopic imagery gives a three-dimensional effect that enhances the features in the photographs and allows interpreters to identify aspects of tree height and elevation as well as land-use and forest-cover types. By combining aerial photo interpretation with surveys and field observations, Canadian cartographers were able to plot comprehensive topographical maps for most of the country's inhabited regions. These had various incarnations but became known as the National Topographic System (NTS) in 1927.[28] The most detailed of these were called "one-mile maps" because of their scale of one inch to one mile (1:63,360). Don Thomson explained how the NTS was "indispensible to any extended, efficient mapping program in this country … and to the development of Canadian air navigation charts."[29] The NTS maps are an important source for environmental historians as they identified features in the built environment such as roads, dams, residential, community, and industrial buildings, and they also offer a view of the natural environment, including features such as surface hydrology, wetlands, coastlines, and a basic breakdown of forest-cover types.

Comparing these documents to more recent data in a GIS allows historians to identify patterns in land-cover and land-use change, and often they reveal the origins of disturbances ranging from deforestation and stream siltation to brown fields and other residual forms of pollution.

Historical maps like the NTS are a valuable source to environmental historians, and GIS has presented a new way to read, visualize, and analyze the documents. However, aerial photographs and the subsequent forest inventories carried out using the photography present a much finer level of detail and greater potential for research. These data effectively fill in the spaces on the map – spaces that to most historians in the twentieth century were simply cadastres with names, property lines, and occasional topographical features. Now historians can identify the built environment, field and forest outlines, land-use and forest-cover types, and even some tree species. This analysis is possible at the township, watershed, or county level and, in the unique case of Prince Edward Island, for the entire province.[30] The documents represent a staggeringly large body of information, even for a small province. Just as computing technology in the 1970s allowed historians and other scholars to process large datasets like the censuses in new ways, increased processing power and the efficiency of GIS allow scholars to better understand and leverage these early remotely sensed data. If the NTS maps were a sort of enumeration of the Canadian environment in certain years, then the original aerial photos were the manuscripts behind the census, and historians and foresters are just beginning to query the data they contain.[31]

Measuring the extent and health of the forest has been a relatively recent endeavour in Prince Edward Island. During the early twentieth century, when industry and the federal government were so focussed on creating forest surveys and managing forest resources, the government of Prince Edward Island took little interest in either surveying or protecting its forests. The province passed an act to establish a "forestry commission" in 1904 and two forest fire prevention acts in the 1930s, but the first *Forestry Act*, which placed restrictions on clear-cutting and burning, was not implemented until 1951. The province's first tree nursery was built the following year.[32] Thus, for the first half of the twentieth century, the province passed on the responsibility of mapping and monitoring its forests to Ottawa. In the 1980s, the province's Department of Forestry began to commission decennial inventories, starting with a forest biomass inventory using the 1980 photographs and field surveys, authored by Dendron Resources in Ottawa.[33] The 1990–1992 Prince Edward Island Forest Biomass Inventory was created by the Prince Edward Island Forest Division using the 1990 false-colour infra-red photographs (scale 1:17,500) and 1991 field surveys of 1,200 ground flora and tree species sampling points. In the 1990s the Department of Environment, Energy and Forestry also used the earlier historical photographs to create forest-cover maps using the 1935/36 and 1958 photography.[34]

The Prince Edward Island forest and land-use inventories were made possible, in part, because the province contains a dense and regular network of roads. Roads and railways were used as control features for georeferencing aerial photographs to the base map. The delineated features on the aerial photographs were transferred to the base map using a light table. This was only possible due to the high density of control features noted above. In the 1958 inventory, land-use categories were created for forest cover, clearcut, partial cut, reverting land, cleared land, roads, railways, and several smaller categories. Reverting land consisted of 5–10-year-old stands of trees growing on cleared land that were large enough to identify but small enough that they could be ploughed under. In other words, reversion was not necessarily long-term and it did not necessarily result in full forests, but the land had clearly not been used for crops or heavy grazing in a decade or more. The availability of more resources for the creation of the 1935/36 inventory meant that the forest-cover category was subdivided into five generalized types (identifying the mix of hardwood and softwood) and classes for reverting land and harvested forest. Specific species were also identified in the 1935/36 photos where possible, such as alder, black spruce, cedar, larch, white spruce, and poplar. Finally, the origin of the forest cover was coded wherever it could be identified; this was often the case with old fields that had grown up in white spruce or larch.[35] Figure 10.1 shows a simplified map of the forest outline based on the 1935 inventory, but what is not visible at this scale are the forest types and land uses within the forest outline. Land that is not identified as forest in this image was either wetland, cleared, or otherwise developed. Figure 10.2 shows the same data from the 2000 inventory.

The 1935/36 inventory is a benchmark for this study, but there are two ways to use historical inventories to estimate land use and forest cover at other points in the early twentieth century. First, because the rural population of Prince Edward Island was in decline for the

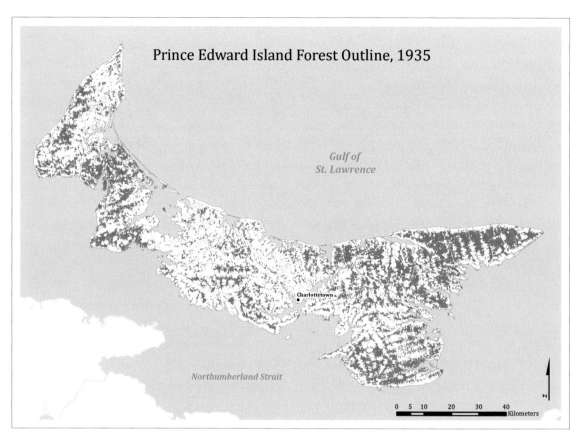

Fig. 10.1. *Prince Edward Island Forest Outline, 1935. (Prince Edward Island, Department of Environment, Energy & Forestry, Forests, Fish & Wildlife Division, 1935 Forest Outline, Coastline, Last modified 31 December, 2003. Software: ESRI ArcGIS 10.1, Adobe CS 4 Illustrator.)*

first six decades of the twentieth century, agricultural land declined and the forest experienced a general expansion rising from 32 per cent to 49 per cent of the area of the province between 1935 and 1990. Therefore researchers can use simple straight-line interpolation (estimating data points between two known data points) when the rate of forest reversion is known in order to estimate the general forest cover and the amount of cleared land. Second, by coding the origin of old fields, the 1935/36 inventory allowed researchers to create a map of the Prince Edward Island forest at close

to its smallest point. Far more accurate than extrapolation, this process involved removing those stands that had regenerated on old fields and the areas of new reversion from the 1935/36 mapping. This process resulted in the earliest forest outline, *circa* 1900, showing forest on only 31 per cent of the province's land mass. Some new clearing might have occurred in isolated areas of agricultural expansion in the period between 1900 and 1935/36, but for the most part this is an accurate estimate of the province's forest at its lowest point of coverage.[36]

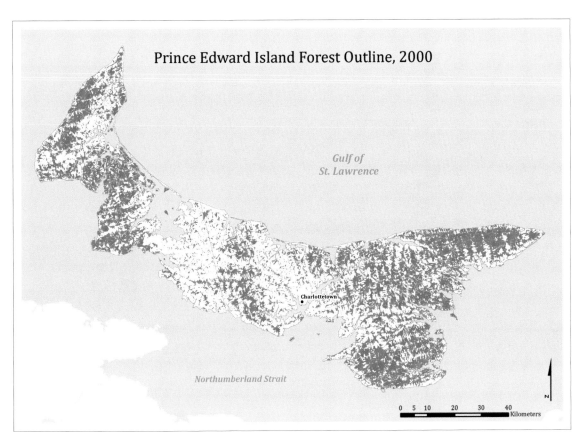

Prince Edward Island Forest Outline, 2000

Gulf of St. Lawrence

Charlottetown

Northumberland Strait

0 5 10   20   30   40
Kilometers

Fig. 10.2. Prince Edward Island Forest Outline, 2000. (Prince Edward Island, Department of Environment, Energy & Forestry, Forests, Fish & Wildlife Division, 2000 Forest Outline, Coastline, Last modified 30 November, 2010. Software: ESRI ArcGIS 10.1, Adobe CS 4 Illustrator.)

PROBLEMS WITH THE TRADITIONAL SOURCES

The Prince Edward Island forest and land-use inventories identify significant problems with the traditional sources used to map changing environments. The province offers an exceptionally useful case study of historical forests in private hands precisely because it has completed inventories of all land regardless of ownership. Early nineteenth-century surveys usually focussed on cadastral characteristics of

private land such as property dimensions and occupancy, but, later in the century, bureaus in agriculture, natural resources, geology, meteorology, and other scientific interests of the state began to survey large tracts of the country's public land on regular intervals. Thanks in large part to the efforts of William Edmond Logan and subsequent officials in the Geological Survey of Canada, the Dominion mapped more of its difficult terrain than most other places on Earth.[37]

For the most part, natural resource surveys in the past were preoccupied with public lands.[38] There were a few exceptions to this norm in the Maritimes. In 1895, Robert Chalmers created maps for the GSC that included a forest outline and origin classification ("old forest growth" and "recent forest growth") for New Brunswick and the western half of Prince Edward Island. However, Chalmers' sources are unknown, and, when compared to the *circa* 1900 outline, his map of western Prince Edward Island proves to be much more of a general estimate than a detailed survey.[39] Other estimates of the forest cover were prepared by provincial officials such as O. L. Loucks and by academics such as F. A. Stilgenbauer and Andrew H. Clark, but these should also be used carefully and compared to the historical inventories.[40] In 1912, the region's first provincial forest inventory was mapped for Nova Scotia by Bernhard Fernow and his University of Toronto students, including C. D. Howe. This survey was impressive, considering the modest budget and "often very inaccurate" base maps the foresters had to work with, but ultimately the source data were simply estimates volunteered by local landowners and lumbermen. Even though Fernow was impressed by the "unusual number of intelligent and well informed men throughout the country," he admitted that "it is only the grand total or the average that is approximately correct and of value."[41] The most comprehensive survey of land use on private properties at this point remained the *Census of Agriculture*.

The census account of human activity in areas of agricultural settlement has been an invaluable source for understanding the economic and ecological impacts of farming. However, in Maritime Canada, the census does not show an accurate trajectory of changes in either cleared land or forest areas over the twentieth century. In the late twentieth century, the

censuses recorded only cropland, and, although cropland has been the main historical variable used by geographers such as Ramankutty, it gives only a partial, and we argue incorrect, picture of land use and obviously says nothing about forest-cover types. Most macro-historical studies of global land use tend to use data for cropland and improved land with relatively few checks for the accuracy or comprehensiveness of the data.[42] It is not possible to use the censuses to trace the changing size of the forest, even in a province such as Prince Edward Island, which had practically no public lands before the Second World War. This province's *Censuses of Agriculture* suggested that woodlots and other "unimproved land" occupied a relatively consistent surface area through the twentieth century. The forest and land-use inventories present a much different story.

The census figure for cropland, furthermore, says nothing about the total area cleared for transportation, industry, housing, recreation, and commerce, and due to the omission of abandoned farms and many subsistence operations it presents only a partial image of land used for growing. In places like the Maritimes, these small operations can represent a significant portion of cleared land. Statistics Canada sometimes reports cropland as a comprehensive measure of human disturbances. "Total cropland in Canada now stands at almost 89 million acres or 53.1% of all land," it claims on its website.[43] This, of course, refers to all land in farms; cropland itself represents less than 6 per cent of the total land area of Canada. By way of comparison, the total land area in the United States is over 23 per cent cultivated and Eastern Europe is well over half in crops.[44] For a country with less than 6 per cent of its land in crops, the Canadian *Census of*

Table 10.1. Two Descriptions of Deforested Land in PEI, ca. 1900–2000.

Area (ha)	1900	1910	1930	1940	1955	1960	1980	2000
Census:	improved	improved	improved	improved	improved	improved	improved	improved
PEI	293,917	309,144	*311,690*	*300,152*	261,222	232,467	190,620	200,375
Prince		109,148	*110,700*	*108,774*	100,657	88,975	74,292	81,746
Queens		125,504	*127,460*	*122,505*	109,351	100,379	82,326	83,537
Kings		74,492	*73,531*	*68,874*	51,214	43,113	33,615	35,095
Inventory:	cleared	cleared	cleared	cleared	cleared	cleared	cleared	cleared
PEI	359,577	359,577	327,765	316,227	292,899	285,625	248,414	236,935
Prince		131,516	114,854	112,928	108,509	107,164	93,904	88,814
Queens		140,317	132,142	127,188	116,682	113,437	103,158	99,893
Kings		87,744	80,769	76,111	67,708	65,025	51,352	48,229
Difference (as % of the inventory)								
Low-end estimate								
PEI	−18.3%	*−14.0%*	*−4.9%*	*−5.1%*	−10.8%	−18.6%	−23.3%	−15.4%
Prince		*−17.0%*	*−3.6%*	*−3.7%*	−7.2%	−17.0%	−20.9%	−8.0%
Queens		*−10.6%*	*−3.5%*	*−3.7%*	−6.3%	−11.5%	−20.2%	−16.4%
Kings		*−15.1%*	*−9.0%*	*−9.5%*	−24.4%	−33.7%	−34.5%	−27.2%
Upper-end estimate								
PEI		n/a	*−11.6%*	*−11.8%*	−13.8%	−21.6%		
Prince		n/a	*−13.4%*	*−13.5%*	−8.7%	−18.5%		
Queens		n/a	*−7.7%*	*−7.8%*	−9.7%	−14.9%		
Kings		n/a	*−15.5%*	*−16.1%*	−29.2%	−38.6%		

1. Italics indicate where abandoned farms were included in "improved land."

2. Improved land in 1980 is estimated through interpolation.

Table 10.1. Two Descriptions of Deforested Land in P.E.I., ca. 1900–2000. (Census data: Census of Canada, 1901, 1911, 1931, 1941, 1956, 1960, 1980, 2000. Inventory data estimated by interpolation using the following datasets: Prince Edward Island, Department of Environment, Energy & Forestry, Forests, Fish & Wildlife Division, 1900 Forest Outline, Last modified 4 November, 2010, 1935 Forest Inventory, Last modified 25 November, 2010, 1958 Forest Outline, Last modified 28 February, 2008, 1980 Forest Outline, Last modified 4 November, 2010, 1990 Forest Outline, Last modified 25 November, 2010, 2000 Forest Outline, Last modified 30 November, 2010. PEI Watershed Boundaries, Last modified 6 April, 2005. Software: ESRI ArcGIS 10.1, Microsoft Excel 2011, Adobe CS 4 Illustrator.)

Agriculture is actually of limited use to historians interested in land-use changes over broad areas. It tells historians nothing about land not in farms, relatively little about farmland not in crops, and it reports incorrect data for cleared land on census farms. In the discussion that follows, we suggest that farmers reporting land in the census routinely, perhaps subconsciously, under-reported the amount of cleared land in their possession.

It is not unusual to find isolated errors in census variables in certain years. In Canada, for example, the 1980 figure for unimproved land was flawed and has been ignored by recent studies.[45] But a discrepancy in cleared land in the Prince Edward Island census and inventories points to something more systemic. The discrepancy existed in all districts in the early twentieth century and widened in later years (Table 10.1).

The earliest forest inventory, *circa* 1900, indicated that sometime around the end of the nineteenth century, farmers in Prince Edward Island had reached the maximum cleared area in the province's history. Almost 360,000 hectares, or 64 per cent of the land in the province's rural townships, had been ploughed at some point in that early period. However, the most improved land the census ever reported in one year (1930) was almost 48,000 hectares short of that amount. It is not possible to compare the *circa* 1900 inventory estimate to a specific census year since the stands of white spruce on old fields had grown up at different times, but, by 1930, we can estimate the extent of the province's forest and cleared land by extrapolating backward five years from the 1935/36 inventory.

The inventory data for 1930 reveal an amount of cleared land much closer to the census figure, but still our most conservative estimate puts the census at approximately 4.9 per cent lower than the cleared land reported in the forest inventory. An upper-end estimate of cleared land in the inventory would suggest a level of under-reporting in the census closer to 12 per cent because of the amount of cleared land reverting to forest. It is likely that farmers reported these areas as unimproved land because the land was currently incapable of growing crops; technically, the land was improved, and, depending on the degree of reversion, it could be ploughed without cutting and stumping the trees.[46] We provide two estimates whenever possible to show how reverting land influenced the *Census of Agriculture*, a document that was not designed to account for this form of land-use change. The low-end estimate shows that at the very least the census figure for improved land was off by 5 per cent; the high-end estimate (12 per cent) suggests that farmers were quick to remove their reverting lands from the record, as well as from active production.

The county-level breakdown shows that the discrepancies varied significantly by region; the low-end estimates appeared much worse in Kings County (9 per cent) than Queens County, which was only below the inventory by 3.5 per cent in 1930. Figure 10.3 shows the spatial variation in the errors between the census and the forest inventory for 1931. Cleared land in most townships was at least 3 per cent higher than the census, but an obvious cluster of townships along the north shore in eastern Kings County under-reported cleared land by more than 10 per cent. If reverting land is added to the figure for cleared land, the discrepancies are even wider.

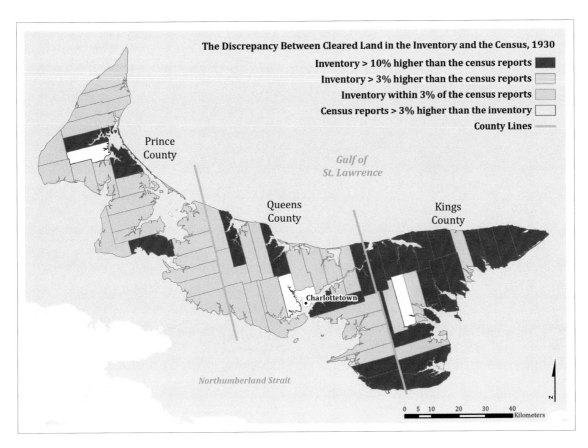

Fig. 10.3. The discrepancy between cleared land in the inventory and the census, 1930. (Prince Edward Island, Department of Environment, Energy & Forestry, Forests, Fish & Wildlife Division, PEI Lot / Townships, 2005. Software: ESRI ArcGIS 10.1, Adobe CS 4 Illustrator.)

Under-reporting cropland and improved land was likely a common phenomenon throughout Canada, and several scholars have argued that it is a regular part of agricultural statistics worldwide. In China, for instance, official estimates for agricultural land are understood to be up to 50 per cent lower than the actual area as determined by remote sensing.[47] The Dominion Bureau of Statistics studied the quality of Canadian agricultural data extensively in 1961, and they formed new estimates of improved land in sample areas of the Maritime provinces that were 16 per cent higher than the

census reports for the 1960 crop year.[48] Table 10.1 shows that the 1958 inventory identified slightly more cleared land, at least in Prince Edward Island, and that under-reporting had been a common part of agricultural statistics in earlier decades, as well.

The land on census farms became an even less useful indicator of environmental change in the late twentieth century. Not only did the census reports for cleared land fall significantly short of the 1935/36 and 1958 forest inventories, but the discrepancy grew to about 23 per cent in the 1980 and 1990 censuses. A large part of the problem is simply that the *Census of Agriculture* did not record information about land use on all private properties or on any public land. The total area of occupied census farms in Prince Edward Island declined over the twentieth century, from 87 per cent of the province's land in 1911 (including abandoned farms) to almost half that proportion (44 per cent) in 2006. After 1955, this category did not include land on abandoned farms and subsistence farms, two important property classes in the Maritime provinces. In most years, the census definition of a "farm" was based on the rather arbitrary cutoff of agricultural production yielding $50 or more.[49] In 1961, this meant the census ignored the equivalent of two whole townships (18,074 hectares) of "subsistence" farms in Prince Edward Island, or over 4 per cent of occupied farmland across the province. In Nova Scotia and New Brunswick, it meant ignoring human activity on 371,157 hectares – over twice the area of farms excluded by the census in Ontario.[50] Information on abandoned farms was also not recorded after 1941, making that growing class of land use impossible to follow.

It may be that the census data for 1930 and 1940 are the most reliable for comparison to the inventory because they included improved land on both census farms and abandoned farms. But even in these two years there is still the problem of land that was not covered by the census (13 and 14 per cent of the total land mass, respectively). We account for the problem of land excluded by the census in two ways. First, almost all of the province's roads, railways, wetlands and inland water would not have been included by the census, and they represented over forty thousand hectares or half of the land not classified as occupied or abandoned land

in these townships. Developed land in villages also accounted for a small amount of land not in farms. Second, we know that farmland that was not developed, cultivated, or grazed intensively quickly reverted to forest in Prince Edward Island, usually to white spruce. It is highly unlikely that other land remained clear if it was of such poor quality that it did not appear on either occupied or abandoned farms. Wetland, urban development, and roads were already removed from the forest inventories; therefore, we conclude that the discrepancies of between 5 and 12 per cent were errors in the enumerated amount of improved land.

Perhaps farmers' main incentive for underreporting the amount of cleared land to enumerators was a fear that accurate reporting would increase the land's assessed value and by extension the amount of property taxes paid. Taxation and production quotas were apparently the primary reason for under-reporting cropland on both communal and privatized farms in China,[51] but in Canada the rationale was not as obvious. Farmers had no reason to lie to the *Census of Agriculture*, as it was not a taxation instrument. However, if land owners were accustomed to giving lower figures to municipal and provincial tax assessors, there would have been no reason to come up with a different amount for enumerators. The province's revenue from real property tax was initially a relatively small proportion of indirect taxation, but it grew significantly in the period of this study. In the 1880s, the land-tax was temporarily cancelled, and in 1914 it represented a fraction of other revenue sources. For example, income tax generated twice the amount of property taxes, and the lucrative "fox tax" brought in ten times that amount from the island's fox farmers. The intensely local nature of property taxation may

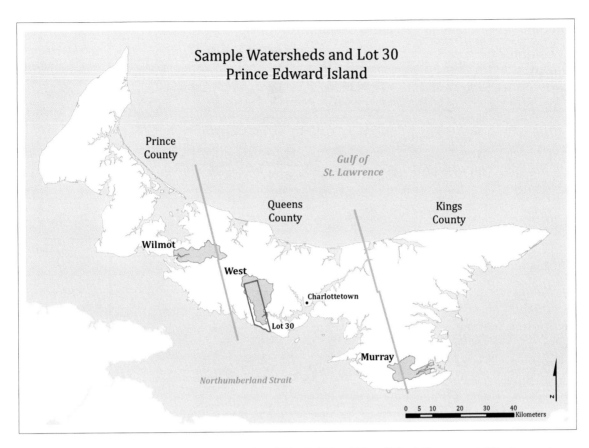

Fig. 10.4. Sample watersheds and Lot 30, Prince Edward Island. (Prince Edward Island, Department of Environment, Energy & Forestry, Forests, Fish & Wildlife Division, PEI Lot / Townships, 2005, PEI Watershed Boundaries, Last modified 6 April, 2005. Software: ESRI ArcGIS 10.1, Adobe CS 4 Illustrator.)

help explain the regional differences visible in Fig. 10.3. J. E. Lattimer has shown that when the province's school taxes were assessed and levied independently by over six hundred local school districts, both the rates and the assessments varied widely from year to year and from district to district.[52]

NEW RESEARCH POSSIBILITIES USING FOREST INVENTORIES AND GIS

In North America, the predominant land-use narrative of the twentieth century was the end of agricultural clearing and the acceleration of forest regrowth. Large tracts of farmland in the Northeast states were reverting to forest, and Prince Edward Island was experiencing similar patterns. The white spruce (*Picea gluaca*)

Table 10.2. Sample Watershed Summaries, ca. 1900–2000.

Watershed	Area (ha)	ca. 1900	1935	1958	1980	1990	2000
West River	high forest	3,018	3,449	4,433	5,511	5,443	5,046
	softwood	140	571	–	1,379	1,131	829
	mixedwood	1,899	1,899	–	1,955	1,304	1,057
	hardwood	978	979	–	2,178	3,008	3,159
	other forest	488	742	540	226	295	775
	total forest	**3,506**	**4,190**	**4,973**	**5,737**	**5,737**	**5,821**
	cleared land	7,686	7,002	6,209	5,343	5,381	4,983
	developed land	203	203	203	300	270	515
	wetlands	18	18	28	33	25	93
	Total Area	**11,413**	**11,413**	**11,412**	**11,413**	**11,414**	**11,412**
Wilmot River	high forest	855	973	1,189	1,109	1,025	822
	softwood	224	343	–	280	134	106
	mixedwood	569	569	–	377	363	271
	hardwood	61	61	–	452	529	446
	other forest	140	312	42	183	125	146
	total forest	**994**	**1,285**	**1,231**	**1,292**	**1,151**	**968**
	cleared land	7,146	6,855	6,843	6,644	6,730	6,600
	developed land	191	191	190	332	380	574
	wetlands	30	30	75	71	80	197
	Total Area	**8,361**	**8,360**	**8,338**	**8,339**	**8,340**	**8,339**
Murray River	high forest	2,700	2,825	–	4,722	4,898	4,191
	softwood	730	853	–	1,598	870	847
	mixedwood	1,628	1,628	–	1,863	2,828	1,970
	hardwood	343	343	–	1,261	1,200	1,375
	Other forest	984	1,239	4,491	292	213	621
	total forest	**3,684**	**4,065**	**4,491**	**5,014**	**5,111**	**4,813**
	cleared land	2,990	2,613	2,146	1,546	1,461	1,519
	developed land	158	158	161	276	308	443
	wetlands	278	278	296	254	216	317
	Total Area	**7,111**	**7,114**	**7,095**	**7,091**	**7,097**	**7,091**

Table 10.2. Sample Watershed Summaries, ca. 1900–2000. (Prince Edward Island, Department of Environment, Energy & Forestry, Forests, Fish & Wildlife Division, 1900 Forest Outline, Last modified 4 November, 2010, 1935 Forest Inventory, Last modified 25 November, 2010, 1958 Forest Outline, Last modified 28 February, 2008, 1980 Forest Outline, Last modified 4 November, 2010, 1990 Forest Outline, Last modified 25 November, 2010, 2000 Forest Outline, Last modified 30 November, 2010. PEI Watershed Boundaries, Last modified 6 April, 2005. Software: ESRI ArcGIS 10.1, Microsoft Excel 2011, Adobe CS 4 Illustrator.)

was the first species to capitalize on the new habitat, making its way across old fields from self-seeding hedgerows. L. M. Montgomery framed the language of reversion as a land-use decision in *Jane of Lantern Hill*. "Nothing had changed really," claimed Jane on returning to Prince Edward Island, "though there were surface changes.… Big Donald had repainted his house … the calves of last summer had grown up … Little Donald was letting his hill pasture go spruce. It was good to be home."[53] But just as scholars like Ramankutty have used remote sensing and revised historical statistics to show the local patterns within the larger trend of forest regeneration in North America, Prince Edward Island forest inventories show that "going spruce" was only one part of a complex suite of land-cover and land-use changes taking place in the twentieth century.[54]

The inventories and aerial photographs also present the ability to examine land use at multiple geospatial scales. Examining land-use data within boundaries such as township lines can be precarious. Communities were not formed along such arbitrary boundary lines, and business did not stop at census subdivisions.[55] The data derived from aerial photograph interpretation allow large areas to be broken up according to geophysical features such as watershed boundaries. Forest inventories and aerial photo data allow for analysis of land-cover and land-use change for all watersheds on Prince Edward Island. For this study, we focussed on three sample watersheds, selected for their similar sizes (7,100 to 11,500 hectares) and their diverse geographies and land-use histories (Fig. 10.4). The Wilmot River in eastern Prince County flows through some of the province's best agricultural land. The West River watershed is found in central

Queens County on steep slopes and shallow soils. In Kings County, the Murray River watershed drains sandy infertile soils.

Table 10.2 shows the forest cover in these three watersheds at various points throughout the twentieth century. The West River watershed experienced the largest regrowth of forest with increases in every inventory; Murray River forest cover increased steadily but experienced a small loss of mixed-wood forest in the 1990s; forest cover in the Wilmot watershed increased only slightly and then experienced a net loss in the 1980s and 1990s. The forest regrowth and abandonment of cleared land in all three areas characterizes the second act in the Maritime historiography of farm expansion into inferior soils and slopes.[56] By examining land-use changes and the amount of reverting land, we can see how the limits of agricultural expansion were reached at different times in different places. For example, the amount of cleared land in each watershed declined as a percentage of total area, but the decline was most severe in West River and quite minor in Wilmot. Reverting land points to a prolonged process of abandonment in West River and suggests that abandonment was most pronounced in Murray River in the early 1900s when almost 9 per cent of its cleared area was reverting to forest.

Developed land took a rapidly increasing share of land use in these watersheds toward the end of the twentieth century. The single largest threat to forests in terms of land-use change in the United States is urban development. Between 1982 and 1992, that country's urban land increased by 14 million acres, and about 5.4 million acres were at a net loss to the forest.[57] A similar trend is occurring in Prince Edward Island, where a significant amount of cleared and forest land is being lost to housing.

The forest-cover type also changed in different ways for each watershed. Murray River saw a modest increase in softwoods, mixedwoods, and hardwoods, but the other two watersheds saw hardwoods rise dramatically and overtake softwoods and mixedwoods cover types. The biggest winner in each watershed was hardwood, and this fits the general pattern for the province.[58]

By combining cadastral maps such as the Cummins Atlas (1928) with the forest inventories at either the township or watershed levels, we can also analyze forest-cover and land-use change on any number of individual properties.[59] We created a land-use database for twenty-seven randomly selected properties from the Cummins cadastral map of Lot 30 (Fig. 10.4). These properties were 54 per cent cleared in 1935/36, which compares favourably with the figure of 57 per cent cleared land in the entire township.

When we examine the forest cover on even these relatively early farms, we see forests shaped in large part by human hands. The extent of clearing on farms in 1935 influenced not only the size of the forest but also the type of forest that remained. As properties in the sample began to approach total clearing, their remaining woodlots were more likely to contain hardwood and hardwood/softwood cover types (Fig. 10.5). These farms were also the most likely to contain harvested parcels and the least likely to contain any reverting land. Thus, harvested forest areas tended to favour the growth of hardwood, and the absence of reverting old fields tended to discourage new stands of white spruce. Woodlot owners placed pressure on the forest for a variety of reasons, and these changed with the market for agricultural and forest products. Clear-cutting was

not only the first step to land-clearing, but it was often a reflection of the value of the wood harvested. In the 1950s pulpwood was a significant source of income in Prince Edward Island, and a significant strain on softwood stands. In the 1980s, fuelwood was important, and in the 1990s there was a strong market for studwood (small logs).

It is not possible to identify abandoned farms with any certitude from one set of aerial photographs, but the forest-cover type and the presence of clear-cuts and reverting parcels all point to farms that were strong candidates for downsizing in this period. The most intensively farmed parcels in the 1930s contained very little softwood and practically no reverting stands, and what forest did remain was predominantly hardwood and harvested land, presumably to supply the family's fuel and lumber needs. Conversely, properties with the lowest proportion of farmland contained the largest stands of reverting land and softwood stands on old fields. The combined forest inventory and cadastral map data suggest that land-use trends visible on these properties in the 1935/36 aerial photographs had clearly been in motion for decades.

We can also use the property boundaries from 1935 as a footprint for future land use on those parcels. Each one of these records in the sample represented the hopes and challenges of real historical actors. Figure 10.6 shows that Thomas McDougald's farm on New Argyle Road was mostly regrown and in the process of reversion in 1935, but it was probably not due to the quality of the land. Later inventories show that this land was mostly cleared and regularly farmed in the late twentieth century. A number of social and economic pressures could have caused the McDougalds to allow reversion on

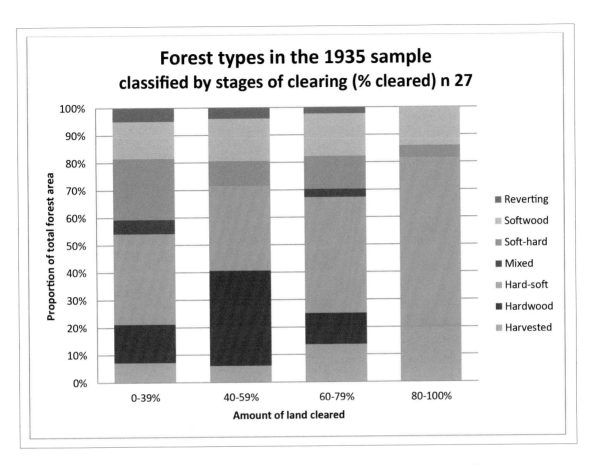

Forest types in the 1935 sample
classified by stages of clearing (% cleared) n 27

Proportion of total forest area

Legend:
- Reverting
- Softwood
- Soft-hard
- Mixed
- Hard-soft
- Hardwood
- Harvested

Amount of land cleared

Fig. 10.5. Forest types in the 1935 inventory sample, Lot 30. (Prince Edward Island, Department of Environment, Energy & Forestry, Forests, Fish & Wildlife Division, 1935 Forest Inventory, Last modified 25 November, 2010. Property boundaries from Atlas of Province of Prince Edward Island, Canada and the World, 1927 (Toronto: Cummins Map Company, 1928). Software: ESRI ArcGIS 10.1, Adobe CS 4 Illustrator.)

such a large portion of their farm, including a death or a gap in the family labour supply or simply the economic difficulties experienced in this period of Maritime history.

Another property in Green Bay, a farm belonging to Frank Costello on the corner of Eliot River Road and Riverdale Road, consisted of hilly farmland showing the edge effect between the hardwood uplands of the Appin Road area and the farmland of Emyvale (Fig. 10.7). This property was almost entirely cleared

in 1935 despite the hilly terrain and substantial wetland. It was undergoing more clearing in 1935 and does not appear to have reverted at any point since. Finally, the next property north of Costello's on Riverdale Road, toward Emyvale (not pictured here), was used mainly for its forest products over the twentieth century. Owned by Frank Dougherty in 1935, it was half-forested, mainly in hardwood with some softwood growth on old fields and with clear evidence of harvesting either for fuel or

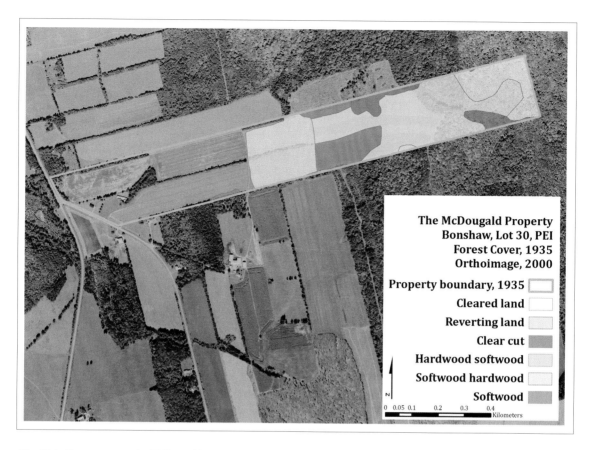

The McDougald Property
Bonshaw, Lot 30, PEI
Forest Cover, 1935
Orthoimage, 2000

Property boundary, 1935
Cleared land
Reverting land
Clear cut
Hardwood softwood
Softwood hardwood
Softwood

0   0.05 0.1   0.2   0.3   0.4
Kilometers

Fig. 10.6. Forest cover on the McDougald property, Bonshaw, Lot 30. (Prince Edward Island, Department of Environment, Energy & Forestry, Forests, Fish & Wildlife Division, 1935 Forest Inventory, Last modified 25 November, 2010, PEI 2000 Orthomap, Last modified 10 December, 2010. Property boundaries from Atlas of Province of Prince Edward Island, Canada and the World, 1927 (Toronto: Cummins Map Company, 1928). Software: ESRI ArcGIS 10.1, Adobe CS 4 Illustrator.)

lumber. By 1958, most of the farmland had reverted, and today the entire property is fully forested or in reversion with a small parcel of land for a home and outbuildings.[60] Costello's and McDougald's properties exemplified the two extremes of the spectrum shown in Fig. 10.5. Costello's small woodlot consisted of mostly hardwood, and McDougald's large and expanding woodlot was under less pressure and had become home to a variety of softwoods.

The image we see of the province *circa* 1900 was a landscape completely transformed by human activity. Pre-contact ecosystems had all but vanished in certain areas, wild life was at an all-time low, and forests had been profoundly altered. Unlike some provinces, however, Prince Edward Island farmland began to revert to forest for a variety of reasons, and the local environment began another form of change. The forest inventories show that farm abandonment and forest reversion were already well in motion by 1935. However, reversion did not occur evenly. We see, for instance, that townships with 75 per cent or more of their

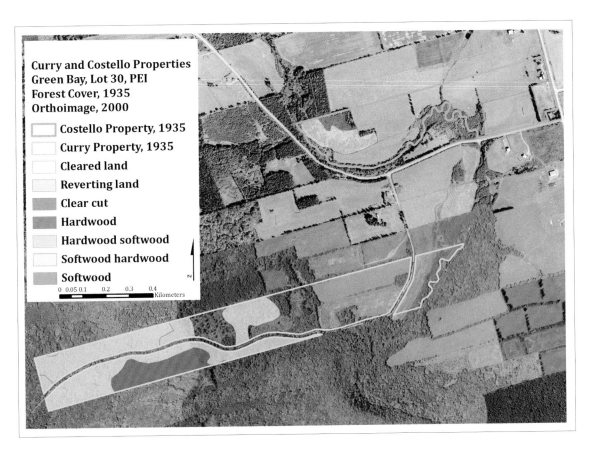

Fig. 10.7. Forest cover on the Curry and Costello properties, Green Bay, Lot 30. (Prince Edward Island, Department of Environment, Energy & Forestry, Forests, Fish & Wildlife Division, 1935 Forest Inventory, Last modified 25 November, 2010, PEI 2000 Orthomap, Last modified 10 December, 2010. Property boundaries from Atlas of Province of Prince Edward Island, Canada and the World, 1927 (Toronto: Cummins Map Company, 1928). Software: ESRI ArcGIS 10.1, Adobe CS 4 Illustrator.)

total area cleared in 1935 only saw about 4 per cent of that area return to forest by 1958. By comparison, the rest of the province was reverting to forest at three times that rate. The inventories are not the only way to measure these trends; the censuses also show us this in "improved acreage" figures. However, as we have shown, problems with the way cleared land was identified in the census result in exaggerated figures for improved land. Thus, the same townships with 75 per cent or more of their land "improved" actually appeared to lose

13 per cent of that land by 1961, and farms in the rest of the province relinquished over a quarter of their improved land to the forest. The forest inventories show a more even decline in cleared land because they captured all properties across the province, including those termed "abandoned" or "subsistence" farms by the *Census of Agriculture*.

None of our sample properties was in a complete state of reversion in either 1935 or 1958. Thus, we must be careful not to equate reverting land with abandoned land. Our sample suggests that reversion was a land-use decision followed by a relatively small number of farms; mainly it was a decision to reduce crop production. In our Lot 30 sample, reverting land occurred on only one quarter of properties. Most other areas in Queens County experienced lower reversion rates, so we can assume that well over three quarters of farmers in the county did not take farmland out of crop rotation long enough for reversion to occur. In townships with a higher proportion of cleared land in reversion, it is certainly possible that more farmers were downsizing, but this practice was far from ubiquitous in the 1930s. Some properties experienced relatively rapid reversion from average farms to almost completely treed plots. On Frank Dougherty's property, only a small parcel of cleared land remained at the end of the twentieth century for residential dwellings and small business activity. The 11.5 hectares of reverting land on this property, in 1935, was the largest parcel of new forest in our sample; the average property with reverting land contained only 3.2 hectares. The Dougherty's forest history shows how entire farms could "grow over" in less than half a century.

Using GIS we can query the parcels of cleared land in 1958 that had been in the process of reversion in 1935. The result shows that 3,664 hectares were reverting in 1935 but had been cleared a second or perhaps even a third time before 1958. Clearing on Prince Edward Island was not always a single event that led to a long-term land-use activity. Instead, the residents of this declining agricultural region explored a variety of options including part-time farming, renting, and allowing some fields to revert to forest.

CONCLUSIONS

If, as James Scott has argued, cadastral maps and scientific censuses made local information available to the state, then GIS and aerial photographs attach a time and a place to remotely sensed data and make it available to historians. After all, each cell of remotely sensed information was part of a local narrative. As we interpret the aerial photographs, we are gazing down on hundreds of human agents who probably glanced upwards from farmyards, fishing boats, and woodlots at this marvel of aeronautical engineering and symbol of a changing world. The story of their worlds has mainly been told by the census, the most extensive historical record for the majority of rural, and usually voiceless, Canadians. But, the twentieth century censuses of agriculture missed large tracts of land in Atlantic Canada, even in Prince Edward Island where a vast majority of land was privately owned and used for agricultural production.

Historians can now see this new side of the Canadian environment in many regions, especially the forested areas that attracted aerial photographers from government or industry. GIS allows historians to use the census in conjunction with new sources such as forest inventories and aerial photographs; on its own the census simply cannot provide an accurate portrayal of land-use activity. The *Census of Agriculture* provides an excellent starting point toward an understanding of environmental change in Canada, but, as we have shown, it has its weaknesses. Environmental historians should be attentive to the proportion of land enumerated in any given census year before using the census to make estimates of the size of the forest and the amount of deforestation. Even in areas where commercially farmed private land made up over 85 per cent of total land mass, as it did in Prince Edward Island in 1930 and 1940, historians should expect some degree of under-reporting of cleared land in the *Census of Agriculture.*

To economic historians who use improved land as an indicator of farm value, the revised estimates based on aerial photography suggest that farmland has been undervalued. The amount of improved acreage was critical for estimating the value of land and the costs of farm-making.[61] Furthermore, historians commonly use acreage to measure the productivity of farming as an economic activity since, as Marvin McInnis states, land is "an obvious starting point" for analysis as it "is the most directly measurable input."[62] To environmental historians, the census over-estimated the total biomass available on farms, and it ignored forest and other land cover on a growing number of properties in the late twentieth century. Historians need to understand these standard units of production if they are to ask questions about human interactions with the environment. Recent research emphasizes material and energy flow analysis as a way to measure the effects of human activity on natural ecosystems, and the symbiotic effects of environmental change on quality of life.[63] Land-cover and land-use change are fundamental variables in this research, and the forest inventories help to establish more spatially explicit land-use models.[64] In the Maritime provinces, and other areas that experienced agricultural decline, these sources also introduce the category of reverting land. Reversion occurred only on a small percentage of farms, but it was capable of returning entire properties to forest in a matter of a few decades.

The land-use and forest inventories and the aerial photographs they were built from are a powerful resource for historians and geographers. By interpreting the data in a GIS, users can investigate the land-cover and land-use change at any level, from the individual property to the townships, municipalities, watersheds, and larger jurisdictions such as the census district or the province as a whole. Where they exist, these sources would also allow for a more extensive project that performs sample inventories from aerial photographs across the country – a kind of "environmental census" for representative regions. Although the *Censuses of Agriculture* offer a limited view of human activity in forest and agro-ecosystems, a more comprehensive picture may be created by combining inventory samples with cadastral data and linking them to digital censuses of population and other geospatial databases. Air photos are some of the best time-slices of Canada's vast and diverse environments, and dynamic systems such as GIS allow historians to extract, manipulate, and link the information to other geospatial databases. These linkages present a clearer picture of past land use in order to better understand environmental change and changing land use practices over time.

NOTES

1   Michael Williams, *Americans and their Forests: A Historical Geography* (Cambridge: Cambridge University Press, 1992): 119–20; J. David Wood, *Making Ontario: Agricultural Colonization and Landscape Re-Creation before the Railroad* (Montreal: McGill-Queen's University Press, 2000).

2   N. Ramankutty and J. A. Foley, "Estimating Historical Changes in Land Cover: North American Croplands from 1850 to 1992," *Global Ecology and Biogeography* 8, no. 5 (1999): 382.

3   For more on the literature surrounding abandonment, see Mark Lapping, "Stone Walls, Woodlands, and Farm Buildings: Artifacts of New England's Agrarian Past," in Blake A. Harrison and Richard William Judd, eds., *A Landscape History of New England* (Cambridge, MA: MIT Press, 2011), 130, 132; N. Ramankutty, E. Heller, and J. Rhemtulla, "Prevailing Myths about Agricultural Abandonment and Forest Regrowth in the United States," *Annals of the American Association of Geography* 100, no. 3 (2010): 502–12.

4   For example, most studies of agriculture and agro-ecosystems in projects from the *Historical Atlas of Canada* to *L'Atlas historique du Québec* used improved land or land in crops as the primary indicator of farm activity.

5   See, for example, Table II, Land Occupied According to Tenure and Condition, in Canada, *Census of Canada, 1911. Agriculture*, vol. 4 (Ottawa: J. de L. Taché, 1914).

6   R. M. McInnis, "Perspectives on Ontario Agriculture, 1815–1930," *Canadian Papers in Rural History* 8 (1992): 17–127, 103; see also R. M. McInnis, "Output and Productivity in Canadian Agriculture, 1870–71 to 1926–27," in Stanley L. Engerman and Robert E. Gallman, eds., *Long-Term Factors in American Economic Growth* (Chicago: University of Chicago Press, 1992 [reprint]), 751.

7   O.F.G. Sitwell, "Difficulties in the Interpretation of the Agricultural Statistics in the Canadian Censuses of the Nineteenth Century," *Canadian Geographer* 13, no. 1 (1969): 75. See also the danger of misinterpretation, given the exclusion of grass crops in pasture land in the "crops" category, identified by Andrew H. Clark, *Three Centuries and the Island: A Historical Geography of Settlement and Agriculture in Prince Edward Island, Canada* (Toronto: University of Toronto Press, 1959), 158; Alan A. Brookes, "'Doing the Best I Can': The Taking of the 1861 New Brunswick Census," *Histoire sociale – Social History* 9, no. 17 (1976): 70–91.

8   Early censuses such as the Prince Edward Island *Census of 1841* asked farmers for the "Number of acres of *arable* Land held by each family," and in 1911 census makers had returned to this classification of improved land as "land which has been brought under cultivation and has been cropped and *is fitted for* producing crops."

9   Donald H. Akenson, *The Irish in Ontario: A Study in Rural History* (Montreal: McGill-Queen's University Press, 1984), 385.

10   Ruth W. Sandwell, "Rural Households, Subsistence and Environment on the Canadian Shield, 1901–1940," paper delivered at *Bringing Subsistence Out of the Shadows*: *An Environmental History Workshop on Subsistence Relationships*, Nipissing University, North Bay, Ontario (3 October 2009).

11   Peter A. Russell, "Forest into Farmland: Upper Canadian Clearing Rates, 1822–1839," *Agricultural History* 57, no. 3 (1983): 338; Catharine A. Wilson, *A New Lease on Life: Landlords, Tenants and Immigrants in Ireland and Canada* (Montreal: McGill-Queen's University Press, 1994), 33–34.

12   James C. Scott, *Seeing Like a State: How Certain Schemes to Improve the Human Condition Have Failed* (New Haven, CT: Yale University Press, 1998), 44.

13   Douglas McCalla, *Planting the Province: The Economic History of Upper Canada, 1784–1870* (Toronto: University of Toronto Press, 1993), 165; Catharine Wilson shows how the tax burden for Irish immigrants was much lighter in Canada than in Ireland, and she points to the "wild land tax" implemented to discourage absentee speculators from holding undeveloped land. Wilson, *A New Lease on Life*, 47, 186.

14   Harold M. Groves, "The Property Tax in Canada and the United States," *Land Economics* 24, no. 1 (1948): 24.

15   Suzanne Zeller, *Inventing Canada: Early Victorian Science and the Idea of a Transcontinental Nation* (Toronto: University of Toronto Press, 1987).

16   Public Archives and Records Office (Prince Edward Island), "Report of the Superintendent of Census Returns, Charlottetown, August 23, 1871;" Manitoba, *Report of the Minister of Agriculture of the Province of Manitoba*, 1880–83 (Montreal: Gazette Printing, 1881).

17   Manitoba, *Report of the Minister of Agriculture of the Province of Manitoba*, 1880–83, Table xiii; Canada, *Census of Population and Agriculture of the Northwest Territories, 1886*, Table XIV. Capturing agricultural data for two growing seasons was a unique characteristic of the Canadian censuses, and the 1921 and 1911 Censuses of Canada were each conducted in June to allow "ascertaining

the areas sown to field crops for the harvest years 1921 and 1911 as well as the area and production of crops for the preceding year." Canada, *Sixth Census of Canada, 1921*, vol. v – Agriculture (Ottawa: F. A. Acland, 1925), ix.

18   Geoff Cunfer, *On the Great Plains: Agriculture and Environment* (College Station: Texas A&M University Press, 2005), 30–31, 35–36.

19   S. B. Carter, S. S. Gartner, M. R. Haines, A. L. Olmstead, R. Sutch, and G. Wright, eds., *Historical Statistics of the United States: Earliest Times to the Present*, vol. 4 (New York: Cambridge University Press, 2006); Ramankutty et al., "Prevailing Myths," Fig. 2; M. L. Liu and H. Q. Tian, "China's Land Cover and Land Use Change from 1700 to 2005: Estimations from High-Resolution Satellite Data and Historical Archives," *Global Biogeochemical Cycles* 24, no. 3 (2010).

20   L. E. Philpotts, *Aerial Photo Interpretation of Land Use Changes in Fourteen Lots in Prince Edward Island*, Economics Division, Canada Department of Agriculture, 1958; C. W. Raymond and J. A. Rayburn, "Land Abandonment in Prince Edward Island," *Geographical Bulletin* 19 (1963): 78–86; W. M. Glen, *Prince Edward Island 1935/1936 Forest Cover Type Mapping*, Forestry Division, P.E.I. Department of Agriculture and Forestry, Charlottetown, Prince Edward Island, 1997; S. M. McDonald and W. M. Glen, *1958 Forest Inventory of Prince Edward Island*, P.E.I. Department of Environment, Energy and Forestry, 2006; D. G. Sobey and W. M. Glen, "A Mapping of the Present and Past Forest-types of Prince Edward Island," *Canadian Field-Naturalist* 118, no. 4 (2004): 504–20.

21   Don W. Thomson, *Men and Meridians: The History of Surveying and Mapping in Canada, vol. 2, 1867 to 1917* (Ottawa: Queen's Printer, 1967), 255.

22   Jay Sherwood, *Furrows in the Sky: The Adventures of Gerry Andrews* (Victoria: Royal BC Museum, 2012), 57; "Aviator to Detect Forest Fires," *American Forestry*, September 1915, 914–15; Hugh A. Halliday, "The Forest Watchers: Air Force, Part 35," *Legion Magazine* 20 (October 2009); Thomson, *Men and Meridians*, 289–90.

23   Richard A. Rajala, *Feds, Forests, and Fire: A Century of Canadian Forestry Innovation*, Canadian Science and Technology Museum, Transformation Series 13 (2005), 40–41. See also Peter Gillis

and Thomas Roach, *Lost Initiatives: Canada's Forest Industries, Forest Policies, and Forest Conservation* (New York: Greenwood, 1986), 211; "A Nation-Wide Inventory of our Forest Resources," *Natural Resources Canada* 8 (March 1929): 2; H. E. Seeley "The Use of Air Photographs for Forestry Purposes," *Forestry Chronicle* 12 (December 1935): 287–93; and Roland D. Craig, "Forest Surveys in Canada," *Forestry Chronicle* 11 (September 1935): 31.

24   The small gaps created by spaces where the flight lines were too far apart were corrected by using aerial photography from the subsequent survey in 1958.

25   The pilot's altitude actually ranged from 9,910 to 10,410 feet. Glen, *Prince Edward Island 1935/1936 Forest Cover Type Mapping*, 2–3.

26   McDonald and Glen, *1958 Forest Inventory*, 2–3, 4.

27   See, for example, Canada Bureau of Geology and Topography, Topographical Survey, "St. Andrews, Charlotte County, New Brunswick," Geological Survey of Canada, *"A" Series*, Map 523a (1939). The National Air Photo Library website may be accessed at http://airphotos.nrcan.gc.ca. http://www.nrcan.gc.ca/earth-sciences/products-services/satellite-photography-imagery/aerial-photos/search-air-photos/920 (accessed May 2013).

28   Several of the precursors to NTS maps surveyed privately owned farms and woodland in the early twentieth century, but these are only available in limited areas. They were surveyed by the Geographical Section of the Department of National Defence. See, for instance, Department of National Defence, "Topographic Map, Nova Scotia: Halifax Sheet, Number 201, Surveyed in 1920" [map], 1:63,360 (n.p., 1923).

29   Thomson, *Men and Meridians*, 298.

30   This form of research is also underway in other parts of the country, although more collaboration is needed. For example, Diane Saint-Laurent has used similar sets of aerial photographs to identify land-use and zoning changes in Sherbrooke and Richmond, Quebec, between 1945 and 2000. Stephane Castonguay and Diane Saint-Laurent, "Reconstructing Reforestation: Changing Land-Use Patterns along the Saint-Francois River in the Eastern Townships," in Alan MacEachern and William Turkel, *Method and Meaning in Canadian*

Environmental History (Toronto: Nelson, 2009), 286, 287.

31 The aerial photos are now stored mainly in the National Air Photo library in Ottawa, although many sets of historical air photos are also available at university map libraries and in municipal collections. The Prince Edward Island forest inventories are available for download at the provincial GIS website: http://www.gov.pe.ca/gis.

32 Colin MacIntyre, "The Environmental Pre-History of Prince Edward Island, 1769–1970: A Reconnaissance in Force," Master's thesis, University of Prince Edward Island, 2010, 126, 161–62.

33 Maritime Resource Management Service Inc., *Proposal for a Prince Edward Island Inventory Pilot Study: Prepared for P.E.I. Dept. of Energy and Forestry* (Amherst: Maritime Resource Management Service, 1987).

34 William Glen was the project manager for all but the 1980 inventories, and, within the resources available, the classification used in each of the inventories was comparable.

35 Glen, *Prince Edward Island 1935/1936 Forest Cover Type Mapping*, 4.

36 Variables such as tree species or agricultural land use are different in some inventories, due to the quality of the photographs and availability of resources. For example, a range of crop types and land uses were identified in 2000, but in 1980 and earlier years all contiguous areas of cleared land were included in a single polygon.

37 Don W. Thomson, *Men and Meridians: The History of Surveying and Mapping in Canada, vol. 1, Prior to 1867* (Ottawa: Queen's Printer, 1966).

38 Elsbeth Heaman, *The Inglorious Arts of Peace: Exhibitions in Canadian Society during the Nineteenth Century* (Toronto: University of Toronto Press, 1999), 160.

39 Robert Chalmers, "Report on the surface geology of eastern New Brunswick, north-western Nova Scotia and a portion of Prince Edward Island," *Annual Report*, Geological Survey of Canada, 8 (1895): Part M; Douglas G. Sobey, *Early Descriptions of the Forests of Prince Edward Island: A Source-Book, vol. II, The British and Post-Confederation Periods – 1758–c.1900* (Charlottetown: 2006), Part B: The Extracts, 222.

40 Clark, *Three Centuries and the Island*, 19; F. A. Stilgenbauer, *The Geography of Prince Edward Island*, PhD dissertation, University of Michigan, Ann Arbor, 1929; O. L. Loucks, "A Forest Classification for the Maritime Provinces," *Proceedings of the Nova Scotia Institute of Science* 25 (1962): 85–167.

41 Stephen J. Pyne, *Awful Splendour: A Fire History of Canada* (Vancouver: UBC Press, 2007), 223; Bernhard E. Fernow, *Forest Conditions of Nova Scotia* (Ottawa: Commission of Conservation, Canada, and Department of Crown Lands, Nova Scotia, 1912), 2–5.

42 Williams relied on a rather general interpretation of "improved land" for his early estimates, and according to Ramankutty there were flaws in some of the census statistics on which his estimates for forest areas were based. N. Ramankutty and J. A. Foley, "Characterizing Patterns of Global Land Use: An Analysis of Global Croplands Data," *Global Biogeochemical Cycles* 12 (1998): 667–85.

43 "Snapshot of Canadian agriculture," http://www.statcan.gc.ca/ca-ra2006/articles/snapshot-portrait-eng.htm, (accessed May 2013).

44 N. Ramankutty, J. A. Foley, and N. J. Olejniczak, "Land-Use Change and Global Food Production," in Ademola K. Braimoh and Paul L. G. Vlek, eds., *Land Use and Soil Resources* (Dordrecht: Springer, 2008), 28.

45 Canada, *Census of Agriculture*, 1986, xvi; Donald C. E. Robinson, Werner A. Kurz, and Christine Pinkham, *Estimating the Carbon Losses from Deforestation in Canada*, prepared for the National Climate Change Secretariat, Canadian Forest Service (Vancouver: ESSA Technologies, 1999), 4, Table 2.1; G. Kornelis Van Kooten, "Bioeconomic Evaluation of Government Agricultural Programs on Wetlands Conversion," *Land Economics* 69, no. 1 (1993): 29; P. Sellers and M. Wellisch, *Greenhouse Gas Contribution of Canada's Land-Use Change and Forestry Activities: 1990–2010*, final draft, prepared by MWA Consultants for Environment Canada, July 1998.

46 For the 1930 and 1940 upper-end estimates, we added the percentage of reverting land (6.7 per cent of cleared land) to our interpolated figures, and for 1955 and 1960, we used the amount of reverting land in 1958 (3 per cent of cleared land).

47 V. Smil, "China's Agricultural Land," *The China Quarterly* 158 (June 1999): 414, 417; Frederick W. Crook, "Underreporting of China's Cultivated Land Area: Implications for World's Agricultural Trade," *China Situation and Outlook Series* (Washington, D.C.: Department of Agriculture, 1993): 33–39; Stefan Siebert, Felix T. Portmann, and Petra Döll, "Global Patterns of Cropland Use Intensity," *Remote Sensing* 2 (2010): 1639; Ramankutty et al., "Land-Use Change and Global Food Production," 24.

48 The first "Agriculture Quality Check" performed by the Dominion Bureau of Statistics was in 1956, although it only investigated total farm area and not the condition of farm land. Canada, *Census of Canada 1956, Agriculture*, Bulletin 2–11 (Ottawa: Edmond Cloutier, Queen's Printer, 1957), xii; Canada, *1961 Census of Canada, vol. 5, Agriculture*, Bulletin 5.1 (Ottawa: Roger Duhamel, Queen's Printer, 1963), Table 4, xviii; Don W. Thomson, *Skyview Canada: A Story of Aerial Photography in Canada* (Ottawa: R.B.W. Ltd., 1975), 167.

49 The 1951 and 1956 censuses defined farms differently, incorporating more farms into the total, and as a result they made the drop from 1955 to 1960 appear more drastic. Canada, *Eighth Census of Canada, 1941, vol. VIII, Agriculture*, Part I (Ottawa: Edmond Cloutier, King's Printer, 1947), xx; Canada, *1961 Census of Canada, vol. 5*, vii.

50 According to the 1961 Census, "a farm is defined as an agricultural holding of one acre or more with sales of agricultural products during the past 12 months of $50 or more." Canada, *1961 Census of Canada: Agriculture, Bulletin 5.1–1* (Ottawa: Roger Duhamel, Queen's Printer, June 1963), Table 1, vi–vii.

51 Smil, "China's Agricultural Land," 417.

52 Prince Edward Island, *Journal of the Legislative Assembly of the Province of Prince Edward Island*, 1915, 109, "Collectable;" J. E. Lattimer, *Taxation in Prince Edward Island, A Report* (Charlottetown: Department of Reconstruction, 1945), 28.

53 L. M. Montgomery, *Jane of Lantern Hill* (Toronto: McClelland & Stewart, 1937), chap. 34.

54 Ramankutty et al., "Prevailing Myths," 503.

55 In some cases the agricultural production data in the census did not stop at subdivision boundaries, either. At least two townships in Prince Edward Island (Lot 65 and Lot 6) contained properties where the house, and therefore head of household, was in one subdivision and large portions of the property belonged to the adjacent subdivision. It appears in this case that the farm land and its products were entirely attributed to the subdivision containing the house.

56 Kris Inwood and Jim Irwin, "Land, Income and Regional Inequality: New Estimates of Provincial Incomes and Growth in Canada, 1871–1891," *Acadiensis* 31, no. 2 (2002): 157–58; Julian Gwyn, "Golden Age or Bronze Moment? Wealth and Poverty in Nova Scotia: the 1850s and 1860s," *Canadian Papers in Rural History* 8 (1992): 195–230.

57 Robinson et al., *Estimating the Carbon Losses*, 6, Table 2.2.

58 D. G. Sobey and W. M. Glen, *The Forest Vegetation of the Prince Edward Island National Park and the Adjacent Watershed Area Prior to European Settlement as Revealed from a Search and Analysis of the Historical and Archival Literature*, Phase 2; Analysis and Discussion (Contracted Report for Prince Edward Island National Park, 2002).

59 *Atlas of Province of Prince Edward Island, Canada and the World, 1927* (Toronto: Cummins Map Co., 1928).

60 It should be noted that the reverting section in 1958 included a significant strip of the neighbouring property's farm land. By 2000, the neighbouring farm had re-cleared the encroaching forest along the property lines, indicating the dynamic processes of reversion and clearing along parcel edges.

61 Jeremy Atack, "Farm and Farm-Making Costs Revisited," *Agricultural History* 56 (October 1982): 673.

62 McInnis, "Perspectives on Ontario Agriculture," 103.

63 Marina Fischer-Kowalski and Helmut Haberl, eds., *Socioecological Transitions and Global Change: Trajectories of Social Metabolism and Land Use* (Cheltenham: Edward Elgar, 2007).

64 Liu and Tian, "China's Land Cover and Land Use Change from 1700 to 2005."

11

The Irony of Discrimination: Mapping Historical Migration Using Chinese Head Tax Data

Sally Hermansen and Henry Yu

For over a century, from the late nineteenth through the late twentieth century, Chinese migrants to North America were the targets of racial discrimination and immigration exclusion and control. As targets of government surveillance, Chinese migrants were usually detained before being allowed to pass borders. They were forced to give details about their origins and destinations and to be physically described and measured. Chinese were given identification papers and tracked in government data sets long before other migrants came under the same regime of documentation. In comparison, European migrants passed through ports into Canada and the United States with relative ease, leaving much less paperwork. It is a great irony of the history of migration to North America that Chinese migrants – those who were most unwelcome – have given historians more detailed government data than the more readily welcomed trans-Atlantic migrants from Europe.

In 2004, this ironic consequence made the Chinese Migration Project possible. In 1885, the Canadian government imposed a $50 head tax on all Chinese entering Canada. A substantial proportion of a year's wages as a labourer, the head tax had been deliberately designed by the British Columbia provincial government and the Canadian federal government to discourage Chinese migration.[1] After imposing the head tax in 1885, the federal government created a detailed register that tracked not only who had paid but also a variety of other details such as age, height, village of birth, county of birth, last place of residence, occupation, port of origin, place and date of arrival and registration, and ship's name.

These data, and particularly the data documenting registrants' height, caught the interest of Peter Ward, a historian of health at the University of British Columbia (UBC). Because the 97,123 registrants were spread out over half a century, Ward believed that statistically analyzing their heights could provide a health measure reflecting the effects of changing childhood diets in China. A conversation with UBC colleague Henry Yu, a specialist in Chinese migration history, led to a SSHRC grant to create a digital database of each of the nineteen columns of information for each of the Chinese migrants. Yu was particularly interested in the geographical data. Beginning in 1910, the register began recording the migrant's destination in Canada – in addition to their geographic origins in China – creating the possibility for a detailed analysis of both origin and destinations for over a third of the migrants (35,731 between 1910 and 1923).

The SSHRC project involved a team of student researchers spending almost two years in laborious data entry, deciphering the often difficult-to-read handwriting of the register line by line. Led by Dr. Feng Zhang, then a doctoral candidate in sociology, the research team of Jason Chan, Mary Chan, Judy Maxwell, Alyssa Pultz, Denise Wong, and Lucy Lihong Zhang performed an amazing feat of alchemy, transforming over four thousand microfilm images of the original pages of the Head Tax Register – each with roughly twenty-five lines and twenty columns of handwritten data – into an enormously detailed digital database with nearly three million individual entries of data. This was no mean feat, since the longhand cursive writing was painstaking to read and decipher.

In 2009, the resulting database was featured on the Library and Archives Canada website with a streamlined search function for genealogical purposes.[2] In 2012, a fully searchable database with access to all of the original columns of data was made publicly available through a joint UBC and *Simon Fraser University* Library project.[3] The main use of these databases has been for genealogical purposes, and so one of the first yields of the hard (almost blinding) work of the research team was a valuable public history resource that allows descendants of the nearly 100,000 Chinese migrants who came to Canada in the late nineteenth and early twentieth century to search electronically for their ancestors. Rather than having to pore through the same difficult-to-decipher microfilm images of cursive handwriting that the research team used, family history researchers can now find their relatives with the ease of typing a name into their computer at home.

There was much more of value in the newly created database, however. One of the first research analyses applied to the database utilized the detailed height of each migrant. Ward used statistical analysis techniques to find that there had indeed been a steady and significant rise in average height for the Chinese migrants over the period. These increases, he hypothesized, resulted from improvements in diet that were themselves the result of financial remittances from the very migrants captured in the database. The economic effects of the earlier generations of migrants on those who followed from the same villages were significant, not only in terms of better food and housing but also in inspiring aspirations for wealth that created generation after generation of chain migration.[4]

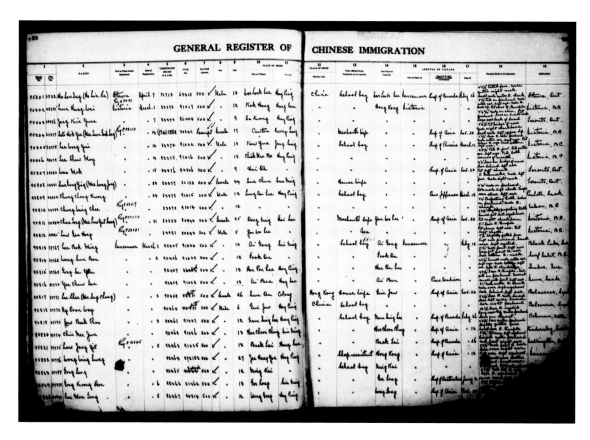

Fig. 11.1 Sample page from General Register of Chinese Immigration. (Source: Library and Archives Canada Website: http://www.collectionscanada.gc.ca/chinese-canadians/021022-1000-e.html. Government of Canada text on The Early Chinese Canadians 1858–1947 Website © Government of Canada. Reproduced with the permission of the Minister of Public Works and Government Services Canada [2011].)

It was this "imagined geography" of aspirations through mobility that we hoped to capture through applying GIS methods to the database. Worldwide, roughly 100 million people outside of China are descended from ethnic Chinese migrants who left China over the last five centuries, and the vast majority of them came from just two southern coastal provinces – Guangdong and Fujian. Within those two provinces, villages in a handful of counties were the main sending regions for century after century of continual out-migrations that created elaborate communication and transportation networks connecting these regions to an array of destinations around the globe. Beginning with short distance networks and then expanding to southeast Asia and around the Pacific, by the late nineteenth and early twentieth century, migrants were establishing and following circuits that took them as far as Africa and South America. These networks were generally circular in nature, with ideas, trade goods, and people moving in multiple directions and with complex and constantly adjusting flows.

Eight particular counties in Guangdong province dominated Chinese migrations to

Canada, the United States, Australia, and New Zealand. Beginning with the "gold rushes" to California and up and down the North American coast in 1849, and continuing around the other side of the Pacific to the Australian colonies of the British, migrants fuelled their journeys with dreams of gold. Even after the gold rushes diminished, aspirations for wealth continued as the organizing metaphor for migration. Leaving for North America, Australia, or New Zealand was called, in the Cantonese dialects that the migrants spoke, going to "Gum San" – literally Gold Mountain (金山). "Gold Mountain" was not the name of a specific place (the Cantonese had separate terms for "Canada," the "United States," and "Australia"). Rather, it named an imagined geography created by a life cycle of aspiration involving mobility and long distance linkages. Stories of wealth came to young children from men overseas or recently returned, creating desires within younger men to follow the same paths. Loans and other forms of support in finding jobs and housing smoothed the journeys, and those already established made money by organizing labour contracts for newcomers, or selling shares in businesses and ventures they had established overseas.

"Gold Mountain dreaming" involved imagining a lifetime ahead inspired by exemplary tales passed back to rural villages from far overseas. The life cycle idealized working hard and saving towards marriage – perhaps an arranged marriage to a woman back in the village who would raise children and take care of the household, or to a woman in "Gold Mountain" whose connections to local communities could help in establishing a business – and the education of children for eventual success. The dispersion of the local sites for this life cycle

– small villages in China, seaports and migration nodes such as Hong Kong, Singapore, San Francisco, Sydney, Honolulu, Victoria, and Vancouver, and rural and urban spaces all around the Pacific and the Caribbean – meant that geographically split families were common, with women and children on one side of the ocean and mobile men in sites around the world. Remittances built houses, hospitals, and schools in home villages, but there were also investments in businesses and education in North America and other local sites. An economics of relative location developed out of the continual calculus of wages, prices, and the comparative currency values for labour, goods, and services in multiple locations. What to trade, when to move, and where to go – all were a product of the powerful geographic imaginary created by "Gold Mountain" dreams.

What Yu hoped to gain from an analysis of the General Register of Chinese Immigration to Canada was a detailed yet aggregate interpretation of the patterns of flow and geographic networks of these almost 100,000 migrants. Moving beyond the "head tax" and other discriminatory anti-Chinese acts as the main historiographic issue that made Chinese in Canada interesting to historians, the goal was to understand what Chinese Canadians were themselves doing rather than only what was done to them. If the detailed surveillance of Chinese migrants provided the raw material for re-imagining the role of Chinese Canadians in Canada's history, there was an elegance and poignancy to bringing a new life to the dead data scribbled in the longhand of the clerks of Canadian racism.

Once the head tax database was complete, the question was whether we could use it to help map this geographic imaginary. Historians are adept at creating narratives of change and continuity over time; most are also familiar with the basics of working with spreadsheets, databases, and statistical software, and the challenges of visualizing numeric data in graphs. The initial research team on this project used these skills to produce graphs of the immigration data. By plotting numbers against arrival dates, they were able to see the peaks and troughs of immigration and how it mimicked federal immigration policy and legislation. But to create a map of the destination cities or a map of the counties in China from which the immigrants came, or to visualize links between origin cities in China and destinations cities in Canada over time – thus visualizing the historical process of chain migration – went far beyond tinkering with Excel or SPSS.

The need for more powerful tools to analyze the data and to create maps and other visualization brought Henry Yu to the UBC geography department, and to GIS specialist Sally Hermansen. Hermansen was teaching an upper-division undergraduate-level geo-visualization class and recognized that this incredibly rich database would be perfect for a student group project. Hermansen's students created a subset of the overall dataset focussing on origins and destinations. Because destination data was only available for migrants after 1910, Hermansen and her students decided to refine the 97,123 records down to only those between the years 1910 and 1923 (when Chinese exclusion laws essentially cut off further migration), resulting in 38,410 individuals. These records were further refined to include only those that contained destination information (35,731).

The destination data had been transcribed from handwritten entries for those migrants who landed at the port with known destinations.

Mistakes had been made in the original registering of the destinations, as well as in the coding of this data into the digital database. For example, there were often two or more variants in spelling of the same place name. After a process of analysis and data cleaning, a unique destination list of 522 place names was created to cross-reference with the current DMTI Spatial Inc.(a geospatial data provider) Canadian place name database. When the unique destination table was matched to the place name spatial file with the GIS software, 150 of the destination names did not match, many the result of spelling variants. When the spelling variants were corrected, only thirty-six destinations remained unmatched. The final database of immigrants contained 35,680 records in 460 unique destination cities and towns in Canada. In summary, after cleaning up the data, there was a remarkable 99 per cent match of the destination city of Chinese immigrants between 1910 and 1923 to current place names of cities and towns.

Combining Yu's research expertise in the history of Chinese migration with Hermansen skills as a GIS researcher allowed us to produce some initial maps of origin and/or destination over various time frames. A variety of Canadian destination maps were created by relating the database of Chinese immigration data to the place name GIS layer. Once the data had been related, any number of maps could be created. The simplest, and possibly most visually effective, map created was a simple dot map plotting all 460 unique destinations.

This map revealed instantly the vast expanse of Canadian geographic space to which Chinese immigrants moved – their destinations were by no means constrained to the obvious port cities of Vancouver, Victoria,

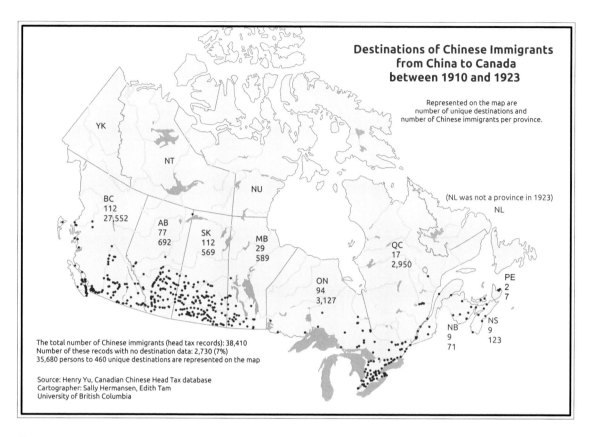

Fig. 11.2. Unique destinations of Chinese immigrants between 1910 and 1923.

and Montreal. (Newfoundland was not part of Canada at the time of this immigration, so there were no cases of Chinese immigrants who recorded Newfoundland as their destination in the General Register, although Newfoundland had its own head tax of $300 enacted in 1906 and a registry database of Chinese immigrants has been created by Dr. Miriam Wright of the University of Windsor.) There were also no recorded immigrants in the Northwest Territories or Yukon, but every other province had recorded destinations. We then created a proportional symbol map of destination cities, where the size of the dot was proportional to the number of immigrants who went to that destination. Subsequent choropleth maps were

also created where the destination data were aggregated by province.

For the creation of the China origin maps, the immigration head tax database contained three origin variables or three levels of origin resolution: province, county, and village or town. Of the 36,000 records at the province level, almost all the immigrants came from one province: Guangdong. Only six came from four other surrounding provinces; 155 came from outside of China; and 850 records were 'unknown.' A choropleth map was created of the provincial origin data by digitizing an 1820 provincial boundary map of China. Drilling down to the county level, the data are mapped to 1911 county boundaries, thereby creating a

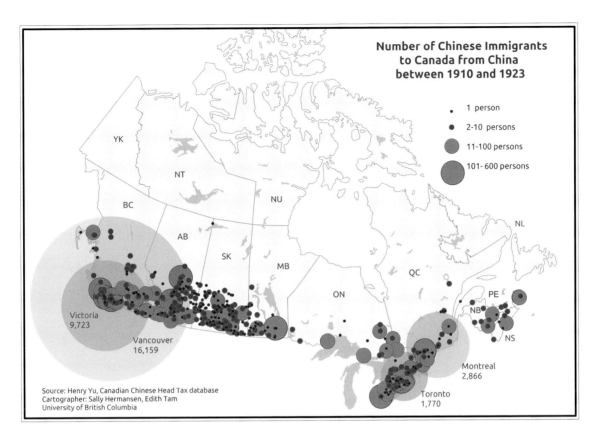

Source: Henry Yu, Canadian Chinese Head Tax database
Cartographer: Sally Hermansen, Edith Tam
University of British Columbia

Fig. 11.3. Number of immigrants per destination between 1910 and 1923.

choropleth map of origin data by county. Drilling down further to village and town proved impossible for this initial project because the data contained too many variations in the phonetic transliteration of Chinese village names into English, and detailed GPS coordinates of Chinese villages were unavailable.

The GIS team continued to create maps that explored the relationship between origins and destinations in Canada, as well as other fields of the database. For example, on an origin county choropleth map, pie charts were created of the destination region in Canada. Further maps were created of different time frames in order to depict temporal patterns of migration, in particular highlighting the effects of anti-Chinese legislation on the numbers of migrants, as well as periods of war and economic depression.

Like all historical GIS projects, the maps and other resultant visuals greatly facilitate narrative explanations about the historical process of trans-Pacific Chinese chain migration. They also do far more. They engage the reader visually, sparking and enhancing spatial perspectives that otherwise might be muted by linear narratives of change and continuity. In doing so, they informed research in a generative manner, leading to other questions that without the visuals may not have become apparent (or were not thought possible). For instance, after the class project was finished and the power of the maps for visualizing the spatial distribution of Chinese Canadian migrants was realized, it became immediately apparent that a project to regularize the English transliterations of Chinese village names and to match them to the actual villages in China might be worth the effort because of the potential to correlate migration patterns between geographic locations on both sides of the Pacific.

One of the students, Edith Tam, began working on a project led by Eleanor Yuen and Phoebe Chow of UBC's Asian Library to map the Chinese villages in the head tax database. Because historical records were of limited use for understanding the specific ways in which migrants speaking highly variant dialects of Cantonese would pronounce their village names, Yuen and Chow devised an important methodological innovation. They engaged with a number of elders from the Chinese Canadian community who had deep familiarity with the local counties in question, as well as fluency in the local dialects, which are not only highly idiosyncratic in both vocabulary and pronunciation compared to both Cantonese and Mandarin, but often unintelligible to other Chinese speakers. Beginning with Hoisan, 台山 (otherwise known as Sunning, 新寧), county where 45 per cent of all migrants to Canada originated, and then moving on to Heungsan, 香山 (later renamed Chungsan, 中山), county near Macau, the Asian Library's mapping project went through twenty rounds of community workshops with Chinese Canadian elders, a painstaking but rewarding process.

The enormous challenge of finding the villages in Hoisan and Heungsan counties that corresponded to the wide range of English transliterations found in the database would have been impossible if the researchers had not engaged local community elders. Over the initial two years of workshops, the phonetic transliterations of village names found in 44,131 entrants from Hoisan county and 5,898 entrants from Heungsan county were analyzed. Using their knowledge of local dialects to "reverse engineer" the original village dialect pronunciation that the English-speaking clerks must have heard, and then matching this hypothesized name to an actual village, these elders were able to match the villages of over 90 per cent of the migrants from those two counties to historical village names on a 1911 map.

This remarkable feat of detective work was only possible because the living memory of these village dialects still exists among small numbers of Cantonese elders overseas. Because of Mandarin language policies and the introduction of a single national dialect in school education following the 1949 Chinese revolution, many local village dialects have fewer and fewer living speakers, and this project may well be impossible to do after another generation. After these two counties had been completed, one of the elders involved in the workshops, Rudy Chiang, undertook a third sending county, Sunwoy (新會), by himself as a labour of respect and devotion to the immigrant ancestors from that county who had paved the way for his own family.[5]

Since these three counties together accounted for over 65 per cent of all Chinese migrants to Canada between 1858 and 1923, it is now possible for a GIS project mapping origin villages to be undertaken if an accurate GPS mapping of the historic villages in these three counties can be created. Unfortunately, many of these villages no longer exist, as suburbanization and industrial development have subsumed rural areas in Guangdong province over the last three decades, but many of the more remote villages still remain much as they were over a century ago. Historical maps give a rough idea of where many of the original villages were located, but this important project remains to be done.

Subsequently, Yu began working on another large-scale project, which aimed to create a portal website for Chinese Canadian history. The "Chinese Canadian Stories" Project,[6] a $1.17 million public history project between 2010 and 2012, involved collaborations between digital librarians and archivists, university researchers and students, and community members from twenty-nine local organizations across Canada. Funded by the Community Historical Recognition Program of the Ministry of Citizenship and Immigration Canada, with in-kind funding from UBC and SFU, the goal of the project was to create a range of web-accessible resources that exemplified the most current historical scholarship, as well as extensive reciprocal partnerships between community organizations and universities. The project also involved the Critical Thinking Consortium, a nation-wide non-profit network of teachers who worked with researchers at UBC to create digital learning resources that could be downloaded by teachers across the country. These resources included a digital historical learning game called "Pages from the Past" that asked social studies students to give advice, using a magical photo album, to historical Chinese Canadian characters as they made important life decisions, at the same time learning about the building of Canada by immigrants in the early twentieth century.[7]

The Chinese Canadian Stories project was ambitious in scope, involving everything from digital oral history recordings to the creation of digital games. The oral history stories of elders across the country were created on a specific model, involving extensive interviews that were several hours long, subsequently logged and digitally preserved by UBC Library, along with short "YouTube" videos that were edited highlights from the interviews. This formula of saving detailed interviews for posterity and future research, as well as creating professionally edited short films that were widely accessible online, came out of the imperative that community participants should be given back versions of the interviews that they could proudly (and easily) show family members and friends. Based upon a model of "photo album" interviews pioneered by the UBC research team, interviewees were encouraged to share memories by flipping through treasured photo albums, discussing meaningful photographs that were subsequently scanned at high resolution for digital preservation. Metadata for these scanned images would include information gleaned from the interviews, preserving memories even as the original photos remained as treasured heirlooms with the families. This model allowed the project to avoid the unnecessary act of alienating important family photographs from descendants, even as important stories surrounding the photos were recorded and preserved in UBC Archives. The project's aim was to create archival collections where photographs with Chinese Canadian subjects would never again have metadata such as "Unknown Chinese, location unknown, date unknown."

The Chinese Canadian Stories project also involved students in all aspects of its work, from the collection of oral histories to the design and construction of mobile museum kiosks that were hosted in 2012–13 in high-traffic locations such as the Vancouver Public Library and the Ottawa Public Library. An online immersive video game called "Gold Mountain Quest," plunging ten- to twelve-year-olds into the small town world of 1910 Canada, was created by digital design company Catstatic and master's students at the Centre for Digital Media's Great Northern Way Campus. "Gold Mountain Quest" used the most up-to-date historical research on the lives of Chinese Canadians in small-town Canada to populate a fictional town of a century ago. The game player helps an array of the townspeople fulfill mini-quests, at the same time collecting historical objects drawn from the over 25,000 objects in the Drs. Wallace and Madeline Chung Collection at UBC Special Collections (one of the best archival collections concerning Chinese Canadian and Canadian Pacific Railroad history).[8] Designed as a fun yet still educational companion to the "Pages from the Past" learning game, the pair show how partnerships between research scholars, teachers, and digital media designers work best when each is involved collaboratively at every step of the way from conception through completion. Working iteratively took longer (each game took over eighteen months), but the results were worth the additional time and effort, in particular in terms of the historical accuracy and authenticity of portrayals of Chinese Canadian life.

The Chinese Canadian Stories project also produced a searchable interface for all of the columns of data within the Chinese head tax database. (The online search function at the Library and Archives Canada site only enabled searching by the name of the individual, year of arrival, or certificate number.) This more fulsome search interface allows more advanced search functions, which should also be more useful for researchers.[9] Chinese-language character search capability for village and county names was also added.

Yu's research team also began working with the Stanford University Spatial History Lab, a project that grew out of multidisciplinary collaborations developed by historian Richard White of Stanford's Bill Lane Center for the American West. As with the earlier stage of work by Sally Hermansen's students on the Chinese head tax database, the continuing work with the Stanford spatial history team emphasizes the generative potential of visualization to highlight and disclose aggregate patterns that otherwise would be nearly impossible to imagine. Using GIS and other visualization tools such as Flash to show temporal changes in spatial patterns, Stanford research students Stephanie Chan and Oliver Khakwani worked with Spatial History Lab Principal Investigator Zephyr Frank, Creative Director Erik Steiner, Spatial Historian Jake Coolidge, and UBC's Chinese Canadian Stories research team to create visualizations that capture the nodal migration pattern in Canadian cities and small towns from the villages and counties of origin in China. By looking for ways to visualize the kinship networks and continually repeating spatial distribution of family chain migration from the same villages, we hope to realize the potential of spatial history approaches to capture both temporal and spatial patterns from large-scale data sets such as the Chinese Head Tax Register.

The first two results of the collaboration between UBC and Stanford's Spatial History Lab are now available online.[10] Figure 11.3 is a screen capture of a Flash visualization that shows the flow of Chinese migrants to five

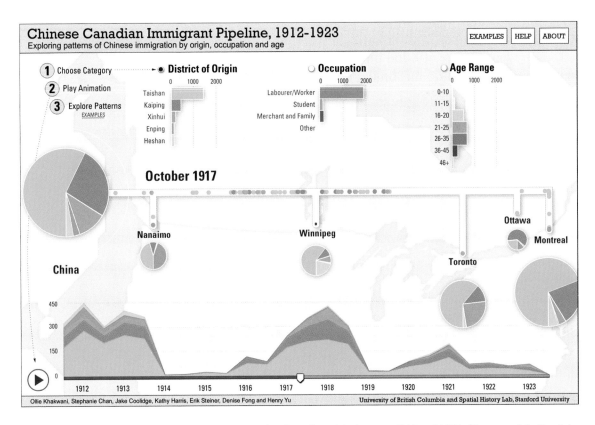

Ollie Khakwani, Stephanie Chan, Jake Coolidge, Kathy Harris, Erik Steiner, Denise Fong and Henry Yu · University of British Columbia and Spatial History Lab, Stanford University

Fig. 11.4. Visualizing the flow of Chinese migrants to five Canadian cities between 1912 and 1923. (Courtesy of the Spatial History Project, Center for Spatial and Textual Analysis [CESTA], Stanford University.)

cities in Canada. Initially named the "gumball machine" in its early conception, the visualization was designed to capture the ebb and flow of migrants by their county origins over a ten-year period, showing each individual migrant, colour-coded to show either their county of origin, their profession, or their age and how each of these three variables was reflected in the migration pattern to the various Canadian cities. We explicitly aimed at creating a visualization that would allow us to see aggregate patterns changing over time that could not otherwise be understood using just static tables. The power of dynamic visualizations combined with GIS data lies in the ability to capture temporal change along with spatial relationships.

As shown by the original visualization from which Fig. 11.4 is drawn, creating a number of datasets representing change over time that can then be played as a dynamic visualization allows a historical researcher to see aggregate patterns shift and to capture relationships between variables that could be imagined through a careful analysis of data but would otherwise be impossible to render visually.

A further series of analyses using a small subset of the head tax database was undertaken by Stanford student Stephanie Chan, using the software program Gephi, originally designed to analyze the relationship between websites in terms of web traffic. Chan and Spatial Lab Creative Director Erik Steiner worked with Yu

Fig. 11.5. Preliminary visualization using Gephi of the migration patterns to Saskatchewan of the Mah family, 1912–23. Using the variables of family name, village origin, and destination in Saskatchewan, Stanford researcher Stephanie Chan used Gephi to produce network patterns for four Chinese family lineages that visualize the weighted correspondence of family name and village origin in creating family chains and connection between destinations. This preliminary visualization describes the Mah family in Saskatchewan. The length of the lines reflect the weight of connection, with shorter lines reflecting a stronger correspondence. (Courtesy of the Spatial History Project, Center for Spatial and Textual Analysis [CESTA], Stanford University.)

to visualize the migration patterns of four different clans to the Prairie province of Sasketchewan (the Yee, Mah, Kwong, and Wong families, which accounted for over 50% of Chinese Canadians migrating to Saskatchewan). Using the variables of family name, village origin, and destination in Saskatchewan, Chan produced network patterns for each of the four families

that reflected visually the weighted correspondence of family name and village origin in creating "family chains" that could be compared to the importance of shared location. Another way to understand this question would be to ask, if I were a Mah family member in Saskatoon in 1920, would I get better information about Swift Current if I asked another member of the Mah family in Saskatoon or if I travelled to Swift Current to see for myself? The aim was to capture the weight of shared family name and village origin in network patterns, and, by extension, the quality of information and intelligence that would pass along family networks. Although the visualizations are still preliminary (Fig. 11.4 shows the visualized pattern for the Mah family), the initial results show that 1) each of the families showed very different patterns, and 2) as the results are analyzed, it will be necessary to use other historical sources, and in particular oral histories and interviews of migrants and their descendants (few of the original migrants are still alive) to discover the meaning of these different patterns.

Perhaps no other lesson from the Chinese Canadian Stories project and other projects at UBC using GIS mapping techniques to visualize historical data has been more clear that the most effective use of aggregate data has been in combination with other methodologies. The discriminatory nature of the original gathering of the data in the Chinese Head Tax Registry created the ironic consequence that those most unwanted by Canada and the target of racial discrimination and exclusionary legislation were also those for whom we had the most useful statistical data. But in order for this irony to have a happy consequence rather than just a perverse twist of history, there needed to be more than just data analysis. The Community Historical Recognition Program that funded the Chinese Canadian Stories project was created by the federal government as part of its official apology in 2006 for the nation's long history of anti-Chinese legislation and discrimination. That the Chinese Canadian Stories project was able to use Canada's meticulous records of surveillance in order to bring alive again the stories of those 97,123 migrants was the result of the database being a part of a larger concerted, collaborative effort to gather life histories and to collect the oft-ignored histories of Chinese Canadians.

NOTES

1   The proceeds were also a significant source of revenue for both governments, and between 1885 and 1923 (when Chinese were formally excluded from Canada), they evenly split over $23 million – the equivalent of well over $1 billion in today's currency.

2   See http://www.collectionscanada.gc.ca/chinese-canadians/021022-3000-e.html.

3   See http://chrp.library.ubc.ca/headtax_search/.

4   W. Peter Ward, "Stature, Migration and Human Welfare in South China, 1850–1930," *Economics and Human Biology*, in press, corrected proof available online: http://www.sciencedirect.com/science/article/pii/S1570677X12001013.

5   See http://asian.library.ubc.ca/files/2012/01/Head-Tax-brochure2.pdf, http://burton.library.ubc.ca/hclmbc/Documents/ZhongshanTaishan_exhibit_finalscreen.pdf, and http://ccs.library.ubc.ca/en/headtax/mapping.html.

6   The project can be accessed at http://chinesecanadian.ubc.ca.

7   http://ccs.library.ubc.ca/game/index.html.

8   http://ccs.library.ubc.ca/en/GMQ/index.html. The Chung Collection is available online at http://digitalcollections.library.ubc.ca/cdm/landingpage/collection/chung.

9   http://chrp.library.ubc.ca/headtax_search/.

10 See at http://www.stanford.edu/group/spatialhistory/cgi-bin/site/project.php?id=1049.

12

Mapping Fuel Use in Canada: Exploring the Social History of Canadians' Great Fuel Transformation

R. W. Sandwell

In the early twenty-first century, as people around the world take in the unwelcome news from scientists about both the limited supply of cheap fossil fuels and the surprisingly virulent impact that large-scale burning of these fuels is having on the global environment, we are being reminded on a daily basis of the deep connections between our everyday lives – particularly the energy systems that provide us with heat, light, power, water, waste disposal, and food – and the larger social and material environments within which we live. At the same time that the limited supply and huge environmental costs of these "new" (in terms of massive human consumption) sources of power become ever clearer in the present, their profound (indeed 'cyclonic'[1]) role in Canada's social, political, economic and environmental history is emerging with more coherence and force. For if, as now seems possible, the twentieth century will appear in the historical record as both the first and the last Age of Abundant Energy, our heightened awareness about the role of fossil fuels past, present, and future is going to have profound implications for the ways in which historians see and analyze what used to be thought about as 'progress,' 'industrialization,' 'urbanization,' and 'modernity.'

This chapter is part of a larger study that seeks to provide the first nation-wide look at the varied relationships between energy use and the practices of everyday life for rural and urban Canadians in the first half of the twentieth century. Much of that research draws on a broad range of sources – oral histories, personal diaries and memoirs, newspapers, educational materials, census data, and advertising and company records (particularly complaint files of utility companies) – to document the ways in which Canadians understood and responded to the new fuels that they were bringing into their homes and farms. Much of that work is devoted to explaining the wide range of supply- and demand-sided problems that influenced the slow adoption of the new fuels in the home. This chapter, however, has a much more specific purpose. It presents my reflections about learning to use spatial data and spatial thinking to explore the relationship between people, place, and energy in Canadian history. More specifically, it offers some suggestions about how and why HGIS might help us understand how new forms of energy and power transformed daily life in place-specific ways. And finally, it draws on the preliminary stages of my own research into electricity to provide some examples of what an HGIS-informed social history of energy might look like.

HGIS has given spatial thinking a boost in recent years, allowing historians to more easily manipulate and compare map-able visual data that is linked to large data sets, such as census data. HGIS has played an important role in allowing geographical data to move beyond the merely descriptive to take on a clearer explanatory edge than is usually found in historians' work, outside of environmental history at least. The attention to space and place has

been particularly welcome in my social history of fuels and energy project, even though, unlike most history GIS projects, my research looks beyond a small geographic focus on a river valley, a city, a county, or a region to examine the entire country over half a century or more.[2] Like other HGIS projects, however, this portion of my research involved a steep learning curve about using the technology, a process made even more difficult because the professionally accepted computer program for creating HGIS maps is definitely not (based on the experience of myself and other historians) user-friendly. To make things worse for this researcher, it is not available for Macintosh computers. Difficult as it was learning to make use of ArcGIS software, I would not have been able to do any HGIS work without guidance and assistance from a collaborative team of students and professionals over a number of months, particularly geography students Jordan Hale (who entered volumes of data into a database) and Sarah Simpkin (who used the data to create the maps that appear in this chapter), and University of Toronto GIS and map librarian Marcel Fortin, who oversaw much of the work. Marcel generously donated his time to this project, and SSHRC provided the funding for my students. And this project would not have been possible without the considerable lengthy and painstaking work carried out by the SSHRC-funded Canadian Century Research Infrastructure Project. This project, via Byron Moldofsky, provided the GIS co-ordinates for 1941 and 1951 Census Divisions that allowed me to map the census-division level data for the entire country.[3] Finally, this project shared with other HGIS research an element of risk; historians characteristically do not know the outcome of their research before they embark on it, and research questions most often emerge through or are clarified in the process of the research itself.[4] HGIS, therefore, requires an unusual 'up front' investment in the time and money attendant on collaborative research, and research dependent on highly sophisticated and expensive technologies. There is no guarantee that it will, in the end, be "worth" the investment.

This chapter will tentatively suggest that for this project the investment has paid off. HGIS has certainly played a role in increasing my historical understanding of the importance of place to the social and the environmental history of fossil fuels and hydroelectricity in Canada. It has done this, on one level, by simply drawing attention to the spatially specific factors involved in energy use across the country. By relating the extraction, processing, transportation or transmission, and the consumption of these fuels to particular places, however, HGIS also sharpens and deepens our understanding of how place, space, and local environments have influenced the profound social transformations of the twentieth century, just as surely as social and cultural changes in fuel use have continued to transform the environment. This research suggests that the mapping and layering of census data over time and through space that HGIS allows can provide an additional, powerful tool of description and analysis for the social historian.

BACKGROUND: TOWARDS A SOCIAL HISTORY OF FOSSIL FUELS AND HYDROELECTRICITY

Historians have long explored the development of new forms of energy, albeit typically with more of an emphasis on the new technologies and the professional, economic, and technological systems within which they developed than on the fuels themselves.[5] Following trends in other countries, Canadian historians are now taking tentative steps towards a more explicit history of energy in Canada.[6] Canadian social histories of the impact of changing energy use in the first half of the century remain scarce, however. An important spate of research and writing in the 1980s and '90s linked women's changing domestic role to larger changes in Canadian society via the 'modernized' household. But aside from Joy Parr's important work on the post-war period, there has been little published recently that explicitly explores the ways in which the new forms of fuel transformed the relationships within families and between families and the larger Canadian society in the 1900 to 1950 period.[7] Evidence is compelling, however, that changing energy consumption in Canadian homes, though occurring long distances from the production of the new fuels and in tiny household-sized consumption units was indeed significant to the 'big picture' of energy consumption: whereas industry consumed the lion's share of energy produced in Canada for many decades, buildings (residential and commercial) are now estimated to be responsible for over a third of all power use and therefore greenhouse gas emissions in Canada.[8] And compelling evidence suggests that changes in domestic energy use were directly and indirectly related to broadly based social, economic, cultural, and political change across the country in the twentieth century.

To give just one example of the relationship between energy use in the home and broad social change, Emanuela Cardia has used econometric data from the 1941 census in the United States to suggest a strong correlation between running water in the home and women's workforce participation; she suggests that women's work within the home was simply too labour-intensive without running water for women to be able to leave the home to take on paid work.[9] My own research into Canadian households suggests a similar correlation between oil heating and women's work outside the home; it was not until thermostatically controlled (automated) oil furnaces replaced wood (and to a lesser extent coal) that houses could be left empty for more than a few hours in the winter without bringing on the fairly serious consequences attendant on the fire going out. While historians may disagree about the pace, origins, and exact consequences of changing energy use within North American homes, there is little disagreement that changes in fuel use were related to the 'modernization' of the family and of Canadian society as a whole throughout the twentieth century.[10]

As Canadians contemplate what the 'post-carbon world' of the future is going to look like, there is good reason for historians to look again, and through the lens of fossil fuel and hydroelectricity use, at areas of social history already explored through the lenses of cultural history, the history of the family, and

gender studies, from the rise of mass consumerism and urbanization to women's emergence in the twentieth century as the non-productive consuming housewife, to the implications of women's full-scale proletarianization by the late twentieth century. For these twentieth century phenomena are inseparable from both the new kinds of energy used at home as well as at work. My current research, "Heat Light and Work in Canadian Homes 1900–1950: A Social History of Fossil Fuels and Hydro-electricity" seeks to address this lacunae in Canadian history.

SPATIALIZING FUEL AND ENERGY

How can the spatial evidence from the past – evidence that HGIS (and of course the electricity that powers this historical method or practice[11]) now makes available to and workable by historians – shed light on the great transformation in domestic energy use that arguably comprises one of the most significant changes of the twentieth century? Fossil fuels and hydro-electricity,[12] after all, are perhaps distinguished most clearly from other forms of household fuel by their near-invisibility, particularly when compared to the materiality of and labours involved in (for example) finding, storing, cutting, and burning wood. Coal more closely resembles wood as a fuel, but the experience of using oil or gas or electricity in the home differs significantly from the household fuels of an earlier time: no one, quite literally, knows exactly where the natural gas that automatically heats our homes and fuels our stoves comes from, and few understand even approximately how it ends up there. Electricity, being pure energy, doesn't even have a substance or form to it, so how can it be said to have a spatial dimension? On a more practical level, householders, previously completely responsible for bringing fuels into the home and managing their use themselves, are now warned against, and sometimes legally prohibited from, customizing or even repairing modern furnaces and stoves; handling the fuels that these appliances consume can be highly dangerous. Within the contours of daily life today, it is almost as if these fuels come from nowhere specific, no particular place at all.[13] If hydroelectricity and some fossil fuels (particularly oil and natural gas) have qualities that render them almost invisible at the level of daily household experience, theorists argue that a defining characteristic of fossil fuels (including coal) is their ability, as uniquely concentrated and extremely valuable sources of power, to be transported long distances. As a result, fossil fuels and hydroelectricity are the first fuels that have allowed us, as a species, to successfully transcend our local environments for the first time.

Allowing people to transcend the local may be the single most important innovation brought about by fossil fuel and hydroelectricity, but it is a spatial anomaly so familiar to twenty-first-century Canadians that its significance is easy to miss. As Christopher F. Jones recently explained, drawing on the work of E. A. Wrigley, there have been two kinds of economies in the world: the remarkable new Mineral Economies (based on fossil fuels, and including hydroelectricity) and the Organic Economies that preceded them.[14] They differ in two significant ways. In pre-fossil-fuel Organic Economies, all of the energy in human lives comes from the direct capture of solar energy: people and their animals live and work

by eating sun-powered plants and animals, and by using the muscle power these sustain to find and burn the wood (also a direct product of solar energy) that provides the energy needed for cooking and heating. Land and energy are directly linked; energy use is limited by the carrying capacity of the land. In more densely populated areas, this means that choices must be made about whether land should be used to grow fuel in the form of wood or be used for food. Furthermore, in an Organic Economy, the energy sources available (like wood) are bulky and typically expensive to transport, while "the power from wind or falling water could not be transported at all; the energy from a water wheel had to be used at the river bank and wind was only useful in those places where it blew regularly."[15] Because of these two factors – the limited carrying capacity of the land and the expense of transporting bulky fuel sources – Organic Economies tend to be self-sufficient in, and relatively small consumers of, energy.

A Mineral Economy has very different characteristics. Even though fossil fuels were originally the product of solar energy, vast amounts of time and geological forces have transformed that organic matter into a uniquely dense, uniquely efficient, form of mineral energy:

> Fossil fuels represent massive stocks of energy available for immediate use rather than the direct capture of solar energy flows. And the high energy density of fossil fuel deposits justifies investing in infrastructure to transport energy long distances, thereby separating sites of energy production and consumption. Energy use is no longer necessarily local or limited.[16]

The implications of this transcendence for Canadian societies in the nineteenth and early twentieth centuries were myriad and profound. At the level of economic development, fossil fuel sources allowed industries to locate where the availability of the labour force, or materials for production, or transportation routes were the most favourable, rather than where sources of energy could be found. Manufacturing and transportation expanded rapidly as a result, and with them the Canadian economy as a whole. Resource extraction also became vastly more extensive and profitable: with hydroelectricity and fossil fuels available, mines became safer and exponentially more productive due to lighting, fans, motors, and pumps. Pulp and paper mills could be located in remote areas of the Canadian Shield closer to the pulp resource (a.k.a. spruce trees) and production boomed. Gasoline-powered bulldozers could quickly build roads and railroads that allowed fossil-fuel-powered trucks and trains to transport Canada's resource wealth out of its industrial rural areas to global markets and to bring workers (seasonal, part-time, or full-time) and the food and other goods needed to support them into extraction zones.

Using HGIS to map the growth of the number and extent of the conduits by which fossil fuels and hydroelectricity were generated and moved through Canadian landscapes could potentially deepen our understanding of the geographies of Canada's economic growth as well as its settlement – and its 'unsettlement' or ghost town – history. Finding ways to map increasing fossil fuel and hydroelectric use opens up an important dimension to our understanding of their impact on economic development and the social and cultural changes associated with that. It also opens up the potential of understanding the impact of the production and transportation of fuels on particular environments or locations.

MAPPING SITES OF PRODUCTION AND TRANSPORT

There are other important spatial dimensions to fossil fuel and hydroelectric generation and transmission resulting from their particularly efficient, easily transportable nature. In allowing people to divorce energy sources from other land uses, the new fuels have also allowed the unprecedented exploitation of land for a single purpose – the extraction of a particular mineral, the flooding of land for a hydroelectric dam, or monocropping of a particular valuable vegetable commodity. With the relationship between land, population, and energy fractured by the import of energy sources, people were free to destroy any particular environment, and with it the possibility of sustainable ecologies and communities. Landscapes of destruction that characterize so much of rural Canada, for example, must be counted as part of the 'fall out' of fossil fuel use. As Joy Parr has already shown, simply mapping these areas sacrificed to industrial development offers an interesting counter-narrative to Canada's journey from 'colony to nation.'[17] Far from transcending place, therefore, or erasing the local, fossil fuels and hydroelectricity have allowed the creation of very particular kinds of places, narrowly focussed on particular kinds of activities that are of benefit to people a long way from the sites of production and often profoundly destructive to the local land and peoples.

Although they may be transported to almost anywhere with relative ease, fuels do in fact come from a particular place, and they do travel through particular places. This is particularly clear to the people who live in or near such places. The extraction of fossil fuels, like the generation of hydroelectricity (as well as most types of mining and forestry) is typically accompanied by more or less catastrophic consequences to those environments. A number of historians are studying the political and technological complexities involved in establishing new systems of power in Canada and are emphasizing the devastating impact that these new forms of power have had on the local environments and the human communities (indigenous and immigrant) from which they originate and which they travel through.[18] They are examining the pollution attendant on the extraction and processing of coal, oil, and gas as well as the resulting habitat destruction and danger to fish and animals, including human beings. The oil sands of Alberta, oil and gas processing plants, and both coal and uranium mines are already providing rich sources of evidence about the environmental and social impact of fuel production on local places.[19] Other historians are examining the massive altering of ecosystems following in the wake of hydroelectric dam construction and the attendant flooding required to get the appropriate year-round flows of water needed to turn the turbines. Others are working to understand the impact of building and maintaining transmission lines on local environments, including, most recently, studies of the impact on local human, animal, bird, and fish populations of Agent Orange and other pesticides used, it turns out, by the hydro companies to control plant growth under the lines.[20] Joy Parr's recent work, noted above, has made significant contributions to our understanding of the social, cultural, and environmental contexts of the systems of modernity in the post-war period, including domestic electrification, nuclear power, the militarization of rural landscapes, and the "lostscapes" created by industrial mega-projects.[21] Mapping the time of the extension of the routes by which fossil fuels were created and transported to, from, and through Canadian landscapes, when linked with time series of data about industrial development, environmental devastation, and settlement, will provide an important addition to our understanding of the link between energy production, economic activity, environmental destruction, and sustainable communities in rural Canada.

MAPPING SITES OF CONSUMPTION

Locating the production, extraction, processing, and the transportation of fuels and energy in space and over time – 'spatializing' fuels and energy – provides a deeper understanding of the intricate and vital relationships among individuals, communities, and particular places or environments, relationships which change drastically with changing fuel use. What can spatializing electricity *consumption* add to our understanding of the social history of the early twentieth century?

In the late nineteenth and early twentieth centuries, the introduction of electricity into homes and public spaces was heralded as a wondrous phenomenon, filled with fantasy, magic, and mystery. From early on, it was predicted to provide labour-saving devices that would be

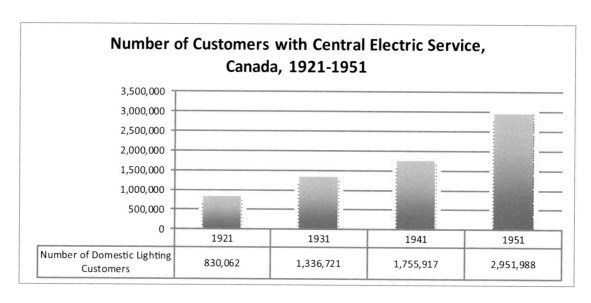

Fig. 12.1. Number of customers with central electric service, Canada, 1921–51. Source: M. C. Urquhart, ed. Historical Statistics of Canada (Toronto: Macmillan, 1968), Series P34–38, 450.

very useful for the homemaker in particular. While historians have been suitably skeptical of early-twentieth-century boosters' extravagant claims about the 'dawn of the new day' that would be ushered in by electricity, they nevertheless widely acknowledge that domestic electrification was an important component of the modernization, most particularly industrialization and urbanization, used to characterize change in Canadian society in the first half of the century.[22] Statistics documenting electricity use confirm this national trend.

As Fig. 12.1 illustrates, the number of Canadian households with electrical service grew substantially between 1921 and 1951, more than tripling from over 830,000 to just under three million households.

Fig. 12.2 confirms the general trend of increasing electrification, but highlights as well the considerable differences amongst the provinces in the number of households

("domestic lighting customers") with lighting. By cross-linking published census data about the number of households in Canada, and Central Electrical Stations' information about the number of residential customers, we see that the differences amongst provinces were not simply the result of total population. While the proportion of households with electric lighting increased in every province, raising the national average from under 50 percent in 1921 to almost 90 per cent in 1951, Figure 12.3 illustrates that the proportion of households with electricity in each province varied widely: while almost half of Canadian homes had electricity by 1921, fewer than a third of those in Prince Edward Island, Nova Scotia, New Brunswick and Saskatchewan had electrical service by that date Even in 1951, when Ontario, Quebec, and British Columbia were approaching 100 per cent domestic electrification, fewer than a third of households in Saskatchewan and Prince

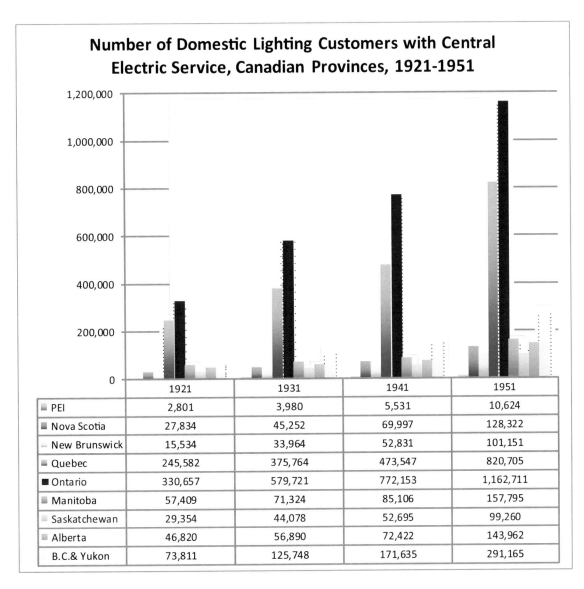

Number of Domestic Lighting Customers with Central Electric Service, Canadian Provinces, 1921-1951

	1921	1931	1941	1951
PEI	2,801	3,980	5,531	10,624
Nova Scotia	27,834	45,252	69,997	128,322
New Brunswick	15,534	33,964	52,831	101,151
Quebec	245,582	375,764	473,547	820,705
Ontario	330,657	579,721	772,153	1,162,711
Manitoba	57,409	71,324	85,106	157,795
Saskatchewan	29,354	44,078	52,695	99,260
Alberta	46,820	56,890	72,422	143,962
B.C.& Yukon	73,811	125,748	171,635	291,165

Fig. 12.2. Number of domestic lighting customers with central electric service, Canadian province, 1921–51. Sources: Canada, Dominion Bureau of Statistics, Census of Industry, 1921, Part I: Statistics Central Electric Stations in Canada, Ottawa, 1923, *Table 8; Dominion Bureau of Statistics, Transportation and Public Utilities Branch, Census of Industry, 1931,* Central Electric Stations in Canada, *Ottawa, 1933, Table 7; Government of Canada, Census of Industry, 1947,* Central Electric Stations in Canada, *DBS Publication [1947], Table 2; Government of Canada,* Central Electric Stations, *1951 (Ottawa, 1953).*

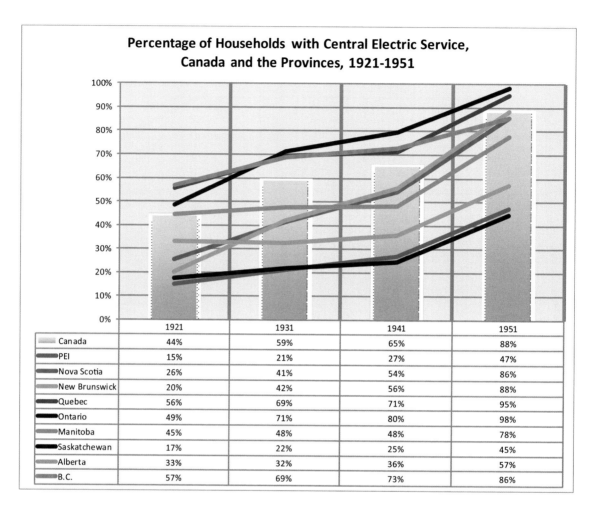

	1921	1931	1941	1951
Canada	44%	59%	65%	88%
PEI	15%	21%	27%	47%
Nova Scotia	26%	41%	54%	86%
New Brunswick	20%	42%	56%	88%
Quebec	56%	69%	71%	95%
Ontario	49%	71%	80%	98%
Manitoba	45%	48%	48%	78%
Saskatchewan	17%	22%	25%	45%
Alberta	33%	32%	36%	57%
B.C.	57%	69%	73%	86%

Fig. 12.3. Percentage of households with central electric service, Canada and the province, 1921–51. Sources: Canada, Dominion Bureau of Statistics, Census of Industry, 1921, Part I: Statistics Central Electric Stations in Canada, Ottawa, 1923, Table 8; Dominion Bureau of Statistics, Transportation and Public Utilities Branch, Census of Industry, 1931, Central Electric Stations in Canada, Ottawa, 1933, Table 7, Number of Customers; Government of Canada, Census of Industry, 1947, Central Electric Stations in Canada, DBS Publication [1947], Table 2; Government of Canada, Central Electric Stations, 1951 (Ottawa, 1953); Census of Canada, vol. 3, Housing and Families, Table 1, Dwellings, households and average number of Persons per dwelling and per household, for Canada and for the provinces, 1881–1951.

Edward Island, and more than a third in Alberta were still without electricity supplied by central power stations.

There were also significant differences within the provinces between rural and urban households. Figure 12.4 documents very low rates of farm household electrification across the country before the 1940's, compared to the national average. Rates of farm electrification grew slowly before the Second World War, with only 19% of farm households equipped with electric lighting by 1941, as compared to 65%

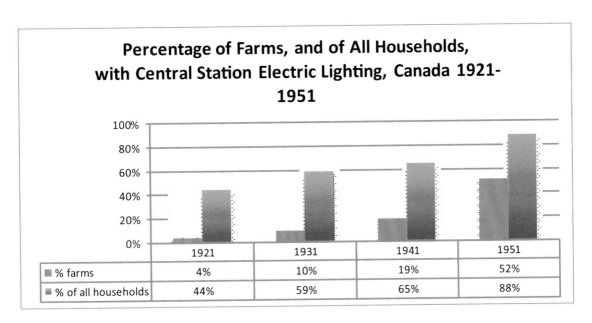

Fig. 12.4. Percentage of farms, and of all households, with central station electric lighting, Canada 1921 to 1951. Sources: 1931 Census of Canada, Agriculture, vol. 8, Table XXVI, p. lxvii "Farm Facilities by Province, 1921–1931"; 1941 Census of Canada, Housing, vol. 9, Table 13 "Lighting facilities in occupied dwellings, 1941"; Farmholdings in Canada, Statistics Canada Series M12-22, http://www.statcan.gc.ca/pub/11-516-x/sectiona/4147436-eng.htm; 1951, Census of Canada, Housing and Families, vol. 3, Table 36.

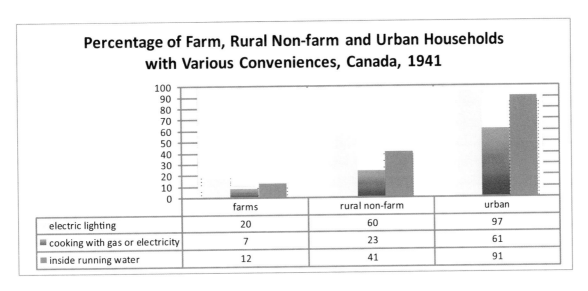

Fig. 12.5. Percentage of Farm, Rural Non-Farm and Urban Households with Various Conveniences, Canada, 1941. Source, 1941 Census of Canada, vol. 9, Housing, Table 18: Occupied Dwellings with Specified Conveniences.

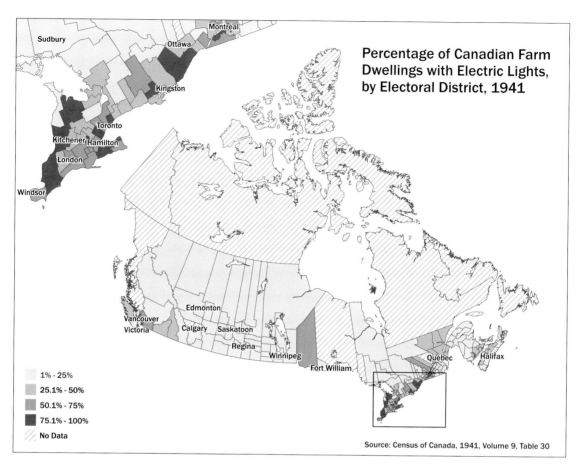

Fig. 12.6. Percentage of Canadian farm dwellings with electric lights by electoral district, 1941. Source: 1941 Census of Canada, vol. 9, Housing, Table 30.

of all households in the country. Though the rate of farm electrification more than doubled in the next decade, by 1951 still only half of farm households were 'electrified'. Fig. 12.5 draws on detailed census comparisons of farm, rural non-farm and urban households available in the 1941 census to illustrate differences in electric lighting, electric and gas cooking, and inside running water available to these different groups.

While graphs do a good job of documenting the regional variations, and the differences that existed amongst farm, non-farm and urban households, the level of complexity that it is possible to convey with a graph has probably been reached when documenting change across nine provinces and four dates. With HGIS, however, it is possible to provide much more detail in a comprehensible way. The three pairs of maps, Figures 12.6 to 12.11 use census data, aggregated by electoral district in 1941 and reported in the Census of Housing that year to map key variations amongst households across the country. The fist pair of

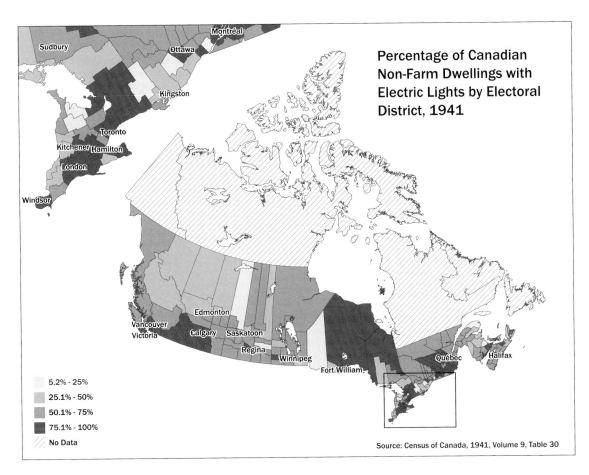

Percentage of Canadian Non-Farm Dwellings with Electric Lights by Electoral District, 1941

Source: Census of Canada, 1941, Volume 9, Table 30

Legend:
- 5.2% - 25%
- 25.1% - 50%
- 50.1% - 75%
- 75.1% - 100%
- No Data

Fig. 12.7. Percentage of Canadian non-farm dwellings with electric lights by electoral district, 1941. Source: 1941 Census of Canada, vol. 9, Housing, Table 30.

maps illustrates striking differences relating to household electrification across the country for each of farm and non-farm households. The comparison between farm (12.6) and non-farm (12.7) households provides a clear, highly detailed, overview of the difference that living on a farm made to household electrification.

Figures 12.8 and 12.9, maps showing the percentage of dwellings with running water, similarly document (in considerably more detail than graphs can accommodate) differences both across the country, and between urban

and rural, in the prevalence of homes with running water.

As late as 1941, most farms were not located on municipal water systems, and lacked the electricity needed to pump water from the local sources of water on which they relied. While some farms had hand pumps inside the house, most did not; instead, they continued to depend on the age-old practice of carrying water into the house in a bucket. This was typically the work of women and children. Figures 12.10 and 12.11 document the different percentages

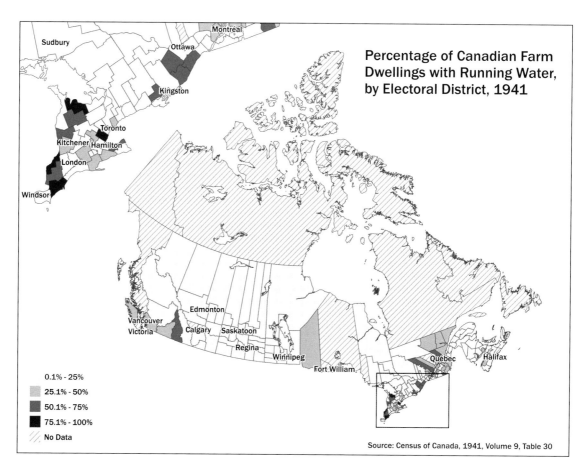

Fig. 12.8. Percentage of Canadian Farm Dwellings with Running Water, by Electoral District 1941. Source: 1941 Census of Canada, vol. 9, Housing, *Table 30.*

of farm and non-farm dwellings that had all four of the 'conveniences' associated with the 'fully modern' house: radio, automobile, telephone, and electric vacuum cleaner. Once again, both the extent of regional variation across the country and the differences between farm and non-farm are striking.

These maps identify in space-specific detail a key element in the history of residential electrification in Canada and one that demands explanation. The explanation lies in the combination of the particular qualities of electricity with the geography of Canadian settlement.

Electricity, whether generated by fossil fuels or water, shares fossil fuel's general, distinctive characteristic of being an energy source efficient enough to warrant transportation over long distances. But electricity has other spatial characteristics that limit where it travels and which have a profound impact on how and why it was (and is) consumed – notwithstanding what seems, at first glance, to be its decidedly insubstantial nature. Perhaps the most unusual of these qualities is that electricity is not only *able to be* transported cheaply, it *must be* transported constantly – the electrons must flow – in

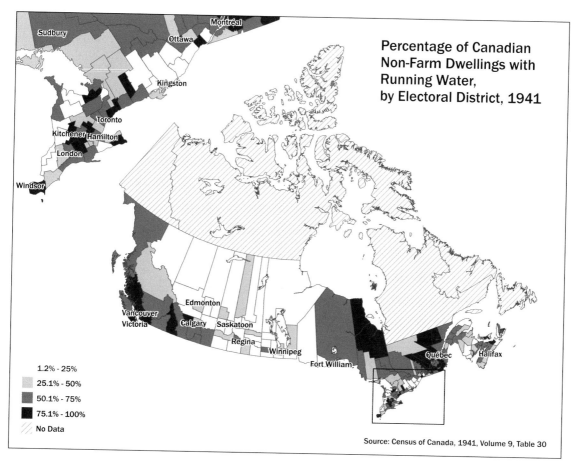

Fig. 12.9. *Percentage of Canadian Non-Farm Dwellings with Running Water, by Electoral District 1941. Source: 1941 Census of Canada, vol. 9,* Housing, *Table 30.*

order for electric energy to exist at all. Because electrical current cannot be stored, it must be consumed as it is generated or be wasted. Creating electricity, particularly when it is generated by means of water power, necessitated the building and maintenance of, and co-ordination amongst, a great many complex machines, wires, and systems. The North American electrical grid, indeed, has been called the world's biggest machine, and it may be the most important for creating the world in which we now live.[23] In spite of all the rhapsodizing of electricity boosters in the first half of the twentieth

century about the modernizing benefits of electrification, one of its most unusual qualities was (and is) the unprecedented scale and integration of construction, generation, and distribution of this form of power, particularly when it comes from water. As historian Harold Platt put it, "Unlike any previous method of supplying light or power, generating electricity involved a highly interdependent and integrated system. Every one of its several complex components had to be in perfect balance with all the others to maintain an electrical current."[24] The light bulb or electrical appliance visible to the

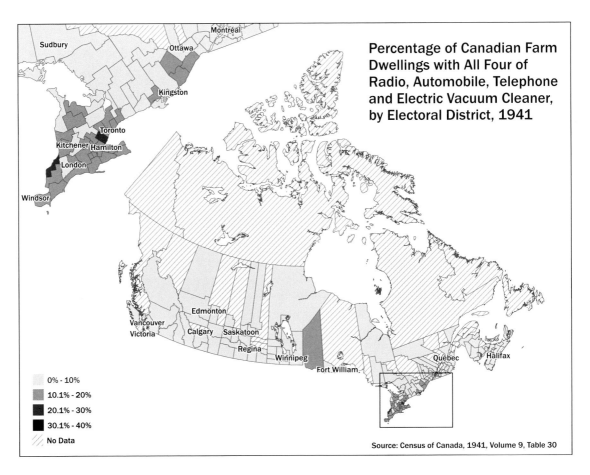

Fig. 12.10. *Percentage of Canadian Farm Dwellings with All Four of Radio, Automobile, Telephone and Electric Vacuum Cleaner, by Electoral District 1941. Source: 1941 Census of Canada, vol. 9,* Housing, *Table 30.*

consumer, therefore, "formed only one part of a symmetrical system that included prime movers, dynamos, regulators, measuring and safety devices, distribution wires, and appliances, all working in instant harmony."[25]

In practice, ensuring "instant harmony" meant that, more so than any other product being consumed in the twentieth century, electrical power was profoundly influenced by economies of scale; without a critical mass, production was uneconomical, and transmission was impossible. When, in 1922, the Dominion Bureau of Statistics noted that British

Columbia had an unusually large proportion of electrical customers, they did not explain this simply by reference to the province's abundant supply of relatively cheap hydroelectric power in itself. Instead, they found the explanation in British Columbia's unusual clustering of population in the Lower Mainland and Vancouver Island.[26] Although British Columbia was not the most urban province, placing well behind Ontario and Quebec in the first half of the century,[27] it did have 41 per cent of its population concentrated in the Lower Mainland and Victoria urban areas. And, significantly, both of

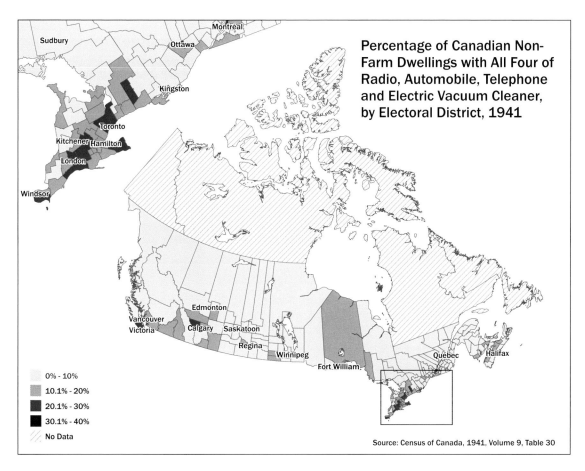

Percentage of Canadian Non-Farm Dwellings with All Four of Radio, Automobile, Telephone and Electric Vacuum Cleaner, by Electoral District, 1941

0% - 10%
10.1% - 20%
20.1% - 30%
30.1% - 40%
No Data

Source: Census of Canada, 1941, Volume 9, Table 30

Fig. 12.11. Percentage of Canadian non-farm dwellings with all four of radio, automobile, telephone and electric vacuum cleaner, by electoral district, 1941. Source: 1941 Census of Canada, vol. 9, Housing, *Table 30.*

these areas were served by the British Columbia Electric Railway Company's urban transit system, which, along with the cities' street lighting systems, demanded the production of massive amounts of hydroelectric power.[28] The other two provinces with the highest rates of electrification were, however, the most urbanized. Three factors, in sum, made Ontario, Quebec, and B.C. 'electrify' earlier and more thoroughly than the other provinces: a large concentration of population, the existence of massive industrial consumers provided by electric street lighting and electric railways within

and between cities, and a power company with a monopoly (or virtual monopoly). These were key supply-related factors in the adoption of electricity by domestic customers.[29]

Before discussing in more detail the difficulties that Canada's vast spaces presented to electricity companies, it is important to note at the outset that the sale of electricity to residential users was dwarfed by sales to such major industrial consumers – not only electric railways and street lighting, but also factories, mills (particularly pulp and paper), mines and smelters – which consumed most of the electricity produced and accounted for most of the profits on a per-customer basis across Canada.[30] In 1933, when national statistics on comparative consumption first became available, the 1.4 million residential customers that year consumed 1.6 million kilowatt hours of electricity, while just under 49,000 industrial customers across the country (3 per cent of the total) were consuming almost ten times the total residential amount, at more than 10 million kilowatt hours.[31] In 1941, residential consumption of electricity comprised only 9 per cent of all electricity consumed, but residential customers generated almost a quarter of the revenue for electrical companies. Even by 1951, when most Canadian households had domestic light, residential customers accounted for over 85 per cent of the customers, consumed only 17 per cent of the total kilowatt hours produced in the country, and continued to generate a disproportionately large portion – 34 per cent – of the revenue.[32]

Even though residential service comprised a minority of customers and total profits, the importance of these small-consumption units, glimpsed in the proportion (or rather the disproportion) of revenues they generated, should not be underestimated. The domestic (i.e., residential or household) market played two very important roles for electrical companies. First, electrical companies did not, by and large, need to make further substantial capital investments to provide electrical lighting for home use: with an electrical company's massive capital costs mostly covered by industrial customers such as electric railways and street lighting, home service provided an important elastic market where returns could be increased simply by persuading more customers to purchase more electricity. For by 1951, the vast

majority of electricity in Canada (over 95 per cent) was being provided by hydroelectricity, which had huge up-front costs in building dams and long-distance transmissions systems, but relatively low running costs compared to fossil-fuel-dependent stations.[33] Because home service did not demand that companies expend much more capital than that already invested to supply industry, residential customers provided the profit icing on the economic cake for many early-twentieth-century electrical producers.

The domestic market, furthermore, provided one more important function. As noted above, one of electricity's most vexing and challenging characteristics is that it cannot be stored. Electrical generating plants and distribution systems had to be created to provide the maximum load that might be needed, even though this maximum might only be reached for a few minutes a day, a week, or even a month. Domestic electrical consumption, with its (at first) tiny but cumulative amounts for cooking, heating, and lighting in the home, was particularly useful in "balancing the load" in off-peak hours, particularly at times when industry did not require massive amounts of energy.[34]

The most significant way that these two related qualities of electricity – the economies of scale and the need to balance production and consumption – influenced Canadian homes in the first half of the century was, as we have seen above, manifested geographically in the stark distinction between farm and non-farm households, or, more specifically, between those who were close enough to an electrical grid to have central station electrical power and those who were not. The electrical grid only serviced areas where a minimum amount of energy consumption could be guaranteed.[35] For although

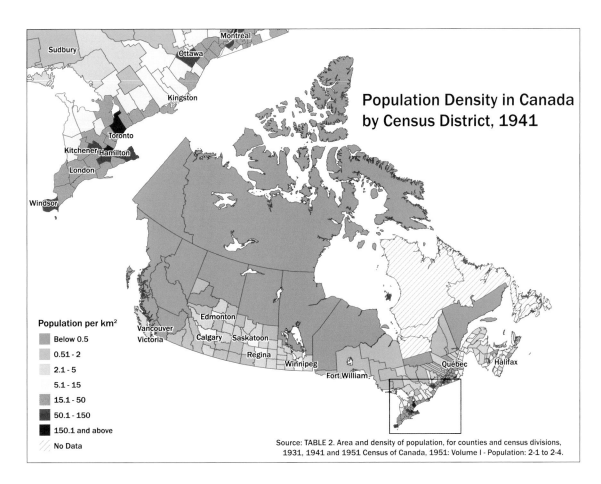

Fig. 12.12. *Population density in Canada by census district, 1941. Source: Table 2. Area and density of population, for counties and census divisions, 1931, 1941, and 1951, Census of Canada, 1951: vol. I – Population, 2-1 to 2-4.*

electricity could theoretically be transported to any place to which a wire could be connected, enough electricity had to be consumed at the end of that wire to make the infrastructure cost-effective to the power provider. Thus, at the same time that electricity was a form of power that transcended the local in some respects, its use was predicated on a very specific kind of local population: high density, and comprised of people who were either highly dependent on electrical energy or could be persuaded to be so. Most regions of Canada, and indeed most

Canadians, met neither of these criteria before 1950. As Figure 12.12 vividly illustrates, the population density of most of Canada was extremely low; indeed Canada, one of the largest countries in the world, still has one of the lowest population densities.

Spatial organization was a vital component in the spread of electrical power in Canada, or, more accurately, in the limitations of its spread. The complexities of electrical generation relating to the economies of scale made it very difficult to provide service to low-density

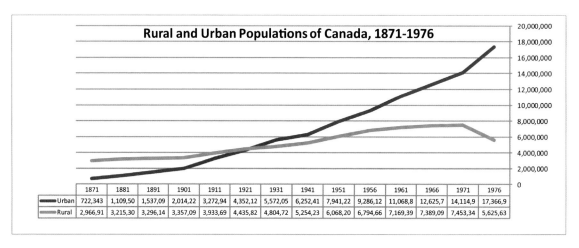

	1871	1881	1891	1901	1911	1921	1931	1941	1951	1956	1961	1966	1971	1976
Urban	722,343	1,109,50	1,537,09	2,014,22	3,272,94	4,352,12	5,572,05	6,252,41	7,941,22	9,286,12	11,068,8	12,625,7	14,114,9	17,366,9
Rural	2,966,91	3,215,30	3,296,14	3,357,09	3,933,69	4,435,82	4,804,72	5,254,23	6,068,20	6,794,66	7,169,39	7,389,09	7,453,34	5,625,63

Fig. 12.13. Rural and urban populations of Canada, 1871–1976. Source: Statistics Canada, Series A67-69, http://www. statcan.gc.ca/pub/11-516-x/sectiona/A67_69-eng.csv.

populations. This was a problem in Canada, where – notwithstanding a national narrative that has identified urban industrialization as the dominant trend in early twentieth century – a majority of the population continued to live in rural areas or in very small towns for the entire first half of the century. Rural Canadians had a strong presence: the number of farms more than doubled between 1871 and 1941, increasing from about 365,000 farms to 733,000.[36] The number of farms began to fall after peaking in 1941, though the acreage of improved farmland continued its increase into the 1970s.[37] While not increasing at the same rate as the urban population, the rural population continued to grow in Canada to 1971, more than doubling during that time from just over three million to almost seven and a half million. As Figure 12.13 illustrates, it was not until 1976 that the rural population of Canada fell for the first time ever.[38] Figure 12.14 evocatively maps the dominance of rural populations across most of the country in 1941.

Although the proportion of people living in rural Canada fell twenty percent in the first half of the century (from 63% in 1901 to 43% in 1951) even as the rural population rose in absolute numbers, census figures tend to overstate the dominance of the urban population; for before 1951, the designation 'rural' or 'urban' had nothing to do with community size or population density. 'Urban' was defined simply as an incorporated municipality, and 'rural' was everything else.[39] This has led to some statistical anomalies, and to a distortion of historians' understanding of Canadians' lived experience in the first half of the twentieth century. While the official statistics designated Canada as more urban than rural from 1921 onwards, It was only in 1941 (when the census confirmed that 46% of the population was rural) that for the first time ever a slight majority of Canadians (51%) lived in communities containing more than a thousand (1,000) people.[40] As Fig. 12.15 indicates, it wasn't until 1961 (when the census confirmed that the rural population had fallen to 39% of Canada's total population) that,

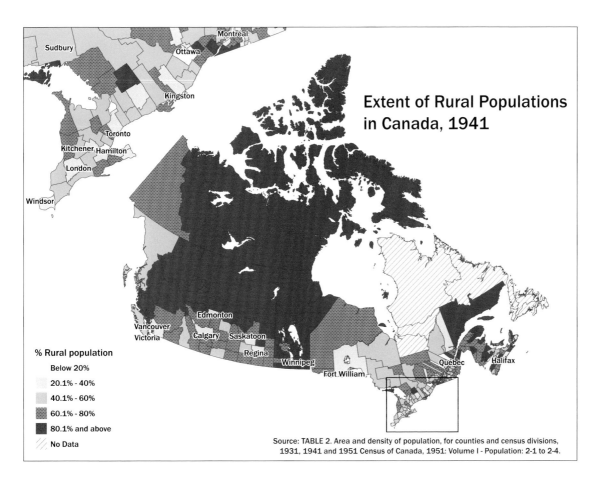

Fig. 12.14. Extent of rural populations in Canada, 1941. Source: Census of Canada, 1951, vol. 1, Population for Counties and Census Divisions, Rural and Urban, 1951 and 1941 (1941 definitions used for both), 14-1 to 14-4.

for the first time ever, a slight majority of Canadians (again, 51%) lived in urban communities larger than five thousand (5,000). Even as late as 1971, when 35% of Canadians were designated rural, 59% still lived in communities smaller than thirty thousand, a population that at least one historian has argued should be the line distinguishing urban from small-town and rural.[41] As the evidence presented here suggests, Canada was a rural and small town place in the first half of the twentieth century, and the country's rurality profoundly influenced the

way Canadian society responded to the changes associated with modernity.[42] This was particularly visible in those elements of modernity closely related to high population densities, and the grids (power, water, sewage) needed to sustain them.

Not only did a smaller percentage of the farm population have access to central electrical service, but, even when they did have it, they used less power. As Fig. 12.16 demonstrates, at the national level, electricity consumption per household was tiny compared to today.

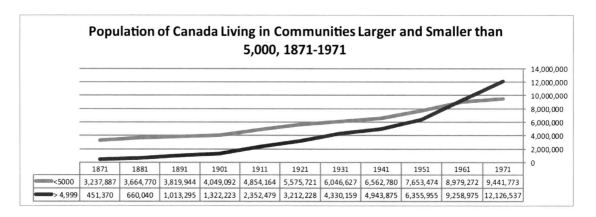

	1871	1881	1891	1901	1911	1921	1931	1941	1951	1961	1971
<5000	3,237,887	3,664,770	3,819,944	4,049,092	4,854,164	5,575,721	6,046,627	6,562,780	7,653,474	8,979,272	9,441,773
> 4,999	451,370	660,040	1,013,295	1,322,223	2,352,479	3,212,228	4,330,159	4,943,875	6,355,955	9,258,975	12,126,537

Fig. 12.15. Population of Canada living in communities larger and smaller than 5,000, Canada, 1871–1971. Source: Statistics Canada, Series A67-69, A70-74, http://www.statcan.gc.ca/pub/11-516-x/sectiona/4147436-eng.htm#1.

But again, as Fig. 12.17 demonstrates, the national trend cloaks huge regional variations. In the case of Manitoba, the variation in consumption was only indirectly related to spatial factors: Manitoba was early in developing its abundant water powers for hydroelectricity, and it dramatically boosted both the number of customers and the volume of their consumption by offering its Winnipeg residents a cheap flat rate for hot water heaters.[43] But low usage in Saskatchewan, Alberta, Nova Scotia, Prince Edward Island, and much of British Columbia was related to distance from urban or industrial areas. When they had any service at all, rural areas were typically served by a disparate collection of rural utilities that did not provide fulltime or full electrical service. Service was subject to frequent interruptions, and was, furthermore, usually limited to when the major industry was not using the service: the evening hours when rural households would require lighting. In British Columbia alone, there were 111 utilities selling electricity outside of the lower mainland area serviced by BC Electric

(which provided 85 per cent of the power in the province), eighty-one of which served fewer than five hundred customers.[44]

The records from the Electric Lighting Departments across the country received hundreds of complaints about rural service when it did exist. While many simply complained about the incomprehensible billing system and improbably high prices, poor service was the target of many letters. In 1928, for example, the Domestic Lighting Department of the Hamilton Cataract Power, Light and Traction Co. reported: "In the last week or two there have been several complaints of low voltage in and around Grimsby.… The Mountain St. residents say they cannot read between the hours of 7 and 10 pm and when it is time to retire they have all kinds of voltage." On July 14, 1930, the company received a complaint from an Aldershot resident, complaining that he "cannot get enough power to run our pump properly and sometimes even the radio will bring in local stations only faintly." Herbert Coates wrote to complain in August 26, 1930, that the voltage

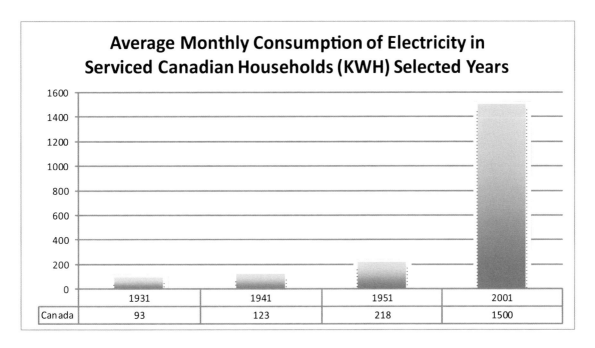

Average Monthly Consumption of Electricity in Serviced Canadian Households (KWH) Selected Years

	1931	1941	1951	2001
Canada	93	123	218	1500

Fig. 12.16. Average monthly consumption of electricity in serviced Canadian households (kWh) in selected years. Sources: Census of Canada, 1951, vol. 3, Housing and Families, *Table 1; Canada, Department of Trade and Commerce, Dominion Bureau of Statistics, Public Utilities Branch,* Index Numbers of Cost of Electricity for Domestic Service and Tables of Monthly Bills for Domestic Service, *Commercial Light and Small Power, 1937: Ottawa, 1938, p. 5; Government of Canada, Census of Industry 1947,* Central Electric Stations in Canada, *DBS Publication [1947] , Table 2; Government of Canada,* Central Electric Stations, *1951 (Ottawa, 1953); 2001 data from Statistics Canada average household size, http:// www12.statcan.ca/english/census01/products/analytic/companion/fam/canada.cfm, and from William J. Hausman, Peter Hertner, and Mira Wilkins,* Global Electrification: Multinational Enterprise and International Finance in the History of Light and Power, 1878–2007 *(Cambridge: Cambridge University Press, 2008), p. 5.*

was so low that he could not run their refrigerator in Hamilton Beach. Receiving yet another complaint about low voltage, the Operating Engineer wrote to the irate customer to explain:

> This district is served by a 40,000 volt line only and for that reason is liable to interruptions for repairs as well as breakdown and the service in consequence is not as reliable in communities served by a duplicate transmission line. I feel sure that the service

compares favourably with rural service furnished by the Commission in other districts.[45]

Complaints from rural residents extended to more than the inconvenience of poor lighting. As Mrs. Robertson complained to BC Electric Co. in 1932 from her home in rural Sooke,

> Out of the 164 pheasants hatched out of the incubator, BC Electric killed 124 by chilling and cutting off the

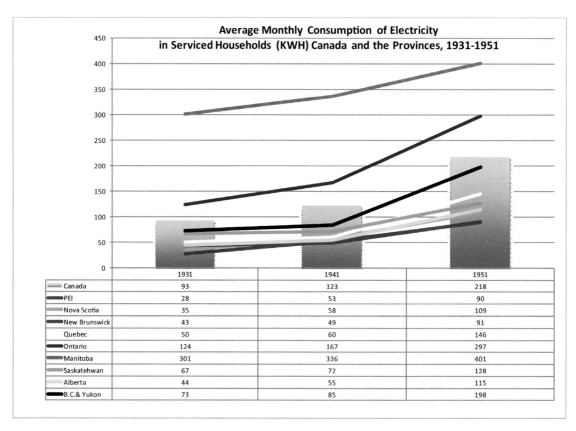

	1931	1941	1951
Canada	93	123	218
PEI	28	53	90
Nova Scotia	35	58	109
New Brunswick	43	49	91
Quebec	50	60	146
Ontario	124	167	297
Manitoba	301	336	401
Saskatchewan	67	72	128
Alberta	44	55	115
B.C.& Yukon	73	85	198

Fig. 12.17. Average monthly consumption of electricity in serviced households (kWh) Canada and the provinces, 1931–51. Sources: Census of Canada 1951, vol. 3, Housing and Families, Table 1; Canada, Department of Trade and Commerce, Dominion Bureau of Statistics, Public Utilities Branch, Index Numbers of Cost of Electricity for Domestic Service and Tables of Monthly Bills for Domestic Service, Commercial Light and Small Power, 1937: Ottawa, 1938, p. 5; Government of Canada, Census of Industry, 1947, Central Electric Stations in Canada, DBS Publication [1947], Table 2; Government of Canada, Central Electric Stations, 1951 (Ottawa, 1953).

current from time to time during the most critical time of their existence.

This means a dead loss to us of $76.20, the cost of the chicks at that age. I had hoped this year to have added quite a lot to a very small income to help with my two little children and the many other overhead expenses in keeping a good home together. But you have ruined all my hopes! I shall never use the brooders again as you will not guarantee electricity at all times. But

I think it is very unfair to coerce us to buy electrical appliances before you can guarantee electrical current at all times.[46]

For their part, electrical companies complained about the reluctance of their rural customers to consume more power -- or to pay their bills on time. The Central Electrical Stations of Canada annual report of 1926 noted with dissatisfaction that across the country, "small plants sell almost entirely to lighting customers, requiring service for a comparatively short period of time each day."[47] Usage was, therefore, limited: "In 90 percent of the distribution areas, the use of electricity is limited to lighting only. In fact, the plant equipment in many cases apparently has been installed for lighting only ... and the operation cannot be regarded as a modern electric utility service"[48]

Electricity in rural areas was not only limited in its extent, therefore, but it was available in only the most rudimentary forms, with the result that, although the "cost of electrical service ... is relatively high, ... the quality of service is relatively poor."[49] People in the largest cities generally had the cheapest rates and consumed the most power, while people in rural areas could pay six or seven times as much for their electric lighting.[50] As the Progress Report for Rural Electrification in 1945 summarized: "rural electrification is not a distinct brand of the utility business; it is an outgrowth of the central station industry – the widening of the areas to which certain average costs are applied."[51] As the authors explained, rural electrification would only succeed if it were "conditional upon urban and suburban service."

It is a simple matter to supply electricity to fifty farms on the borders of a city of 100,000 population; it is a much more difficult matter to supply fifty farms surrounding a town with 2,000 people.... Statistics on service in rural areas will have more meaning, therefore, if accompanied by corresponding statistics on the associated urban services.... Availability, according to modern standards of service, must be general and at the option of the subscriber. The service must be reliable and adequate and it must be offered at rates which will permit the widest possible uses.[52]

Economies of scale were at the heart of the availability of affordable electrical power, and Canada's rural population and vast spaces were disadvantaged. By 1951, Canada's 51% of farms with power line service (over 300,000), compared poorly to about 80 per cent of the farms in the United States.[53] While about 30,000 Canadian farms (less than 10 per cent) generated their own electricity by the use of gasoline engines or windmills, this kind of service was intermittent, inconvenient, and occasional and did not represent what electrical service providers considered the fulltime "modern" service.

CONCLUSION

This chapter has suggested that, in spite of the near invisibility of fossil fuels and hydroelectricity in the contemporary home and in spite of their apparent spatial transcendence of the local, our understanding of Canadian social history can be enhanced by thinking in spatial terms about the production, the transmission/transportation, and the consumption of these new and probably short-lived fuels. A focus on the geographies of where these fuels came from and the places that they travel through and in which they are consumed allows us to tease out key themes, issues, and relationships of power that would otherwise not be apparent. Spatial evidence about fuels can be effectively used to describe the location and the growth of various kinds of fuel; mapping can also be used to illuminate the close *causal ties* between fossil fuel use and changes in the relationships, not only between people and the land, but amongst people as their relationship with the land changed.

The dominance of Canada's rural and small-town population into the 1950s makes the mapping of census district data particularly appropriate at the national level. While mapping data for highly concentrated urban populations poses conceptual problems in a national study,[54] trends within rural areas with low population density emerge clearly from the existing historical record. Data comparing rural and urban, farm and non-farm, places across the country not only highlight regional differences but allow us to 'see' the importance of rural and urban as categories of analysis, along with the more familiar class, race, and gender. Mapping does not replace other sources of evidence about fuel and energy use in Canada, but it does work to illuminate and recontextualize their significance. This kind of research is particularly important in the case of rural Canada, where historians are only just beginning to recognize the unusual rural society that persisted, though in ever-changing relation to urban and industrializing Canada, long after the first large cities began to emerge in the late nineteenth century.

My conclusion, at about the mid-point of my research for this project, is that the economies of scale model so important to modern fuel production, transportation, and consumption is a good one for historians wanting to work with HGIS: the more data we create, the more we can share our data and our questions with other historians, the easier it will be to work more, and more effectively, with historical GIS.

NOTES

1   Harold Innis, cited by Arn Keeling, "'Born in an Atomic Test Tube': Landscapes of Cyclonic Development at Uranium City, Saskatchewan," *The Canadian Geographer* 54, no. 2 (2010): 228–52.

2   For the importance of microhistorical work to environmental history in particular, see R. W. Sandwell, "History as Experiment: Microhistory and Environmental History," in Alan MacEachern and William Turkel, eds., *Method and Meaning in Canadian Environmental History* (Toronto: Thomas Nelson, 2008), 122–36.

3   http://www.canada.uottawa.ca/ccri/CCRI/index.htm.

4   For an interesting discussion of how historians' work differs from social scientists in this regard, see John Lewis Gaddis, *The Landscape of History: How Historians Map the Past* (New York: Oxford University Press, 2002).

5   See, for example, provincial histories of electrical and gas utilities; Christopher Armstrong and H. V. Nelles, *Monopoly's Moment: The Organization and Regulation of Canadian Utilities, 1830–1930* (Toronto: University of Toronto Press, 1986); H. V. Nelles, *The Politics of Development: Forests, Mines and Hydro-Electric Power in Ontario, 1849–1941* (Toronto: Macmillan, 1974); David Breen, *Alberta's Petroleum Industry and the Conservation Board* (Edmonton: University of Alberta Press, 1992).

6   See, for example, Joy Parr, *Megaprojects*, http://megaprojects.uwo.ca/, accessed 21 March 2011; Joy Parr, *Sensing Changes: Technologies, Environments, and the Everyday, 1953–2003* (Vancouver: UBC Press, 2010); Matthew Evenden, *Fish versus Power: An Environmental History of the Fraser River* (Cambridge: Cambridge University Press, 2004); Jean L. Manore *Cross-currents: Hydroelectricity and the Engineering of Northern Ontario* (Waterloo: Wilfrid Laurier University Press, 1999). A number of graduate students are currently working on these issues across the country, and their work will make a welcome addition to this literature.

7   See, for example, Ruth Schwartz Cowan, *More Work for Mother: The Ironies of Household Technology from the Open Hearth to the Microwave* (New York:

Basic Books, 1983); Corrective Collective, *Never Done: Three Centuries of Women's Work in Canada* (Toronto: Canadian Women's Educational Press, 1974); Dianne Dodd, "Women in Advertising: The Role of Canadian Women in the Promotion of Domestic Electrical Technology in the Interwar Period," in Marianne Ainley, ed. *Despite the Odds* (Montreal: Véhicule, 1990); Bettina Bradbury, "Women's Workplaces: The Impact of Technological Change on Working Class Women in the Home and in the Workplace in Nineteenth Century Montreal," in A. Kobayashi, *Women, Work and Place* (Montreal and Kingston: McGill-Queen's University Press, 1994), 27–44; Joy Parr *Domestic Goods: The Material, the Moral and the Economic in the Postwar Years* (Toronto: University of Toronto Press, 1999).

8   "The Commission for Environmental Cooperation (CEC) released a North American report in spring 2008 revealing that buildings (including both commercial and residential) are responsible for 33 per cent of all energy used and 35 per cent of greenhouse (GHG) emissions in Canada." Canadian Environmental Commission (CEC), "Green Building in North America: Opportunities and Challenges," CEC: Montreal, 2008. National Roundtable on the Environment and the Economy, http://www.nrtee-trnee.com/eng/publications/commercial-buildings/section1-commercial-buildings.php, accessed 17 February 2011.

9   Emanuella Cardia, "Household Technology: Was It the Engine of Liberation?" Université de Montréal and CIREQ, current version April 2010, http://www.cireq.umontreal.ca/personnel/cardia.html, accessed 28 March 2011.

10   See, for example, David Nye, *Electrifying America: Social Meanings of a New Technology, 1880–1949* (Cambridge, MA: MIT Press, 1990); Harold L. Platt, *The Electric City: Energy and the Growth of the Chicago Area, 1880–1930* (Chicago: University of Chicago Press, 1991); Ronald C. Tobey, *Technology as Freedom: The New Deal and the Electrical Modernization of the American Home* (Berkeley: University of California Press, 1996); Ronald R. Kline, *Consumers in the Countryside: Technology and Social Change in Rural America* (Baltimore, MD: Johns Hopkins University Press, 2000).

11   For an overview of whether HGIS is a method or a practice, see Anne Kelly Knowles, "GIS and History," in Anne Kelly Knowles, ed. *Placing History: How Maps, Spatial Data and GIS are Changing Historical Scholarship* (Redlands, CA: ESRI Press, 2008), 7–8.

12   Electricity can be generated from a variety of energy forms and fuels, - coal, oil, gas, diesel fuel (fossil fuels) or water. Because the generation, transmission, and waste disposal relating to hydro-electricity differs significantly from fossil fuel-generated electricity. It is, therefore, useful to make this important distinction when discussing electricity in general.

13   "Nowhere-ness" as a characteristic of modernity has been explored by a number of authors, from James Howard Kunstler's *The Geography of Nowhere: The Rise and Decline of America's Man-Made Landscape* (New York: Touchstone, 1993) to Anthony Giddens' *Social Theory and Modern Sociology* (Stanford: Stanford University Press, 1987).

14   Christopher F. Jones, "A Landscape of Energy Abundance: Anthracite Coal Canals and the Roots of American Fossil Fuel Dependence, 1820–1860," *Environmental History* 15 (July 2010): 449–84.

15   Ibid., 451.

16   Ibid., 453.

17   Joy Parr *Megaprojects*, http://megaprojects.uwo.ca/, accessed 2 May 2012.

18   An important study of the impact of another kind of power – hydrogen bombs – on mostly First Nations lands was written by Valerie L. Kuletz, who coined the terms 'zones of sacrifice' to describe lands given over completely to destruction. *The Tainted Desert: Environmental and Social Ruin in the American West* (New York: Routledge, 1998).

19   See, for example, Keeling, "Born in an Atomic Test Tube"; Paul Chastko, *Developing Alberta's Oil Sands: from Karl Clark to Kyoto* (Calgary: University of Calgary Press, 2004); Liza Piper, *The Industrial Transformation of Subarctic Canada* (Vancouver: UBC Press, 2009).

20   Parr, *Sensing Changes*; Evenden, *Fish versus Power*; Manore, *Cross-currents*; John Sandlos, *Hunters at the Margin: Native People and Wildlife Conservation in the Northwest Territories* (Vancouver: UBC Press, 2007).

21 Parr, *Sensing Changes*; Joy Parr, "'Lostscapes': Found Sources in Search of a Fitting Representation," *Journal of the Association for History and Computing* 7, no. 2 (August 2004) http://quod.lib.umich.edu/cgi/t/text/text- [no paginations] idx?c=jahc;view=text;rgn=main;id-no=3310410.0007.101 accessed 2 May 2013.

22 See, for example, Veronica Strong-Boag's *The New Day Recalled: Lives of Girls and Women in English Canada, 1919–1939* (Toronto: Copp-Clark, 1988).

23 Julie Cohn, "Expansion for Conservation: The North American Power Grid," paper presented at the American Society for Environmental History, Phoenix, Arizona, April 2011.

24 Platt, *Electric City*, 7.

25 Ibid., 17.

26 Canada, Dominion Bureau of Statistics, Census of Industry 1922, *Central Electric Stations in Canada, 1922* (Ottawa) , 8–9.

27 In 1931, both Ontario and Quebec had populations that were 41 per cent rural, while British Columbia's was 45 per cent rural. All of the other provinces in that year, as in 1941, were predominantly rural. It was only in 1951 that Nova Scotia and Manitoba joined Ontario, Quebec, and British Columbia in having predominantly urban populations, while Alberta, Saskatchewan, Prince Edward Island, Newfoundland, and New Brunswick remained rural. Table 4, Census Monograph No. 6, *Rural and Urban Composition of the Canadian Population*, Dominion Bureau of Statistics, reprinted from vol. XII, Seventh Census of Canada (Ottawa: J. O. Patenaude, 1938), p. 44. In 1941, see Table 21, *Eighth Census of Canada 1941, vol. II, Population by Local Subdivisions* (Ottawa: Edmond Cloutier, 1944). pp. 232–37. Table 2, Canada. Ninth Census of Canada, 1951, vol. III, pp. 2-1 to 2-4.

28 Canada, *Central Electric Stations in Canada, 1922*, p. 8.

29 By 1931, electric streetcars (invented by Canadian John Joseph Wright in 1883) were running in cities across Canada, including Victoria, Vancouver, Calgary, Edmonton, Regina, Saskatoon, Moose Jaw, Winnipeg, Port Arthur and Fort William, Hamilton, Niagara Falls, St. Catharines, Oshawa, Toronto, Ottawa, Kingston, Montreal, Quebec City, Trois Rivière, Halifax, and St. John's (http://www.ahearn.com/english/contact/history.html, accessed 19 July 2011). For a fuller evaluation of the importance of these factors in the Canadian context, see H. V. Nelles, *Politics of Development*, and, in the American, see Roland Tobey, *Technology as Freedom*.

30 See, for example, Table 4, Dominion Bureau of Statistics, *Census of Industry, 1935, Central Electrical Stations in Canada* (Ottawa, 1937), 16–17.

31 M. C. Urquhart, ed. *Historical Statistics of Canada* (Toronto: Macmillan, 1965), Series P29-33 and P35-38, 450.

32 Urquhart, Historical Statistics, Series P29-33, 34-38, and P39-45, pp. 450–51. In 1951, there were 2.95 million residential customers out of a total of 3.44 million, and they generated about $1.8 million of the total $3.75 million revenue. Industrial customers comprised 3 per cent of the customers, consumed 76 per cent of the electricity, and generated 45 per cent of the revenue.

33 In 1951, only 3.5 per cent of electricity in Canada came from "steam and internal combustion engines," compared to 73 per cent in the United States, which explains why American electricity cost about 70 per cent more than in Canada, on average at 2.8 cents per kWh, compared to 1.6 in Canada. Government of Canada, Central Electric Stations, 1951 (Ottawa: Edmond Cloutier, 1963), 8–9. Hydro-electricity dominated production in Canada, providing more than 90% of electricity to Canadians until 1961, though, by 2007, that proportion had been reduced to 59%. Statistics Canada Series Q85-91,"Electrical generation by utilities and industrial establishments, by type of prime mover, 1919 to 1976," http://www5.statcan.gc.ca/access_acces/archive.action?l=eng&loc=Q85_91-eng.csv accessed 2 May 2013. 2007 figures from Statistics Canada, Electric Power Generation, Transmission and Distribution, Catalogue no. 57-202-X, Table 2, p. 11.

34 On the particular ways in which these two factors were centrally important in the growth of electrification in the United States, Roland Tobey's work is particularly insightful. Tobey, *Technology as Freedom*, chap. 1, 9–40. The necessary integration demanded by the massive grid system presented huge problems when it came to estimating the

bills for residential customers, as domestic customers took years to understand why they should pay less per kilowatt hour when they consumed more, let alone why the electrical companies should charge a flat fee every month for "readiness to supply service." As the Department of Trade and Commerce Branch of the Dominion Bureau of Statistics, Transportation and Public Utilities branch, charged with describing and explaining why everyone across the country was paying such different rates for their electricity, put it, "the cost of electricity is one of the most controversial topics in Canada.… It is seldom that satisfactory explanation is given of the many differences in rates as they exist." *Index Numbers of Cost of Rates for Residence Lighting and Tables of Monthly Bills for Domestic Service, Commercial Light and Small Power* [1931], published by authority of the Hon. H. H. Stevens, MP, Minister of Trade and Commerce (Ottawa, 1932), 1.

35 Ontario, with its large, early and provincially owned system, and with a large rural population crowded into the south-western triangle of the province, had one of the first and best developed rural electrification programs in the world. The minimum number of rural customers needed to put in rural service varied across the country and was a matter of great concern everywhere. See, for example, Keith Fleming, *Power at Cost: Ontario Hydro and Rural Electrification, 1911–58* (Montreal and Kingston: McGill-Queen's University Press, 1992); Frank Dolphin, *Country Power: The Electrical Revolution in Rural Alberta* (Edmonton, Plain Publishing, ca. 1993) and and Nelles, *The Politics of Development.*

36 Series M12-22 Farm holdings, census data, Canada and by province, 1871 to 1971. . http://www.statcan.gc.ca/pub/11-516-x/sectionm/M12_22-eng.csv accessed 2 May 2013.

37 It increased more than 500 per cent between 1871 and 1971, from just over 17 million acres to over 108 million. M34-44 Area of improved land in farm holdings, census data, Canada and by province, 1871 to 1971 (thousands of acres).

38 Statistics Canada, Series A67-69, Rural and Urban Populations of Canada, 1871–1976.

39 See definitions for Series A67-69 at http://www.statcan.gc.ca/pub/11-516-x/sectiona/4147436-eng.htm#Population, accessed 20 July 2011.

40 In 1941, there were 5,653,052 living in urban communities larger than 1,000, and 5,853,603 living in rural communities and urban communities smaller than 1,000. Figures calculated from Statistics Canada Series A67-60 and A70-74. http://www.statcan.gc.ca/pub/11-516-x/section-a/4147436-eng.htm

41 Rex Lucas, *Minetown, Milltown, Railtown: Life in Canadian Communities of Single Industry* (University of Toronto Press, 1971). In 1971, there were 8,812,511 people living in Canadian communities larger than 29,999, and 12,755,799 living in rural communities and urban communities smaller than 30,000. Statistics Canada, Series A67-60 and A70-74.

42 For a discussion of the complex nature of Canada's rural population in the period 1870–1940, and its relation to social and economic change, see R. W. Sandwell, "Missing Canadians: Reclaiming the A-Liberal Past," in Jean-François Constant and Michel Ducharme, eds. *Liberalism and Hegemony: Debating the Canadian Liberal Revolution* (Toronto: University of Toronto Press, 2009), 246–73; "Notes towards a History of Rural Canada, 1870–1940," in John R. Parkins, and Maureen G. Reed, eds. *Social Transformation in Rural Canada: New Insights into Community, Cultures, and Collective Action* (Vancouver: UBC Press, 2013), 21-42; and "Rural Reconstruction: Towards a New Synthesis in Canadian History," *Histoire Sociale/Social History* 27, no. 53 (1994): 1–32.

43 Canada, Department of Trade and Commerce, Dominion Bureau of Statistics, Public Utilities Branch, *Index Numbers of Cost of Rates for Domestic Service and Tables of Monthly Bills for Domestic Service, Commercial Light and Small Power*, 1939 (Ottawa, 1940), 12.

44 Report of the Rural Electrification Committee, p. 25.

45 Correspondence regarding Complaints, Hamilton Cataract Power, Light and Traction Co., RG1-1/1-4; file 3.22 Box 3 [labelled 15-3] RG1-1/1-4, correspondence from W. C. Thomson, 28 August 1928, 14 July 1930, Ontario Hydro Archives.

46 British Columbia Electric Railway fonds, Add Mss 4, vol. 241-539, Mrs. Robertson, 5 September 1932, British Columbia Archives.

47   *Central Electric Stations of Canada,* 1926, p. 12.

48   Province of British Columbia, *Progress Report of the Rural Electrification Committee as of January 24, 1944* (Victoria: Charles Banfield, 1944), 81.

49   *Progress Report,* p. 9.

50   Whereas consumers in Vancouver in 1944 were paying about $2.30 to consume the average 85 kWh per month, consumers in small towns were paying as much as $15.00 per month for the same amount of electricity. The average consumption for B. C. was 85 kWh per month in 1941, and the figures stated here are for 75 kWh per month, which is the closest gradation of consumption for which figures are available. Households in Castlegar were paying $4.05, in Vernon $5.88, in Field $9.00. *Progress Report,* pp. 40–43. In 1935, The Department of Trade and Commerce, Transportation and Public Utilities Branch, *The Index Numbers of Cost of Electricity for Domestic Service and Tables of Monthly Bills for Domestic Service, Commercial Light and Small Power* (Ottawa: 1936) provided a rare comparison of average residential consumption. Their charts of relative urban consumption were organized not by province, or region, but by the size of the city. Cities over 100,000 varied in their monthly consumption between 46 kWh (Montreal) and 385 (Winnipeg), with cities of 50–100,000 consuming between 35

and 213. Few cities between 10 and 50,000 consumed more than 150 kWh, whereas most municipalities "up to 10,000 population" consumed well under 75 kWh (4-6).

51   *Progress Report,* p. 20.

52   *Progress Report,* pp. 40–43.

53   "There are 623,000 occupied farms in Canada [in] 1951, [and] 336,345 farm customers.... Between 1941 and 1951, the number of gasoline engines used for power purposed on Canadian farms increased 9 per cent from 168,225 to 183,041. At the same time the number of electric motors rose 238 per cent from 58,192 to 196,681." Government of Canada, Department of Trade and Commerce, Dominion Bureau of Statistics, *Census of Industry, Central Electric Light Stations in Canada, 1921–1951*; Census of Canada, *Households, 1921–1951* (Ottawa: 1953), 11. The census figures tell a slightly different story: Table 36, *Census of Canada, 1951*, vol. 3, Housing and Families, report that 324,065 farm households out of a total of 629,785 had Power Line Source electricity, with a further 29,995 with power from a Home Generated Source.

54   Ian N. Gregory, "'A Map Is Just a Bad Graph': Why Spatial Statistics Are Important In Historical GIS," in Knowles, *Placing History,* 125.

13

Exploring Historical Geography Using Census Microdata: The Canadian Century Research Infrastructure (CCRI) Project

Byron Moldofsky

Most readers of this volume will be familiar with the historical censuses of Canada and its predecessor colonies and other regional censuses that were conducted in different parts of the country since 1832. Most will have used the published census volumes for research and reference. As in many Western nations, they are a core source of data for historical research for anyone interested in how, when, and where this nation grew. Less well known, however, are the primary source data for the censuses – the manuscript census schedules themselves, which attempted to record every individual man, woman, and child in the country, usually at ten-year intervals. Although the published volumes contain summary tables derived from these records, the manuscripts may be seen as a vast, scarcely tapped resource for accessing the hidden "history of the anonymous." Most of these original census schedules – thousands of pages of handwritten enumerators' log books – have been available only in analog form as images on microfilm in government archives; in recent years, some of the early ones have been scanned and put online for genealogical uses.[1] However, it is still virtually impossible to use these digital pictures of old ledgers for analytical purposes. The "microdata" contained within these manuscript censuses – the information attached to Canadian individuals in a particular place and time – contain enormous potential for spatial history.

The Canadian Century Research Infrastructure (CCRI) project opens a window into this vast repository of information. A collaborative initiative among fourteen academics at seven Canadian universities, as well as private-sector and governmental agencies, the CCRI has produced a database of a sample of manuscript census data for the Canadian decennial censuses from 1911 through 1951.[2] The CCRI team elected to focus its efforts on the census enumerations of the first half of the twentieth century – "Canada's century" – in order to build understanding of the country's transformation from a "sparsely populated, quasi-colony with a highly dependent economy and settlements spread across a vast territory" to "a highly urbanized and industrialized country on the world stage."[3] These resources are being made available to all scholars for research and provide a core framework that can be built upon in the future.

The project takes as its departure point the concept of research infrastructure: the understanding that the resources underpinning major and collaborative research work can often not be built by individual researchers alone but need public support to provide and enable this public good. The establishment of the Canadian Foundation for Innovation (CFI) in 1997 inaugurated a new approach to academic support in Canada by funding this type of research infrastructure. Initially targeting scientific and engineering fields, the CFI awarded its first major grant in the social sciences and humanities sector to the CCRI project. The project has been the subject of several articles, including a 2007 special issue of the journal *Historical Methods*. Additional background information

and user guides are available on the CCRI gateway website.[4] This chapter offers an introduction to the project and the resources it offers for historical GIS research in Canada and outlines some of the considerations in working with the data in a GIS environment.

WHAT IS THE CCRI?

The CCRI project consists of three interrelated components: 1) the database of sample microdata; 2) a contextual database of secondary data sources; and 3) a GIS framework that allows researchers to situate the microdata within their geographic context. The sample microdata database incorporates individual records from the manuscript census schedules as recorded by enumerators for the 1911 through 1951 decennial censuses of Canada. The main census population schedules were sampled at rates of 3–5 per cent by dwelling, depending on year (for example, in 1911 a 5 per cent sample was taken, with all individuals in every twentieth dwelling listed).[5] Every enumeration subdistrict was sampled, which in effect provided geographic stratification. These data were entered via a customized data-entry interface and migrated into a relational database for processing. Standardized coding systems based on internationally recognized models were then applied to allow for both easy and sophisticated data manipulation. The main products are five coded data files, one for each census, suitable for working within a variety of statistical software packages. The original database of about eighty hierarchically linked tables, used for data validation, cleaning, and coding, will also be made available through Statistics Canada for generating more products in the future.

The contextual database includes newspaper and journal writings of the time, documents created by the statistical agencies of the day, and information from the published census volumes. These expressions of public knowledge and opinion about the census reflect the social and cultural context of their times. An important aspect of the project vision was enabling multiple perspectives and views on the censuses from both quantitative and qualitative approaches. Not only the "results" of the census were considered important, but also the "multi-layered political, social, economic and cultural contexts within which enumerations took place."[6] As members of the CCRI team have commented, "whereas the microdata derived from the census manuscripts will enable research into the hidden history of the individual lives of Canadians, the contextual data will make possible inquiries about the making and interpreting of that data and the challenges of the great enumerations of contemporary history."[7] Thematically indexed and coded, the bilingual contextual database is publicly accessible through a user-friendly web interface.[8]

The GIS component of the CCRI is unique among similar census database projects in its mandate to link the historical census microdata to a geographical framework at a level of detail that was as granular as possible.[9] This provides major improvements over the conventional mapping of published aggregate census data by standard census divisions (CDs). First of all, the CCRI sample microdata provides access to individual-level data, allowing sub-populations to be extracted for analysis. Researchers can focus, for example, on children, the elderly, women of child-bearing age, or any other age/gender cohort. Secondly, cross-tabulations of data can be made, such as income levels by ethnic groups. Finally, any of these resulting datasets can then be mapped or analyzed by detailed geographic location. Within the GIS framework, sample data can be aggregated into a variety of geographical "containers," from standard census subdivisions (CSDs) to customized geographical units, according to the needs of the individual researcher. While issues of sample size, data completeness, and locational accuracy limit to some degree the questions that can be addressed, the CCRI data offer immense potential for historical geographic analysis.

The CCRI builds upon similar work that has been done nationally and internationally to tap historical censuses as an underused source of information about the everyday people who worked, lived, and raised families in different places and times. The project is modelled most directly upon the Minnesota Population Center's (MPC) Integrated Public Use Microdata Series project (IPUMS), a publicly accessible repository of individual-level sample data from the U.S. census and American Community Survey, from 1850 to the present.[10] The MPC also coordinates the North Atlantic Population Project, which is producing a harmonized database of complete-count censuses from Canada, Great Britain, Iceland, Norway, and the United States between 1865 and 1900.[11] The only complete-count census microdata freely available for scholarly research, it is the basis for many international comparative research projects. Several of the principals at MPC provided advice and consultation to the CCRI, and the goal of international comparability was always considered in the decision-making process.

Similar projects in Canada were also influential as models for the CCRI, particularly the 1881 Census project based at the Département de Démographie, Université de Montréal, and the Canadian Families project, based at the University of Victoria, which created a similar sample microdata database for the 1901 Census of Canada and from which came several of the CCRI team leaders.[12] Statistics Canada was undoubtedly the most vital external partner on the project, and their Public Use Microdata Files (PUMF) also influenced the data design. Other projects are now underway, which also have commonalities with the CCRI and present the potential for research interaction, including the 1891 census project based at the University of Guelph, and the BALSAC project, jointly operated by four Quebec universities, which links vital statistics records from Quebec.[13]

The CCRI team aspires to make all of the data and documentation as freely available as possible over the web. However, this goal is complicated by the fact that, unlike most of the sister projects referenced above, the CCRI deals with census enumerations whose confidentiality is protected by Statistics Canada's covenant with the Canadian people to respect individual privacy. Historically, an informal tradition developed in Canada whereby individual-level census data were released to the public only a number of decades after their collection. This was formalized in legislation by the *Privacy Act* of 1983, which established the ninety-two-year rule for the release of census enumeration data.[14] In order to compile the CCRI data still protected under the *Privacy Act*, the project established seven high-security data-handling centres across the country,

and all staff were sworn in as deemed Statistics Canada employees.

As a result of these confidentiality rules, and in accordance with different access restrictions, the CCRI employs a range of methods to disseminate its data products:

1. The 1911 sample microdata is freely available publicly as a downloadable microdata file. Full records can be selected and extracted or the full data file downloaded.[15]

2. The 1921–51 sample microdata files are currently available only in Statistics Canada's Research Data Centres (RDCs). Like the anonymized microdata that StatsCan creates for recent censuses, these files contain records of individual people, but stripped of personal identification such as names and addresses. Researchers must apply to the RDC program, justifying their need for the data and explaining the nature of their use. The review process typically takes two months. Once approved, researchers may use data only within these high-security RDC centres and must follow disclosure rules that limit what output can be taken away and what may be published.[16]

3. The CCRI geographic GIS files (in ESRI shapefile format) and reference maps (in PDF) for 1911–51 are freely available for download from the CCRI geography website.[17] Use of GIS files is subject to Statistics Canada licensing restrictions for 2001

Dissemination Area Cartographic boundary files.[18]

4. The contextual database for 1911–51, and its companion user's guide, are publicly available for download on a bilingual website hosted by the Centre interuniversitaire d'études québécoises (CIÉQ).

5. Two supplementary data products are also publicly available on the main CCRI website: 1) the digitized published tables, which include selected tables of aggregate data from the published census volumes 1911–51 in spreadsheet format; and 2) the CCRI user's guide, which documents the data design, variables, and coding used in the CCRI microdata files.

CONSIDERATIONS IN WORKING WITH THE CCRI DATA

The Census of Canada may appear monolithic from afar, but in truth it is a dynamic instrument that has changed every time it has been conducted. These changes are not limited to changes in geographic boundaries and changes in the questions that are asked; they also include the political influences, the administrative organization, and the personnel and methods used to carry out the assembling of data intended to capture a snapshot of the people and activities of the nation. Researchers should be aware of these issues in working with the CCRI data.

In general, the officers of the census were instructed to use the federal electoral divisions and subdivisions at the time to create the enumeration areas for the collection of census data. These adapted electoral divisions were usually termed census *districts* and *subdistricts*. In contrast, the reporting of the census was to be done on the basis of municipal geographic units: typically counties, townships or parishes, cities, towns, and villages; these were usually termed census *divisions* and *subdivisions*.[19] Some literature refers to the collection process as the "census-taking," and the reporting (i.e., the publishing of the results of the census in official printed volumes and bulletins) as the "census-making."

What we learned by delving into the census data and documentation of 1911–51 was how ephemeral this census-taking and census-making infrastructure could be. It was also poorly documented, especially in early years.[20] An advantage of having Statistics Canada as a partner in the CCRI project was that researchers were given access to their historical files. No maps, however, were found of census enumeration areas or even of census subdivision boundaries for the early part of our time period. Some maps appear in the published volumes from 1921 onward, but they vary in their level of detail and are frequently only available at the census-division level. By 1951 census tract maps had been published for the newly established Census Metropolitan Areas, but even in that year census subdivision boundaries in rural areas were still poorly defined.[21]

Also unexpected was the amount of deviation between the system used for census-taking and that used for census-making. From previous experience with census data, specifically 1901, it was assumed that the geographic areas

Fig. 13.1.
Comparison between
census divisions
and subdivisions
in Alberta in
1911 and 1921.
(Source: Canadian
Century Research
Infrastructure. Map
produced by the GIS
and Cartography
Office, Department
of Geography
and Program
in Planning,
University of
Toronto.)

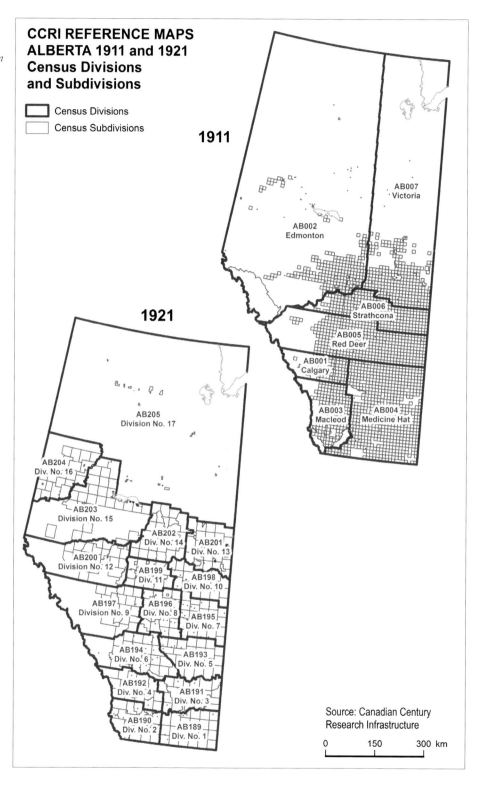

used for these would be coincident, or at least hierarchically nested: individual enumerator's areas and enumeration subdistricts would be grouped together within published census subdivisions and divisions. This turned out to be emphatically not the case. The enumeration areas, subdistricts, and districts used to collect the census data were often significantly different from the subdivisions and divisions used to report the census data.[22]

Another issue that should be understood by users of the CCRI data is the significant differences in the definition of census geography – for example, what comprises a census subdivision in different parts of the country and for censuses taken at different times. Generally, the earlier-settled agricultural areas of the country have a more intensive and persistent system of townships, towns, and villages comprising CSDs. The less-settled northern reaches have more extensive and inconsistent definitions. The most critical concerns about these differences arise when users wish to make comparisons over time. A look at the change in the numbers of census subdivisions between the different census periods puts the magnitude of these changes into perspective. For example, Figure 13.1 shows that in 1911 Alberta had 2,256 CSDs grouped into seven CDs; in 1921 in Alberta the number of CSDs was down to 520, but the number of CDs was up to seventeen.[23] Although the changes between 1921 and later census years were less dramatic, the units of census geography continued to shift and evolve.

Fortunately for users, these incompatibilities are no longer incapacitating. The CCRI geographic team has reconstructed the geographic boundaries of census subdivisions to cover the time period.[24] Using federal electoral atlases published at regular intervals, textual descriptions of census subdivisions in the published census volumes, and contemporaneous historical maps from Library and Archives Canada and other sources, the team created polygons of the components of the CSDs for this period and linked them with the microdata records, allowing researchers to proceed with confidence that data will be matched successfully. The maps and GIS files resulting from this effort, furthermore, can be used to establish compatibility and to enable comparisons between censuses.

WORKING WITH THE CCRI DATA USING GIS

Historical GIS analysis using the CCRI data is similar to other uses of census data in a GIS environment, with a few additional burdens associated with the use of sample data (rather than aggregated data based on the full population).[25] Either selected and grouped microdata or aggregate data from the digitized published tables may be linked to the CCRI GIS files to be mapped or analyzed. It is expected that most users will be selecting and grouping microdata and doing thematic mapping of the results. Mapping the data allows researchers to visualize geographic distributions and patterns and compare these patterns between different variables, places, and times.

To offer a simple example, a researcher might wish to thematically map some of the data from the digitized published tables using the CCRI GIS files. The first step would be to examine the published tables and the documentation describing how they were created.[26]

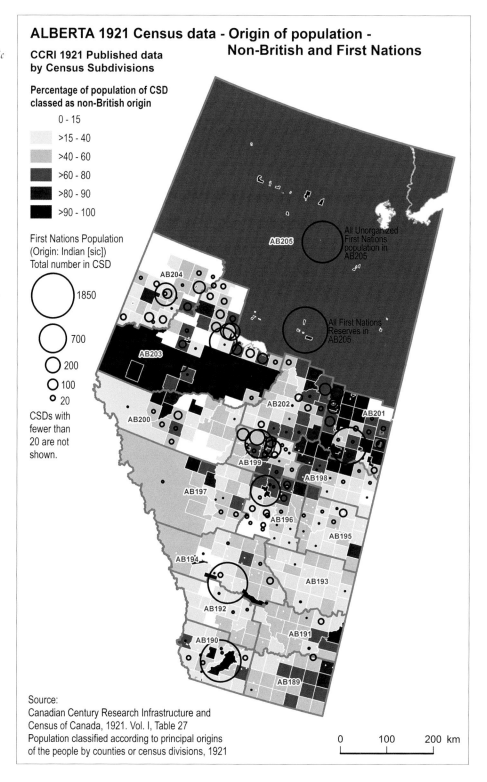

Fig. 13.2.
Example of thematic
map created by
linking CCRI
digitized published
table to CCRI
geographic file.
(Source: Canadian
Century Research
Infrastructure and
Census of Canada,
1921, vol 1, Table
27. Map produced
by the GIS and
Cartography
Office, Department
of Geography
and Program
in Planning,
University of
Toronto.)

ALBERTA 1921 Census data - Origin of population -
Non-British and First Nations

CCRI 1921 Published data
by Census Subdivisions

Percentage of population of CSD
classed as non-British origin

0 - 15
>15 - 40
>40 - 60
>60 - 80
>80 - 90
>90 - 100

First Nations Population
(Origin: Indian [sic])
Total number in CSD

1850

700

200

100

20

CSDs with
fewer than
20 are not
shown.

All Unorganized
First Nations
population in
AB205

All First Nations
Reserves in
AB205

AB205
AB204
AB203
AB202
AB201
AB200
AB199
AB198
AB197
AB196
AB195
AB194
AB193
AB192
AB191
AB190
AB189

Source:
Canadian Century Research Infrastructure and
Census of Canada, 1921. Vol. I, Table 27
Population classified according to principal origins
of the people by counties or census divisions, 1921

0   100   200 km

These aggregate tables present data either at the census-subdivision level or at the census-division level. The CCRI reference maps provide a helpful supplementary reference for reviewing the census geography of the time and place in order to facilitate data selection and download. Both the GIS files and the digitized published tables have a table-specific identification field; for example, "V1T1_1911" is the ID field that appears in the table taken from Volume I, Table 1 of the 1911 published census. This identifier also appears as a field in the map file identifying the corresponding polygon. When the GIS file and the data table are brought into GIS software, the files can be joined using this common variable, and mapping of the aggregate data can proceed. Figure 13.2 presents such a map for Alberta, using the 1921 published data table of "principal origins" of the population, digitized by CCRI. It shows the percentage of the population of "non-British" origins by CSD; it then superimposes proportional circles showing numbers of First Nations people.

Working with the microdata is undeniably more complex. Coding and cleaning of these data has been done, and the documentation for this is available on the CCRI website. Since these tables are based on individual census records, however, all classification and aggregation must be done by the user. Typically, researchers create their own classifications based on their own research needs. For example, to establish a series of household classifications, social historian Gordon Darroch examined each head of household, and the relationship of the other members of the household to him/her (see Table 13.1).[27] This required querying a number of variables in the table – primarily the "Relationship to Head," "Marital Status," and "Gender" fields, for their appropriate codes

– and then creating a new variable to assign a "Household Situation" value. Darroch generated twenty-five of these values, which he then coalesced into eight main "household types," and then combined further into three "household classes." This work may be done interactively in a statistical package like SPSS and then used to generate scripts to document and replicate the process.

The CCRI geographic framework uses as its basic polygon unit the "CCRIUID" – a unique ten-character geographic identifier that forms the key link between the geographic files and the census microdata. It is an alphanumeric variable based on the census division and subdivision numbers as listed in a selected "key" published table. This hierarchical geographical coding system is modelled on Statistics Canada's modern digital boundary file coding. Every record in the CCRI microdata file has been given a CCRIUID, corresponding to the GIS file. Once a researcher's classification has been completed, the data can be aggregated by CCRIUID and then collapsed into any other level of census geography (census subdivision, division, province). The result is a table with, for example, numbers of individuals in households of each of the eight "household types," within each census subdivision. In a GIS package, this table can then be joined to the GIS file of census subdivisions and manipulated to create a map or do spatial analysis, as required. Figure 13.3 illustrates thematic mapping of one adaptation of Darroch's household classifications, which uses four "household classes," developed using the CCRI sample microdata for 1911.

There are a few methodological issues to bear in mind when using the CCRI microdata. Sample size is the most significant issue. The CCRI aimed for a 5 per cent sample in

Table 13.1. Hierarchical classification of household situation based on "Relationship to head."

Level A. Household Situation (25).

Code	Status of head	Living with... Kids*	Kin*	Non-kin*
0.	Married but alone			
0.		no	no	no
	Married			
1.		yes	yes	yes
2.		yes	yes	no
3.		yes	no	yes
4.		yes	no	no
5.		no	yes	yes
6.		no	yes	no
7.		no	no	yes
8.		no	no	no
	Male Single			
9.		yes	yes	yes
10.		yes	yes	no
11.		yes	no	yes
12.		yes	no	no
13.		no	yes	yes
14.		no	yes	no
15.		no	no	yes
16.		no	no	no
	Female Single			
17.		yes	yes	yes
18.		yes	yes	no
19.		yes	no	yes
20.		yes	no	no
21.		no	yes	yes
22.		no	yes	no
23.		no	no	yes
24.		no	no	no

Level B. Household Type (8).

1.	Primary individuals
2.	Single parents
3.	All other lone head with non-kin
4.	Childless couples
5.	Couples with children
6.	All couples with non-kin
7.	Extended household
8.	Extended households with non-kin

Level C. Household Class (3). (Darroch)

1.	Lone head households
2.	Married couple households
3.	Extended households

Level C. Household Class (4). (revised for mapping)

1.	Single individuals
2.	All other single heads
3.	All couples (no kin)
4.	All extended (kin)

Source: Gordon Darroch (personal communication, e-mail, 28 April 2010).

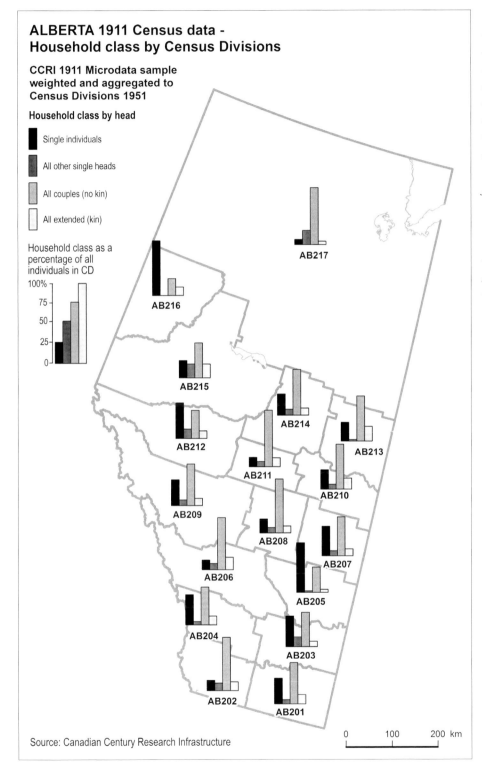

ALBERTA 1911 Census data -
Household class by Census Divisions

CCRI 1911 Microdata sample
weighted and aggregated to
Census Divisions 1951

Household class by head

■ Single individuals

■ All other single heads

■ All couples (no kin)

□ All extended (kin)

Household class as a
percentage of all
individuals in CD

100%
75
50
25
0

AB217
AB216
AB215
AB214
AB213
AB212
AB211
AB210
AB209
AB208
AB207
AB206
AB205
AB204
AB203
AB202
AB201

Source: Canadian Century Research Infrastructure

0   100   200 km

*Fig. 13.3.
Example of thematic
map created by
linking CCRI
classified sample
data for 1911,
weighted and
aggregated to the
CCRI geographic
file of 1951
Census divisions.
(Source: Canadian
Century Research
Infrastructure. Map
produced by the GIS
and Cartography
Office, Department
of Geography
and Program
in Planning,
University of
Toronto.)*

Table 13.2. Sampling rates, sample sizes, and population estimates for each CCRI census year.

Census year	Sampling rate of "Regular dwellings" (by dwelling)	Sampling rates within "Large-dwellings" of single- or multi-unit type		Sample sizes (Ns)		CCRI weighted Canadian population estimate	Published Census population of Canada
		Single-units (by individual)	Multi-units (by unit)	(dwellings)	(individuals)		
1911	5%	10%	25%	73,141	371,373	7,164,970	7,206,643 (+0.6%)
1921	4%	10%	20%	73,505	367,475	8,707,780	8,788,483 (+0.9%)
1931	3%	10%	20%	65,331	348,519	10,277,240	10,376,786 (+0.9%)
1941	3%	10%	20%	80,131	363,935	11,381,887	11,506,655 (+1.1%)
1951	3%	10%	n/a	105,620	443,921	13,915,210	14,009,429 (+0.6%)

Source: Canadian Century Research Infrastructure User's Guide and internal documentation, including Statistics Canada internal certification reports for inclusion of data in Research Data Centres.

1911, 4 per cent in 1921, and 3 per cent for the other three census years. Higher sampling rates were constructed for large dwellings (defined as those with more than thirty residents) to capture the institutional population, so these need to be weighted differently in creating overall population estimates. Sample sizes for all dwellings and the Canadian population are large enough to be adequate for most analyses (see Table 13.2). The standard statistical software packages allow population estimates to be made automatically based on varying sampling rates. Issues will arise, however, when data are extracted for geographic areas with small populations such as census subdivisions, and even for some census divisions. One of the strengths of census microdata is the ability to analyze subgroups of the population.[28] But when a subpopulation is the focus of research, such as a specific age/sex cohort or an occupational group, small sample sizes restrict the analytical capabilities of the data.

Another concern with small sample sizes relates to confidentiality issues for the 1921–51 period. Work may be done on these files at the individual level in the RDC, but results taken out, and especially published results, are subject to disclosure rules designed to maintain confidentiality. Data on individuals cannot be disclosed or mapped; it must first be cross-tabulated by geographic area, and table cell sizes of at least five individuals are required. As a general rule of thumb to ensure representativeness, CCRI guidelines suggest at least twenty sample dwellings or one hundred individuals per geographic unit. Since there are many instances when smaller sample sizes occur, researchers must aggregate cases by geographic areas. Geographic aggregations may be created by standard census geographic areas (e.g., more than one CSD or CD grouped together); by census attributes (e.g., contiguous regions of similar population density grouped together); or by externally defined geography

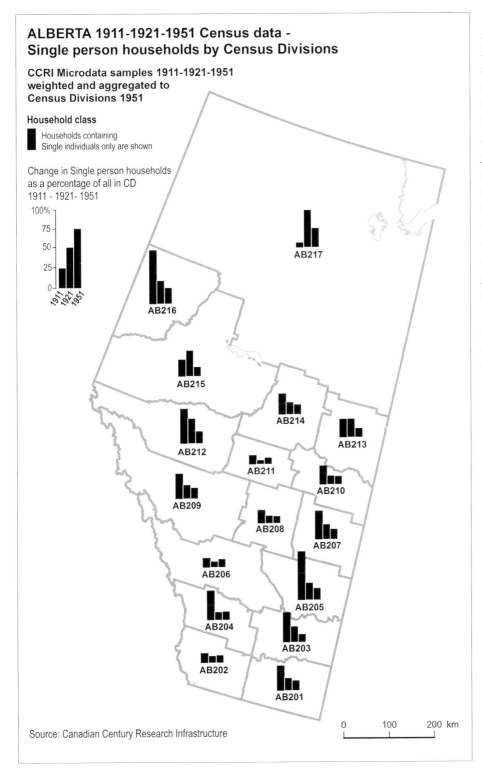

ALBERTA 1911-1921-1951 Census data - Single person households by Census Divisions

CCRI Microdata samples 1911-1921-1951
weighted and aggregated to
Census Divisions 1951

Household class

■ Households containing
Single individuals only are shown

Change in Single person households
as a percentage of all in CD
1911 - 1921- 1951

100%
75
50
25
0

1911 1921 1951

AB217
AB216
AB215
AB214
AB213
AB212
AB211
AB210
AB209
AB208
AB207
AB206
AB205
AB204
AB203
AB202
AB201

0   100   200 km

Source: Canadian Century Research Infrastructure

Fig. 13.4. Example of cross-census comparison map created by linking CCRI classified sample data for three census years, weighted and aggregated to the common base of 1951 Census divisions. (Source: Canadian Century Research Infrastructure. Map produced by the GIS and Cartography Office, Department of Geography and Program in Planning, University of Toronto.)

(areas of similar land characteristics such as urban or rural areas, or ecological zones, may be grouped together).

The decision about how to aggregate will depend upon the framing of the research question, and the amount of aggregation necessary to meet Statistics Canada's disclosure requirements. An example of this is shown in Figure 13.4, which illustrates the change in the proportion of single-person households in Alberta between 1911 and 1951. In order to do this comparison, the classified households for 1911, 1921, and 1951 were aggregated by the census divisions of 1951. Because there were relatively large numbers in the sample for single-person households, the 1921 and 1951 data were approved for disclosure. Other household types such as "single parents" had relatively small numbers; researchers working with these data would be required to combine some of the 1951 CDs in order to meet the Statistics Canada disclosure requirements.

CONCLUSION

Research using the CCRI is a developing story. Since the data are still in the process of being made available to users, this introduction to the CCRI can only hint at its potential. Similar projects have experienced distinct alternating phases of resource creation and methodological innovation.[29] The CCRI is entering such an innovation stage. It invites interested users to follow their own lights in utilizing this resource and to add their voices to the conversation.

The next steps for the CCRI will improve access to the data and other tools. In the long term, online open access to aggregated anonymized data is the goal. In addition to simple access, building tools for online geographic selection, aggregation, and mapping of results will enhance the usefulness of the geographic component of the data. Several additional projects would add considerably to functionality of the CCRI data. One would be to establish cross-census harmonization methods and best practices for these and other historic Canadian census data. Data expansion and enhancement is also a realistic goal; the CCRI geographic data were designed to be extensible and capable of infill. One important objective that could take advantage of other historical GIS work done elsewhere would be to fill in the census geography and supplement microdata geocoding at the intra-urban level for major cities.

The geographic team on the CCRI has always seen the project as part of the larger picture of conducting and enhancing spatial historical studies in Canada. As an infrastructure project, it is only one piece in what should be a shared resource of historical data and mapping made available to scholars and the public. This may be conceived of as part of a "national historical GIS" comparable to those in the United States and Britain. The experience of the CCRI indicates that academics, government. and the private sector can work together to develop such resources. As Gregory and Ell commented on the Great Britain HGIS project, "Although the investment that has been put into this may seem large, compared to the investment in collecting the statistics that they contain, these resources are small, and the potential for reinvigorating [their] use ... is large."[30] Building a partnership to leverage this investment seems like a good place to start.

NOTES

1 Library and Archives Canada holds the early manuscript census schedules on microfilm or fiche and is in the process of scanning them and making them available online, either through their own website or those of genealogical societies (see http://www.collectionscanada.gc.ca/genealogy/022-911.009-e.html). The Automated Genealogy project (http://automatedgenealogy.com/) has made digital images for several censuses available and indexes them by name. Statistics Canada holds all census records for 1921 up to the present securely in its archives.

2 The University of Ottawa was the lead institution for the CCRI project. Other university partners included the Université du Québec à Trois Rivières (Centre interuniversitaire d'études québécoises – CIÉQ); Université Laval (CIÉQ); Memorial University of Newfoundland; University of Toronto; York University; University of Victoria; and the University of Alberta, which continues to host the CCRI archive and distribute its microdata for 1911. Institutional and government partners included: Statistics Canada; Newfoundland and Labrador Statistics Agency; Institut de la statistique Québec; International Microdata Access Group; and Library and Archives Canada. The CCRI was supported by the Canada Foundation for Innovation, the Ontario Innovation Trust, the FCAR Funds (Quebec), and the Harold Crabtree Foundation.

3 Chad Gaffield, "Introduction. Conceptualizing and Constructing the Canadian Century Research Infrastructure," *Historical Methods* 40, no. 2, Special Issue on the Canadian Century Research Infrastructure (2007): 55.

4 See http://ccri.library.ualberta.ca/.

5 Gordon Darroch, Richard D. B. Smith, and Michel Gaudreault, "CCRI Sample Designs and Sample Point Identification, Data Entry, and Reporting (SPIDER) Software," *Historical Methods* 40, no. 2, Special Issue on the Canadian Century Research Infrastructure (2007): 65–75.

6 Chad Gaffield, "Evidence of What? Changing Answers to the Question of Historical Sources as Illustrated by Research Using the Census," in *Building New Bridges: Sources, Methods and*

Interdisciplinarity, ed. Jeff Keshen and Sylvie Perrier (Ottawa: University of Ottawa Press, 2005), 265–74.

7 C. Bellavance, F. Normand, and E. S. Ruppert, "Census in Context: Documenting and Understanding the Making of Early-Twentieth-Century Canadian Censuses," *Historical Methods* 40, no. 2, Special Issue on the Canadian Century Research Infrastructure (2007): 92.

8 See http://ccri-cd.cieq.ca, also accessible through http://ircs1911.cieq.ca.

9 Similarities exist to earlier projects such as the Great Britain Historical GIS project (http://www.gbhgis.org) and to contemporaneous efforts such as the American National Historical Geographic Information System project (http://www.nhgis.org).

10 See http://www.ipums.org.

11 See http://www.nappdata.org.

12 See, respectively, the Programme de recherche en démographie historique (http://www.prdh.umontreal.ca/census) and the Canadian Families project (http://web.uvic.ca/hrd/cfp/). Two publications based on the Canadian Families project were particularly important to the CCRI project and the research examples in this chapter: a special issue on the Canadian Families Project in the Fall 2000 issue of *Historical Methods* (33, no. 4); and Larry McCann, Ian Buck, and Ole Heggen, "Family Geographies: A National Perspective," in *Household Counts: Canadian Households and Families in 1901*, ed. Eric Sager and Peter Baskerville (Toronto: University of Toronto Press, 2007).

13 See the 1891 Census of Canada website (http://www.census1891.ca) and the BALSAC website (http://wordpress.uqac.ca/~balsac-en/) respectively.

14 Gaffield, "Introduction," 57.

15 The NESSTAR data download service for the CCRI database for 1911 is accessible at: http://nesstar2.library.ualberta.ca/webview/.

16 See the Statistics Canada Research Data Centres website at: http://www.statcan.gc.ca/rdc-cdr/index-eng.htm.

17　See the CCRI Geographic Files Access website at: http://ccri.library.ualberta.ca/endatabase/geography/gislayers.

18　See the Statistics Canada Dissemination Area files documentation at: http://www.statcan.gc.ca/bsolc/olc-cel/olc-cel?lang=eng&catno=92F0169X.

19　According to the instructions to the Officers of the 1911 Census, and the Census Act of Canada (Revised Statutes of Canada, 1906, c. 68), *census districts* "shall correspond as nearly as may be with the existing electoral divisions and subdivisions; and the said Census districts may be further divided into such sub-districts as the Minister of Agriculture may direct as units of enumeration for Census purposes." Census sub-districts "shall ordinarily consist of townships, parishes, cities, towns and incorporated villages." In the published census volumes the use of the terms "division" and "district" are used differently in different years, as summarized below:

1911: Census districts and subdistricts used for all published tables; Electoral divisions and subdivisions used in instructions to Officers.

1921: Census divisions and subdivisions used for the majority of published tables; Electoral districts and subdistricts used for some published tables. Census districts and subdistricts referred to for enumeration purposes in instructions to officers.

1931: Census divisions and "municipality, township or subdivision" used for all published tables.

1941: Census divisions and subdivisions used for the majority of published tables; Electoral districts used for some published tables.

1951: Census divisions and subdivisions used for the majority of published tables; Electoral districts used for some published tables.

20　Byron Moldofsky, "The CCRI Geographic Files: Introduction and Examples," in *The Dawn of Canada's Century: Hidden Histories*, ed. Gordon Darroch (Montreal and Kingston: McGill-Queen's University Press, 2013) (forthcoming).

21　Canada. Department of Trade and Commerce, Dominion Bureau of Statistics, *Ninth Census of Canada, 1951* (Ottawa: Queen's Printer, 1953), vol. 1, Appendix A, "Reference Maps."

22　Marc St-Hilaire et al., "Geocoding and Mapping Historical Census Data: The Geographical Component of the Canadian Century Research Infrastructure," *Historical Methods* 40, no. 2, Special Issue on the Canadian Century Research Infrastructure (2007): 80.

23　Ibid., 78.

24　St-Hilaire et al., "Geocoding and Mapping Historical Census Data".

25　A discussion of CCRI sample designs and their implications is included in Darroch, Smith, and Gaudreault, "CCRI Sample Designs." A further explanation is available in the online user's guide at http://ccri.library.ualberta.ca/endatabase/sampling/sampledesign/.

26　The digitized published tables are available in the online user's guide at http://ccri.library.ualberta.ca/endatabase/geography/digitizedpublictables/index.html.

27　Gordon Darroch, personal e-mail communication, 28 Apr 2010. Darroch analyzes numbers of individuals in different household situations rather than numbers of households themselves. So do the examples here. See also Gordon Darroch, "Household Experiences in Canada's Early 20th Century Transformation," in *The Dawn of Canada's Century: Hidden Histories,* ed. Gordon Darroch (Montreal and Kingston: McGill-Queen's University Press, 2013) (forthcoming).

28　Angela Dale, Ed Fieldhouse, and Clare Holdsworth, *Analyzing Census Microdata* (New York: Oxford University Press, 2000), 30.

29　Ian N. Gregory and Paul S. Ell, *Historical GIS: Technologies, Methodologies, and Scholarship* (Cambridge: Cambridge University Press, 2008), 189.

30　Ibid.

Appendix A:
Historical GIS Studies in Canada

BOOKS AND ARTICLES

Beard, Colleen. "Using Google Earth Technologies to Enhance Digital Map Collections: Visualizing the Historic Welland Canals." *Association of Canadian Map Libraries and Archives Bulletin*, 140 (2012): 32–38.

Bellavance, Claude et Marc St-Hilaire, eds. *Le fait urbain au Québec*. Sainte-Foy: Les Presses de l'Université Laval (in press).

Black, Fiona, and Bertrum MacDonald. "Geographic Information Systems: A New Research Method for Book History." *Book History* 1 (1998): 11–31.

———. "HGIS of Print Culture in Canada." *Historical Geography* 33 (2005): 154–56.

———. "Using GIS for Spatial and Temporal Analyses in Print Culture Studies: Some Opportunities and Challenges." *Social Science History* 24, no. 3 (2000): 505–36.

Brassard, Paul, Kevin A. Henry, Kevin Schwartzman, Michele Jomphe, and Sherry Olson. "Geography and Genealogy of the Human Host Harbouring a Distinctive Drug Resistant Strain of Tuberculosis." *Infection, Genetics and Evolution* 8 (2008): 247–57.

Buck, Ian, David Jordan, Shaun Mannella, and Larry McCann. "Reconstructing the Geographical Framework of the 1901 Census of Canada." *Historical Methods* 33, no. 4 (2000): 199–205.

Buck, P. Louise. "Snapshots of Change: Applying GIS to a Chronology of Historic Charts of the St. Marys River, Ontario and Michigan." *Prairie Perspectives* 12 (2009): 1–22.

Courville, Serge and Robert Garon, eds. *Québec, ville et capitale*. Sainte-Foy: Les Presses de l'Université Laval, 2001.

Courville, Serge, Jean-Claude Robert, and Normand Séguin. *Atlas historique du Québec. Le pays laurentien au XIXe siècle. Les morphologies de base*. Sainte-Foy: Les Presses de l'Université Laval, 1995.

Cunfer, Geoff. "Scaling the Dust Bowl." In *Placing History: How Maps, Spatial Data, and GIS are Changing Historical Scholarship*, 95–122. Redlands, CA: ESRI Press, 2008.

Dean, Joanna. "The Social Production of a Canadian Urban Forest." In *Environmental and Social Justice in the City: Historical Perspectives*, edited by Richard Rodger and Genevieve Massard-Guilbaud, 67–88. Isle of Harris, UK: White Horse Press, 2011.

Dufaux, François, and Sherry Olson. "Reconstruire Montréal, rebâtir sa fortune." *Revue de Bibliothèque et Archives nationales du Québec* 1 (2009): 44–57.

Dufaux, François, Matthieu Lachance, Jean Guérette, and Marc-André Bouchard-Fortin. *Le monastère des Augustines de l'Hôtel-Dieu de Québec : une histoire opérationnelle des bâtiments*. Québec: École d'architecture de l'université Laval, Ministère de la culture, des communications et de la Condition féminine du Québec, 2008.

Dunae, Patrick, John Lutz, Donald Lafreniere, and Jason Gilliland. "Making the Inscrutable, Scrutable: Race and Space in Victoria's Chinatown, 1891." *B.C. Studies* 169 (2011): 51–80.

Dunae, Patrick, John Lutz, Donald Lafreniere, and Jason Gilliland. "Race and Space in Victoria's Chinatown, 1891." In *Home Truths: Highlights from BC History*, edited by Richard Mackie and Graeme Wynn, 206–39. Madeira Park, BC: Harbour Publishing, 2012.

Fortin, Marcel, and Janina Mueller. "The Library as Research Partner and Data Creator: The Don Valley Historical Mapping Project." *Journal of Map and Geography Libraries* 9, no. 1–2 (2013): 157–74.

Frenette, Yves, Étienne Rivard, and Marc St-Hilaire, eds. *La francophonie nord-américaine*. Québec: Les Presses de l'Université Laval, 2013.

Gauvreau, Danielle, and Sherry Olson, "Mobilité sociale dans une ville industrielle nord-américaine: Montréal, 1880–1900." *Annales de démographie historique* 1 (2008): 89–114.

Gentilcore, R Louis, ed. *Historical Atlas of Canada: The Land Transformed, 1800–1891*. Vol. 2. Toronto: University of Toronto Press, 1993.

Gilliland, Jason. "The Creative Destruction of Montreal: Street Widenings and Urban (Re)Development in the Nineteenth Century." *Urban History Review / Revue d'histoire urbaine* 31, no. 1 (2002): 37–51.

———. "Fire and Urban Morphogenesis: Patterns of Destruction and Reconstruction in Nineteenth-Century Montreal." In *Flammable Cities: Urban Conflagration and the Making of the Modern World*, edited by Greg Bankoff, Uwe Lubken, and Jordan Sand, 190–211. Madison: University of Wisconsin Press, 2012.

———. "Modeling Residential Mobility in Montreal, 1860–1900." *Historical Methods* 31, no. 1 (1998): 27–42.

———. Muddy Shore to Modern Port: Redimensioning the Montréal Waterfront Time-space." *Canadian Geographer* 48, no. 4 (2004): 448–72.

———. "Redimensioning the Urban Vascular System: Street Widening Operations in Montreal, 1850–1918." In *Transformations of Urban Form: From Interpretations to Methodologies in Practice*, edited by Roberto Corona and Gian Luigi Maffei. Florence: International Seminar on Urban Form, 1999.

Gilliland, Jason, and Mathew Novak. "On Positioning the Past with the Present: The use of Fire Insurance Plans and GIS for Urban Environmental History." *Environmental History* 11, no. 1 (2006): 136–39.

Gilliland, Jason, and Sherry Olson. "Claims on Housing Space in Nineteenth-Century Montreal." *Urban History Review / Revue d'histoire urbaine* 26, no. 2 (1998): 3–16.

———. "Montreal, l'avenir du passé." *GEOinfo* (January–February 2003): 5–7.

———. "Residential Segregation in the Industrializing City: A Closer Look." *Urban Geography* 31, no. 1 (2010): 29–58.

Gilliland, Jason, Sherry Olson, and Danielle Gauvreau. "Did Segregation Increase as the City Expanded? The Case of Montreal, 1881–1901." *Social Science History* 35, no. 4 (2011): 465–503.

Guérard, François. "Les populations hospitalisées à l'Hôtel-Dieu de Québec et à l'Hôtel-Dieu Saint-Vallier de Chicoutimi de 1881 à 1942." *Cahiers québécois de démographie* 41, no. 1 (2012): 55–85.

Hayek, Michael, Mathew Novak, Godwin Arku, and Jason Gilliland. "Building a Brownfield Database with Historical GIS." *Planning Practise & Research* 25, no. 4 (2010): 461–75.

Lafreniere, Don, and Doug Rivet. "Rescaling the Past through Mosaic Historical Cartography." *Journal of Maps* v2010 (2010): 417–22.

Novak, Mathew, and Jason Gilliland. "Buried Beneath the Waves: Using GIS to Examine the Social and Physical Impacts of a Historical Flood." *Digital Studies* 1, no. 2 (2010): 25.

———. "Trading Places: A Historical Geography of Retailing in London, Canada." *Social Science History* 35, no. 4 (2011): 543–70.

Olson, Sherry. "Ethnic Partition of the Work Force in 1840s Montreal." *Labour/Le Travail* 53 (2004): 159–202.

Olson, Sherry, and Patricia Thornton. *Peopling the North American City: Montreal 1840–1900*. Montreal and Kingston: McGill-Queen's University Press, 2011.

Olson, Sherry, Kevin Henry, Michele Jomphe, Paul Brassard, and Kevin Schwartzman. "Tracking Tuberculosis in the Past: The Use of Genealogical Evidence." *Journal of Historical Geography* 36, no. 3 (2010): 327–41.

Rueck, Daniel, "A Common Tragedy: Enclosure in Kahnawá:ke Mohawk Territory 1850–1900," *Canadian Historical Review*, forthcoming.

Schwartzman, Kevin, et al. "Geomatics as a Tool for Bridging the Cultures of Research." In *The Added Value of Scientific Networking: Perspectives from the GEOIDE Network Members 1998–2012*, edited by N Chrisman and M Wachowicz, 75–100. Quebec: GEOIDE, 2012.

Séguin, Normand, Serge Courville and Jean-Claude Robert. *Le Pays laurentien au XIXe siècle: les morphologies de base*. Sainte-Foy: Les Presses de l'Université Laval, 1995.

St-Hilaire, Marc, Byron Moldofsky, Laurent Richard, and Mariange Beaudry. "Geocoding and Mapping Historical Census Data: The Geographical Component of the Canadian Century Research Infrastructure." *Historical Methods* 40, no. 2 (2007): 76–91.

———. "Geocoding and Mapping Historical Census Data: The Geographical Component of the Canadian Century Research Infrastructure." *Historical Methods* 40, no. 2 (2007): 76–91.

St-Hilaire, Marc, Alain Roy, Mickaël Augeron, and Dominique Guillemet, eds. *Les traces de la Nouvelle-France au Québec et en Poitou-Charentes Québec*. Québec: Les Presses de l'Université Laval, 2008.

Sweeny, Robert C.H. "Risky spaces: the Montreal Fire Insurance Company, 1817–20." In *The Territories of Business*, Claude Bellavance and Pierre Lanthier, eds., 9–23. Sainte-Foy: Les Presses de l'Université Laval, 2004.

———. "Property and Gender: Lessons from a 19th-Century Town." *Journal of Canadian Studies* (London) 22 (2006/7): 9–34.

Sweeny, Robert, and Sherry Olson. "MAP: Montréal l'avenir du passé, Sharing geodatabases yesterday, today and tomorrow." *Geomatica* 57, no. 2 (2003): 145–54.

Thornton, Patricia, and Sherry Olson. "Mortality in Late Nineteenth-Century Montreal: Geographic Pathways of Contagion." *Population Studies* 65, no. 2 (2011): 157–81.

WEBSITES

Don River Valley Historical Mapping Project – http://maps.library.utoronto.ca/dvhmp/index.html

GEORIA: Georeferenced Databases for Accessing Historical Data – http://mercator.geog.utoronto.ca/georia/home.htm

Historical Atlas of Canada – http://www.historicalatlas.ca

The Historic Welland Canals: A Virtual Tour – http://brocku.ca/library/collections/maplibrary/welland-canals-exhibit

Imag(in)ing London – http://www.historicalgis.com/

Inventaire des lieux de mémoire de la Nouvelle-France- http://inventairenf.cieq.ulaval.ca:8080/inventaire/index.jsp

Montréal, l'avenir du passé – http://www.mun.ca/mapm/

Population et histoire sociale de la ville de Québec (PHSVQ) – http://www.phsvq.cieq.ulaval.ca/

Regional Environmental History Atlas of South Central Ontario – http://www.trentu.ca/library/madgic/reha.htm

Vieux-Montréal/Old Montreal – http://vieux.montreal.qc.ca/

viHistory (Victoria HGIS) – http://vihistory.ca/

Select Bibliography

ARCHIVAL SOURCES

Bibliothèque et Archives nationales du Québec à Montréal (BAnQ).

Dossiers de grand format (TP11, S2, SS2, SS42), Expropriations.
Fonds Cour supérieure, District judiciaire de Montréal, Dossiers des faillis (TP11, S2, SS10, SSS1).
Fonds Tutelles et curatelles (CC601, S1); Fonds Greffes de notaires (CN601).

Bibliothèque et Archives nationales du Québec à Québec

CA301, S45, no. 3. McCarthy, J. "Plan and Survey of Cap-au-Diable in Kamouraska." 1781.
P407, famille Taché, no. 2. J. Hamel, "Plan de la seigneurie de Kamouraska." 1826.

Bibliothèque générale de l'Université Laval, Centre d'information géographique et statistique

Mosaïque aérienne du Québec, 1929, F82.
Ministère des Ressources naturelles, Gouvernement du Québec, Q74316-94-95-96 and Q74313-133-134-135 (1974); Q85913-143 (1985).

British Columbia Archives.

A/E/Or3/C15, Alexander Campbell, "Report on the Indians of British Columbia to the Superintendent General of Indian Affairs," October 19, 1883.
British Columbia Electric Railway fonds.
GR 428. Vancouver Island. Police and Prisons Department, Esquimalt, 1862–1868.

National Air Photo Library (Ottawa, Canada), A11660-290, 1948.

Ontario Hydro Archives. Correspondence regarding Complaints, Hamilton Cataract Power, Light and Traction Co. RG1-1/1-4.

Presbyterian Church in Canada Archives. Knox Presbyterian Church Toronto collection.

Ville de Montréal. Service des archives. Rôles d'évaluation et feuilles de route 1847–1915.

PUBLISHED PRIMARY SOURCES

Adams, J. Map of the City and Suburbs of Montreal, 1825.
Atlas of Province of Prince Edward Island, Canada and the World, 1927. Toronto: Cummins Map Company, 1928.
"Aviator to Detect Forest Fires." *American Forestry* (September 1915): 914–15.
Blaiklock, Frederick W., E. H. Charles Lionais and Louis-Wilfrid Sicotte, Cadastral Plans, City of Montreal (Montréal, E.H.C. Lionais, 1880).
Cane, James. Topographical and Pictorial Map of the City of Montreal. 1:5100 (Montreal, Robert W. S. Mackay, 1846).
City of Victoria. 1871 Municipal Census. Online database on Patrick Dunae, ed. Vihistory.ca <http://vihistory.ca>.
Goad, C. E. Atlas of the City of Montreal showing all Buildings and Names of Owners (C. E. Goad Ltd., 1881).
Langevin H. L., in Canada. Sessional Papers. 1872. Vol. 6, No. 10: "British Columbia. Report by the Hon. H. L. Langevin, C. B., Minister of Public Works."
Lovell, John, and Sons. City of Montreal Directory. Lovell's Business and Professional Directory of the Province of Quebec, 1902–03. Montreal: Lovell, 1847–1901.
Powell, I. W. in Canada. Sessional Papers. Department of Indian Affairs. Annual Report, 1877. 32–34.
Williams, R. T., ed. The British Columbia Directory for the year 1882–83. Victoria, 1882.
———. Williams' illustrated official British Columbia directory, 1892. Victoria, B. C.: Colonist Printers, 1892.

Adams, Peter, and Colin Taylor. *Peterborough and the Kawarthas*, 3d ed. Peterborough: Trent University, 2009.

Agarwal, Pragya. "Operationalising 'Sense of Place' as a Cognitive Operator for Semantics in Place-Based Ontologies." In *Spatial Information Theory: International Conference, COSIT 2005: Proceedings*, edited by A. G. Cohn and D. M. Mark, 96–114. New York: Springer, 2005.

Akenson, Donald H. *The Irish in Ontario: A Study in Rural History*. Montreal and Kingston: McGill-Queen's University Press, 1984.

Anderson, Kay. *Vancouver's Chinatown: Racial Discourse in Canada, 1875–1980*. Montreal and Kingston: McGill-Queen's University Press, 1991.

Angus, James. *A Respectable Ditch: A History of the Trent-Severn Waterway, 1833–1920*. Montreal and Kingston: McGill-Queen's University Press, 1988.

Annual Report of the Trustees and Deacons Court of Knox Church For Congregational Year 1883. Toronto: Globe Printing Co., 1884.

Aporta, Claudio, and Eric Higgs. "Satellite Culture: Global Positioning Systems, Inuit Wayfinding, and the Need for a New Account of Technology." *Current Anthropology* 46, no. 5 (2005): 729–53.

Apple, J. D., and P. D. Manion. "Increment Core Analysis of Declining Norway Maples, Acer platanoides." *Urban Ecology* 3, no. 4 (1986): 309–21.

Armstrong, Christopher, Matthew Evenden, and H. V. Nelles. *The River Returns: An Environmental History of the Bow*. Montreal and Kingston: McGill-Queen's University Press, 2009.

Armstrong, Christopher, and H. V. Nelles. *Monopoly's Moment: The Organization and Regulation of Canadian Utilities, 1830–1930*. Toronto: University of Toronto Press, 1986.

———. *Revenge of the Methodist Streetcar Company: Sunday Streetcars and Municipal Reform, 1888–1897*. Toronto: P. Martin Associated, 1977.

"At Caughnawaga, P. Q." *Catholic World* 37, no. 221 (1883): 607.

Atack, Jeremy. "Farm and Farm-Making Costs Revisited." *Agricultural History* 56 (October 1982): 663–76.

Bakhtin, Mikhail. "Forms of time and of the chronotope in the novel; notes towards a historical poetics." In *The Dialogic Imagination*, edited by M. Olquist. Austin: University of Texas Press, 1981.

Barman, Jean. "Taming Aboriginal Sexuality: Gender, Power, and Race in British Columbia, 1850–1900." *BC Studies* 115/116 (1997/98): 237–66.

Beard, Colleen. "The Three 'Old' Lock Ones – A history in maps." *Dalhousie Peer – Port Dalhousie's Community News Magazine* 10 (2000): 6–8.

Bellavance, C., F. Normand, and E. S. Ruppert. "Census in Context: Documenting and Understanding the Making of Early-Twentieth-Century Canadian Censuses." *Historical Methods: A Journal of Quantitative and Interdisciplinary History* 40, no. 2 (2007): 92–103.

Benedickson, Jamie. *The Culture of Flushing: A Social and Legal History of Sewage*. Vancouver: UBC Press, 2007.

Bentham, Christie and Katharine Hooke. *From Burleigh to Boschink: A Community Called Stony Lake*. Toronto: Natural Heritage Books, 2000.

Berland, Adam. "Long-term urbanization effects on tree canopy cover along an urban-rural gradient." *Urban Ecosystems* 15, no. 1 (2012): 721–38.

Bernatchez, P., and J.-M. M. Dubois. "Bilan des connaissances de la dynamique de l'érosion des côtes du Québec maritime laurentien." *Géographie physique et Quaternaire* 58, no. 1 (2004): 45–71.

Bernatchez, P., and C. Fraser. "Evolution of Coastal Defence Structures and Consequences for Beach Width Trends, Québec, Canada." *Journal of Coastal Research* 28, no. 6 (2012): 1550 – 1566.

Blomley, Nicholas. "Law, Property, and the Geography of Violence: The Frontier, the Survey, and the Grid." *Annals of the Association of American Geographers* 93, no. 1 (2003): 121–41.

Boas, Franz. *The Ethnography of Franz Boas: Letters and Diaries of Franz Boas Written on the Northwest Coast from 1886 to 1931*, edited by Ronald P. Rohner. Chicago: University of Chicago, 1969.

Boast, Robin, Michael Bravo, and Ramesh Srinivasan. "Return to Babel: Emergent Diversity, Digital Resources, and Local Knowledge." *The Information Society* 23 (2007): 395–403.

Bocking, Stephen. "Constructing Urban Expertise: Professional and Political Authority in Toronto, 1940–1970." *Journal of Urban History* 33, no. 1 (2006): 51–76.

Bodenhamer, David. "History and GIS: Implications for the Discipline." In *Placing History: How Maps, Spatial Data and GIS are Changing Historical Scholarship*, edited by Anne Kelly Knowles, 222–30. Redlands, CA: ESRI, 2008.

Bonnell, Jennifer. *Reclaiming Toronto's Don River Valley: An Environmental History of an Urban Borderland*. Toronto: University of Toronto Press (forthcoming).

———. "A Social History of a Changing Environment: The Don River Valley, 1910–1931." In *Reshaping Toronto's Waterfront*, edited by Gene Desfor and Jennefer Laidley, 123–50. Toronto: University of Toronto Press, 2011.

———. "An Intimate Understanding of Place: Charles Sauriol and Toronto's Don River Valley, 1927–1989." *Canadian Historical Review* 92, no. 4 (2011): 607–36.

Boone, C. G., M. L. Cadenasso, J. M. Grove, K. Schwartz, and G. L. Buckley. "Landscape, Vegetation Characteristics and Group Identity in an Urban and Suburban Watershed: Why the 60s Matter." *Urban Ecosystems* 13 (2010): 255–71.

Bouchard, V., F. Digaire, J.-C. Lefeuvre, and L.-M. Guillon. "Progression des marais salés à l'ouest du Mont-Saint-Michel entre 1984 et 1994." *Mappemonde* 4 (1995): 28–33.

Boulton, Jeremy. *Neighbourhood and Society: A London Suburb in the Seventeenth Century*. Cambridge: Cambridge University Press, 1987.

Bourdieu, Pierre. *Esquisse d'une théorie de la pratique*. Paris: Droz, 1972.

Bradbury, Bettina. "Women's Workplaces: The Impact of Technological Change on Working Class Women in the Home and in the Workplace in Nineteenth Century Montreal." In *Women, Work and Place*, edited by A. Kobayashi, 27–44. Montreal and Kingston: McGill-Queen's University Press, 1994.

Brealey, Kenneth G. "First (National) Space: (Ab)original (Re)mappings of British

Columbia." PhD thesis, University of
British Columbia, 2002.

Breen, David. *Alberta's Petroleum Industry and
the Conservation Board.* Edmonton:
University of Alberta Press, 1992.

British Columbia, Province of. *Progress Report
of the Rural Electrification Committee as
of January 24, 1944.* Victoria: Charles
Banfield, 1944.

Brody, Hugh. *Maps and Dreams: Indians and the
British Columbia frontier.* Vancouver:
Douglas & McIntyre, 1981.

Bromberg, K. D., and M. K. Bertness.
"Reconstructing New England Salt
Marsh Loss Using Historical Maps."
Estuaries 26, no. 6 (2005): 823–32.

Brookes, Alan A. "'Doing the Best I Can': The
Taking of the 1861 New Brunswick
Census." *Histoire sociale / Social History* 9,
no. 17 (1976): 70–91.

Brunger, Alan. "Early Settlement in Contrasting
Areas of Peterborough County, Ontario."
In *Perspectives on Landscape and Settlement
in Nineteenth Century Ontario,* edited
by David J. Wood, 117–40. Toronto:
McClelland & Stewart, 1975.

———. "The Cultural Landscape." In *Peterborough
and the Kawarthas,* 3d ed., edited by
Peter Adams and Colin Taylor, 119–54.
Peterborough: Trent University, 2009.

Byrnes, Giselle. *Boundary Markers: Land Surveying
and the Colonisation of New Zealand.*
Wellington: Bridget Williams Books,
2001.

Campanella, Thomas J. *Republic of Shade: New
England and the American Elm.* New
Haven, CT: Yale University Press, 2003.

Campbell, Claire. *Shaped by the West Wind: Nature
and History in Georgian Bay.* Vancouver:
UBC Press, 2005.

Canada. *Census of Canada, 1911. Agriculture,* Vol. 4
(Ottawa: J. de L. Taché, 1914).

———. *Census of Canada 1956, Agriculture,*
Bulletin 2-11. Ottawa: Edmond Cloutier,
Queen's Printer, 1957.

———. *Census of Canada 1961,* Vol. 5, *Agriculture,*
Bulletin 5.1. Ottawa: Roger Duhamel,
Queen's Printer, 1963.

———. *Census of Population and Agriculture of the
Northwest Territories, 1886.*

———. Census office. Fourth census of Canada,
1901. Ottawa: Printed by S. E. Dawson,
1902–06.

Canada. Bureau of Geology and Topography,
Topographical Survey. "St. Andrews,
Charlotte County, New Brunswick."
Geological Survey of Canada, *"A" Series,*
Map 523a, 1939.

———. Department of Trade and Commerce,
Census and Statistics Office. *Census of
Canada, 1911.* Ottawa: King's Printer,
1912.

———. Department of Trade and Commerce,
Dominion Bureau of Statistics. *Census of
Industry, Central Electric Light Stations in
Canada, 1921–1951; Census of Canada,
Households, 1921–1951.* Ottawa: 1953.

———. *Sixth Census of Canada, 1921.* Ottawa:
King's Printer, 1924.

———. *Sixth Census of Canada, 1921,* Vol. V –
Agriculture. Ottawa: F. A. Acland, 1925.

———. *Seventh Census of Canada, 1931.* Ottawa:
King's Printer, 1933.

———. *Eighth Census of Canada, 1941.* Ottawa:
King's Printer, 1950.

———. *Eighth Census of Canada, 1941,* Vol. VIII,
Agriculture, Part I. Ottawa: Edmond
Cloutier, King's Printer, 1947.

———. *Ninth Census of Canada, 1951.* Ottawa:
Queen's Printer, 1953.

———. *Census of Industry, Central Electric Stations
in Canada, 1922.* Ottawa: F. A. Ackland,
1924.

———. *Census of Industry, 1935, Central Electrical Stations in Canada*. Ottawa, 1937.

———. *Central Electric Stations, 1951*. Edmond Cloutier: Ottawa, 1963.

———. Census Monograph No. 6, *Rural and Urban Composition of the Canadian Population*. Reprinted from Vol. XII, Seventh Census of Canada. Ottawa: J. O. Patenaude, 1938.

Canada. Department of Trade and Commerce, Transportation and Public Utilities Branch. *The Index Numbers of Cost of Electricity for Domestic Service and Tables of Monthly Bills for Domestic Service, Commercial Light and Small Power*. Ottawa: 1936.

———. Department of National Defence. "Topographic Map, Nova Scotia: Halifax Sheet, Number 201, Surveyed in 1920" [map], 1:63,360, 1923.

———. Public Utilities Branch. *Index Numbers of Cost of Rates for Domestic Service and Tables of Monthly Bills for Domestic Service, Commercial Light and Small Power, 1939*. Ottawa, 1940.

Canada. Statistics Canada. *Census of Agriculture, 1986*. Ottawa: Minister of Supply and Services Canada, 1987.

———. Series A67–69, Rural and Urban Populations of Canada, 1871–1976.

———. Series M12-22, Farm holdings, Canada and by province, 1871 to 1971.

Caniggia, G., and G. L. Maffei. *Interpreting Basic Building: Architectural Composition and Building Typology*. Firenze: Aliena Editrice, 2001.

Cannadine, David. "Residential Differentiation in Nineteenth Century Towns: From Shapes on the Ground to Shapes in Society." In *The Structure of Nineteenth Century Cities*, edited by J. H. Johnson and C. G. Pooley, 235–52. London: Croom Helm, 1982.

Cardia, Emanuella. "Household Technology: Was it the Engine of Liberation?" Université de Montréal and CIREQ, current version April 2010, http://www.cireq. umontreal.ca/personnel/cardia.html.

Careau, Chrystian. "Les marais intertidaux du Saint-Laurent : Complexités et dynamiques naturelles et culturelles." MS thesis, Dépt. de géographie, Université Laval, 2010.

Careless, J.M.S. *Toronto to 1918: An Illustrated History*. Toronto: James Lorimer, 1984.

Carter, Sarah. *Lost Harvests: Prairie Indian Reserve Farmers and Government Policy*. Montreal and Kingston: McGill-Queen's University Press, 1990.

Carter, S. B., S. S. Gartner, M. R. Haines, A. L. Olmstead, R. Sutch, and G. Wright, eds. *Historical Statistics of the United States: Earliest Times to the Present*. Vol. 4. New York: Cambridge University Press, 2006.

Castonguay, Stéphane, and Diane Saint-Laurent. "Reconstructing Reforestation: Changing Land-Use Patterns along the Saint-François River in the Eastern Townships." In *Method and Meaning in Canadian Environmental History*, edited by Alan MacEachern and William Turkel, 273–92. Toronto: Nelson, 2009.

Cataldi, G. "From Muratori to Caniggia: The Origins and Development of the Italian School of Design Typology." *Urban Morphology* 7, no. 1 (2003): 19–34.

Catton, H. A., S. St George, and W. R. Humphrey. "An Evaluation of Bur Oak (*Quercus macrocarpa*) Decline in the Urban Forest of Winnipeg, Manitoba, Canada." *Arboriculture and Urban Forestry* 33, no. 1 (2007): 22–30.

Chalmers, Robert. "Report on the Surface Geology of Eastern New Brunswick, North-western Nova Scotia and a Portion of Prince Edward Island." *Annual Report*, Geological Survey of Canada, 8 (1895).

Champagne, P., R. Denis, and C. Lebel. *Établissement de modèle caractérisant l'équilibre dynamique des estrans de la rive sud du moyen estuaire du Saint-Laurent.* Rapport manuscrit canadien des sciences halieutiques et aquatiques no. 1711. Quebec: Ministère des Pêches et des Océans, 1983.

Chapman, V. J. *Salt Marshes and Salt Deserts of the World.* New York: Interscience Publishers.

Chastko, Paul. *Developing Alberta's Oil Sands: From Karl Clark to Kyoto.* Calgary: University of Calgary Press, 2004.

Chudacoff, Howard. *Mobile Americans: Residential and Social Mobility in Omaha, 1880–1920.* New York: Oxford University Press, 1972.

Cimino, Richard and Don Lattin. *Shopping for Faith: American Religion in the New Millennium.* San Francisco: Jossey-Bass, 1998.

Clark, Andrew H. "Field Work in Historical Geography." *Professional Geographer* 4 (1946): 13–23.

———. *Three Centuries and the Island: A Historical Geography of Settlement and Agriculture in Prince Edward Island, Canada.* Toronto: University of Toronto Press, 1959.

Cohn, Julie. "Expansion for Conservation: The North American Power Grid." Paper presented at the American Society for Environmental History. Phoenix, Arizona. April 2011.

Comité ZIP de la Rive Nord de l'Estuaire. *Forum citoyen 2007 sur l'érosion des berges et l'occupation du territoire en Côte-Nord.* Baie Comeau, QC: Webcréation, 2007.

Connerton, Paul. *How Societies Remember.* Cambridge: Cambridge University Press, 1989.

Cook, Kakwiranó:ron. "Kahnawà:ke Revisited: The St. Lawrence Seaway." Kakari:io Pictures, 2009.

Corrective Collective. *Never Done: Three Centuries of Women's Work in Canada.* Toronto: Canadian Women's Educational Press, 1974.

Cover, Robert M. "Violence and the Word." *Yale Law Journal* 95, no. 8 (1986): 1601–29.

Cowan, Ruth Schwartz. *More Work for Mother: The Ironies of Household Technology from the Open Hearth to the Microwave.* New York: Basic Books, 1983.

Cracroft, Sophia. *Lady Franklin Visits the Pacific Northwest, February to April 1861 and April to July 1870.* Edited by Dorothy Blakey Smith. Victoria: Provincial Archives of British Columbia, 1974.

Craib, Raymond B. *Cartographic Mexico: A History of State Fixations and Fugitive Landscapes.* Durham, NC: Duke University Press, 2004.

Craig, Roland D. "Forest Surveys in Canada." *Forestry Chronicle* 11 (September 1935): 31.

Cronon, William. *Changes in the Land: Indians, Colonists, and the Ecology of New England* New York: Hill & Wang, 1983.

Crook, Frederick W. "Underreporting of China's Cultivated Land Area: Implications for World's Agricultural Trade." In *China Situation and Outlook Series*, 33–39. Washington, D.C.: Department of Agriculture, 1993.

Cummings, H. R. *Early Days in Haliburton.* Toronto: Ontario Department of Lands and Forests, 1963.

Cunfer, Geoff. *On the Great Plains: Agriculture and Environment*. College Station: Texas A&M University Press, 2005.

Dale, Angela, and Ed Fieldhouse, and Clare Holdsworth. *Analyzing Census Microdata*. New York: Oxford University Press, 2000.

Darroch, Gordon. "Constructing Census Families and Classifying Households: 'Relationship to Head of Family or Household' in the 1901 Census of Canada." *Historical Methods* 33, no. 4 (2000): 206–10.

Darroch, Gordon, ed. *The Dawn of "Canada's Century": The Hidden Histories*. Montreal and Kingston: McGill-Queen's University Press, forthcoming.

Darroch, Gordon, Richard D. B. Smith, and Michel Gaudreault. "CCRI Sample Designs and Sample Point Identification, Data Entry, and Reporting (SPIDER) Software." *Historical Methods* 40, no. 2 (2007): 65–75.

Dean, Joanna. "'Said tree is a veritable See:' Ottawa's Street Trees, 1869–1939." *Urban History Review / Revue d'histoire urbaine* 34, no. 1 (2005): 46–57.

———. "The Social Production of a Canadian Urban Forest." In *Environmental and Social Justice in the City: Historical Perspectives*, edited by Richard Rodger and Genevieve Massard-Guilbaud. Isle of Harris, UK: White Horse Press, 2011.

Dear, Michael, Jim Ketchum, Sarah Luria, and Doug Richardson. *Geohumanities: Art, History, Text at the Edge of Place*. London: Routledge, 2011.

Dennis, Matthew. *Cultivating a Landscape of Peace: Iroquois–European Encounters in Seventeenth-Century America*. Ithaca, NY: Cornell University Press, 1993.

Desfor, Gene. *Nature and the City: Making Environmental Policy in Toronto and Los Angeles*. Tucson: University of Arizona Press, 2004.

Desfor, Gene. "Planning Urban Waterfront Industrial Districts: Toronto's Ashbridge's Bay, 1889–1910." *Urban History Review* 17, no. 2 (1988): 77–91.

Desfor, Gene, and Jennifer Bonnell. "Socio-ecological Change in the Nineteenth and Twenty-first Centuries: The Lower Don River." In *Reshaping Toronto's Waterfront*, edited by Gene Desfor and Jennefer Laidley, 305–25. Toronto: University of Toronto Press, 2011.

Desfor, Gene, and Roger Keil. "Every River Tells a Story: The Don River (Toronto) and the Los Angeles River (Los Angeles) as Articulating Landscapes." *Journal of Environmental Policy and Planning* 2, no. 1 (2000): 5–23.

Desfor, Gene, and Jennefer Laidley, eds. *Reshaping Toronto's Waterfront*. Toronto: University of Toronto Press, 2011.

Desplanque, C., and D. Mossman. "Tides and Their Seminal Impact on the Geology, Geography, History, and Socio-Economics of the Bay of Fundy, Eastern Canada." *Atlantic Geology* 40, no. 1 (2004): 1–118.

Dionne, J.-C. "Âge et taux moyen d'accrétion verticale des schorres du Saint-Laurent esturaien, en particulier ceux de Montmagny et de Sainte-Anne-de-Beaupré, Québec." *Géographie physique et Quaternaire* 58, no. 1 (2004): 74–75.

Dodd, Dianne. "Women in Advertising: The Role of Canadian Women in the Promotion of Domestic Electrical Technology in the Interwar Period." In *Despite the Odds*, edited by Marianne Ainley, 134-151. Montreal: Véhicule, 1990.

Dolphin, Frank. *Country power: The Electrical Revolution in Rural Alberta*. Plain Publishing, ca. 1993.

Donahue, Brian. "Mapping Husbandry in Concord: GIS as a Tool for Environmental History." In *Placing History: How Maps, Spatial Data, and GIS are Changing Historical Scholarship*, edited by Anne Kelly Knowles, 151–77. Redlands, CA: ESRI Press, 2008.

Driver, Felix and David Gilbert, eds. *Imperial Cities: Landscape, Display and Identity*. Manchester: Manchester University Press, 1999.

Duchesne, Julie. "Le Canadien Pacifique et la transformation de l'espace est-montréalais : le cas de la gare-hôtel Viger, 1891–1901." In *The Territories of Business*, edited by C. Bellavance and P. Lanthier, 47–62. Sainte-Foy: Les Presses de l'Université Laval, 2004.

Duerden, Frank, and Richard G. Kuhn. "The Application of Geographic Information Systems by First Nations and Government in Northern Canada." *Cartographica* 33, no. 2 (1996): 49–62.

Dufaux, François. "A New World from Two Old Ones: The Evolution of Montreal's Tenements, 1850–1892." *Urban Morphology* 4, no. 1 (2000): 9–19.

———, and Sherry Olson. "Reconstruire Montréal, rebâtir sa fortune." *Revue de Bibliothèque et Archives nationales du Québec* 1 (2009): 44–57.

Dunae, Patrick A. "Making the 1891 Census in British Columbia." *Histoire sociale / Social History* 31, no. 62 (1998): 234–36.

Dunae, Patrick A., John S. Lutz, Donald J. Lafreniere, and Jason A. Gilliland. "Making the Inscrutable Scrutable: Race and Space in Victoria's Chinatown, 1891." *BC Studies* 169 (Spring 2011).

Edmonds, Penelope. *Urbanizing Frontiers: Indigenous Peoples and Settlers in 19th Century Pacific Rim Cities*. Vancouver: UBC Press, 2010.

Einhorn, Robin L. *Property Rules, Political Economy in Chicago, 1833–1872* (Chicago: University of Chicago Press, 2001).

Elson, Jeremy, Jon Howell, and John R. Douceur. "MapCruncher: Integrating the World's Geographic Information." Microsoft Research Redmond, April 2007. (http://research.microsoft.com/pubs/74210/OSR2007-4b.pdf, accessed February 24, 2011)

Ennals, Peter. "Cobourg and Port Hope: The Struggle for Control of 'The Back Country.'" In *Perspectives on Landscape and Settlement in Nineteenth Century Ontario*, edited by David J. Wood, 182–95. Toronto: McClelland & Stewart, 1975.

Escobedo, F. J., T. Kroeger, and J. E. Wagner. "Urban Forests and Pollution Mitigation: Analyzing Ecosystem Services and Disservices." *Environmental Pollution* 159 (2011): 2078–87.

Escobedo, Francisco and Jennifer Seitz. "The Costs of Managing an Urban Forest." University of Florida, IFAS Extension, FOR 217 (2009) available at http://edis.ifas.ufl.edu/pdffiles/FR/FR27900.pdf.

Ethington, Philip. "Placing the Past: 'Groundwork' for a Spatial Theory of History." *Rethinking History* 11, no. 4 (2007): 465–94.

Evenden, Matthew. *Fish versus Power: An Environmental History of the Fraser River*. Cambridge: Cambridge University Press, 2004.

Evenden, Matthew, and Graeme Wynn. "54, 40 or Fight: Writing within and across Borders in North American Environmental

History." In *Nature's End: History and the Environment*, edited by Sverker Sörlin and Paul Warde, 215–46. New York: Palgrave Macmillan, 2009.

Fawcett, Edgar. *Some Reminiscences of Old Victoria*. Toronto: William Briggs, 1912.

Feinburg, Jonathan. "Wordle." In *Beautiful Visualization*, edited by Julie Steele and Noah Illinsky, 37-58. Sebastopol, CA: O'Reilly Media, 2010.

Fernow, Bernhard E. *Forest Conditions of Nova Scotia*. Ottawa: Commission of Conservation, Canada, and Department of Crown Lands, Nova Scotia, 1912.

Fischer-Kowalski, Marina, and Helmut Haberl, eds. *Socioecological Transitions and Global Change: Trajectories of Social Metabolism and Land Use*. Cheltenham: Edward Elgar, 2007.

Fitch, William. *Knox Church Toronto: Avant-garde Evangelical Advancing*. Toronto: John Deyell, 1971.

Fleming, Keith. *Power at Cost: Ontario Hydro and Rural Electrification, 1911–58*. Montreal and Kingston: McGill-Queen's University Press, 1992.

Forkey, Neil S. *Shaping the Upper Canadian Frontier: Environment, Society, and Culture in the Trent Valley*. Calgary: University of Calgary Press, 2003.

Fortin, Marcel, and Janina Mueller. "The Library as Research Partner and Data Creator : The Don Valley Historical Mapping Project." *Journal of Map and Geography Libraries* 9, no. 1–2 (2013): 157–74.

Foucault, Michel. "Space, Knowledge and Power." In *The Foucault Reader*, edited by Paul Rabinow<PAGES?>. New York: Pantheon, 1984.

Fraser, E.D.G., and W. A. Kenney. "Cultural Background and Landscape History as Factors Affecting Perceptions of the Urban Forest." *Journal of Arboriculture* 26, no. 2 (2000): 106–13.

Gabriel-Doxtater, Brenda Katlatont, and Arlette Kawanatatie Van den Hende. *At the Woods' Edge: An Anthology of the History of the People of Kanehsatà:ke*. Kanesatake, Québec: Kanesatake Education Center, 1995.

Gaddis, John Lewis. *The Landscape of History: How Historians Map the Past*. New York: Oxford University Press, 2002.

Gaffield, Chad. "Evidence of What? Changing Answers to the Question of Historical Sources as Illustrated by Research Using the Census." In *Building New Bridges: Sources Methods and Interdisciplinarity*, ed. J. Keshen and S. Perrier, 265–74. Ottawa: University of Ottawa Press, 2005.

———. "Introduction: Conceptualizing and Constructing the Canadian Century Research Infrastructure." *Historical Methods* 40, no. 2 (2007): 54–64.

Gauvreau, Danielle, and Sherry Olson. "Mobilité sociale dans une ville industrielle nord-américaine: Montréal, 1880–1900." *Annales de Démographie Historique* 1 (2008): 89–114.

Giddens, Anthony. *Social Theory and Modern Sociology*. Stanford: Stanford University Press, 1987.

Gilliland, Jason. "The Creative Destruction of Montreal: Street Widenings and Urban (Re)Development in the Nineteenth Century." *Urban History Review* 31, no. 1 (2002): 37–51.

———. "Modeling Residential Mobility in Montréal, 1860–1900." *Historical Methods* 31 (January 1998): 27–42.

———. "Redimensioning the Urban Vascular System: Street Widening Operations in Montreal, 1850–1918." In *Transformations of Urban Form*, edited by G. Corona and

G. L. Maffei, FK2.7–11. Firenze: Alinea Editrice, 1999.

Gilliland, Jason, Don Lafreniere, Sherry Olson, Pat Dunae, and John Lutz. "Residential Segregation and the Built Environment in Three Canadian Cities, 1881–1961." Paper presented to the European Social Science History Conference, Ghent Belgium, April 2010.

Gilliland, Jason, and Mathew Novak. "On Positioning the Past with the Present: The Use of Fire Insurance Plans and GIS for Urban Environmental History." *Environmental History* 10, no. 1 (2006): 136–39.

Gilliland, Jason, and Sherry Olson. "Claims on Housing Space in Nineteenth-Century Montreal." *Urban History Review / Revue d'histoire urbaine* 26, no. 2 (1998): 3–16.

———. "Montréal, l'avenir du passé." *GEOinfo* (January–February): 5–7.

———. "Residential Segregation in the Industrializing City: A Closer Look." *Urban Geography* 33 (January 2010): 29–58.

Gilliland, Jason, Sherry H. Olson, and Danielle Gauvreau. "Did Segregation Increase as the City Expanded? the Case of Montreal, 1881–1901." *Social Science History* 35, no. 4 (2011): 465–503.

Gillis, R. Peter. "Rivers of Sawdust: The Battle over Industrial Pollution in Canada, 1865–1903." *Journal of Canadian Studies* 21 (1986): 84–103.

Gillis, Peter, and Thomas Roach. *Lost Initiatives: Canada's Forest Industries, Forest Policies, and Forest Conservation*. New York: Greenwood, 1986.

Glen, W. M. *Prince Edward Island 1935/1936 Forest Cover Type Mapping*. Charlottetown, PEI: Forestry Division,

P.E.I. Department of Agriculture and Forestry, 1997.

Gould, S. J. *Time's Arrow, Time's Cycle: Myth and Metaphor in the Discovery of Deep Geological Time*. Cambridge: Harvard University Press, 1987.

Gourde, G. *Les aboiteaux : comté de Kamouraska*. Québec: Ministère de l'agriculture, des pêcheries et de l'alimentation, 1980.

Gregory, Ian N. "'A map is just a bad graph:' Why Spatial Statistics Are Important in Historical GIS." In *Placing History: How Maps, Spatial Data and GIS are Changing Historical Scholarship*, edited by Anne Kelly Knowles, 125. Redlands, CA: ESRI Press, 2008.

Gregory, Ian, and Paul Ell. *Historical GIS: Technologies, Methodologies and Scholarship*. Cambridge: Cambridge University Press, 2007.

Groves, Harold M. "The Property Tax in Canada and the United States." *Land Economics* 24, no. 1 (1948): 23–30.

Gwyn, Julian. "Golden Age or Bronze Moment? Wealth and Poverty in Nova Scotia: The 1850s and 1860s." *Canadian Papers in Rural History* 8 (1992): 195–230.

Hall, Stuart, and Paul du Gay, eds. *Questions of Identity*. London: Sage, 1996.

Halliday, Hugh A. "The Forest Watchers: Air Force, Part 35." *Legion Magazine* 20 (October 2009).

Hamel, A. "La récupération et la mise en valeur des alluvions maritimes du St-Laurent." *Agriculture* 20, no. 3 (1963): 77–83.

Hanson, Julienne. *Decoding Homes and Houses*. Cambridge: Cambridge University Press, 1998.

Harney, Robert, ed., *Gathering Place: Peoples and Neighbourhoods of Toronto, 1834–1945*. Toronto: Multicultural History Society of Ontario, 1985.

Harris, Cole. "How Did Colonialism Dispossess? Comments from an Edge of Empire." *Annals of the Association of American Geographers* 94, no. 1 (2004): 165–82.

———. *Making Native Space: Colonialism, Resistance, and Reserves in British Columbia.* Vancouver: UBC Press, 2002.

———. *The Reluctant Land: Society, Space, and Environment in Canada before Confederation.* Vancouver: UBC Press, 2008.

Hatvany, Matthew. *Marshlands: Four Centuries of Environmental Change on the Shores of the St. Lawrence.* Sainte-Foy: Les Presses de l'Université Laval, 2003.

———. "The Origins of the Acadian *Aboiteau*: An Environmental Historical Geography of the Northeast." *Historical Geography* 30 (2002): 121–37.

———. *Paysages de marais : Quatre siècles de relations entre l'humain et les marais du Kamouraska.* La Pocatière, QC: Société historique de la Côte-du-Sud et Ruralys, 2009.

———. "Wetlands." *Encyclopedia of American Environmental History* 4 (2011): 1380–83.

———. "Wetlands and Reclamation." In *International Encyclopedia of Human Geography*, vol. 12, edited by R. Kitchen and N. Thrift, 241–46. Oxford: Elsevier, 2009.

Head, C. Grant. "The Forest Industry, 1850–1890." In *Historical Atlas of Canada*, edited by R. Louis Gentilcore, Plate 38. Toronto: University of Toronto Press, 1987.

———. "An Introduction to Forest Exploitation in Nineteenth Century Ontario." In *Perspectives on Landscape and Settlement in Nineteenth Century Ontario*, edited by J. David Wood, 78–112. Toronto: McClelland & Stewart, 1975.

Heaman, Elsbeth. *The Inglorious Arts of Peace: Exhibitions in Canadian Society during the Nineteenth Century.* Toronto: University of Toronto Press, 1999.

Hertzog, Stephen, and Robert D. Lewis. "A City of Tenants: Homeownership and Social Class in Montreal, 1847–1881." *The Canadian Geographer* 304 (1986): 316–23.

Heynen, Nick. "The Scalar Production of Injustice within the Urban Forest." *Antipode: A Journal of Radical Geography* 35, no. 5 (2003): 980–98.

———. "Green Urban Political Ecologies: Toward a Better Understanding of Inner City Environmental Change." *Environment and Planning A* 38 no. 3 (2006): 499–516.

Heynen, Nick, H. A. Perkins, and P. Roy. "The Political Ecology of Uneven Urban Green Space the Impact of Political Economy on Race and Ethnicity in Producing Environmental Inequality in Milwaukee." *Urban Affairs Review* 42, no. 1 (2006): 3–25.

Hillier, Amy. "Redlining in Philadelphia." In *Past Time, Past Place: GIS for History*, edited by Anne Knowles, 79–92. Redlands, CA: ESRI Press, 2002.

———. "Spatial Analysis of Historical Redlining: A Methodological Exploration." *Journal of Housing Research* 14, no. 1 (2003): 137–67.

Hillier, Bill. *Space Is the Machine: A Configurational Theory of Architecture.* New York: Cambridge University Press, 1996.

Hillis, Peter. *The Barony of Glasgow: A Window into Church and People in Nineteenth-Century Scotland.* Edinburgh: Dunedin Academic Press, 2007.

Hinson, Andrew. "A Hub of Community: The Presbyterian Church in Toronto and its Role Among the City's Scots." In *Ties of Bluid Kin and Countrie: Scottish*

Associational Culture in the Diaspora, edited by Tanja Bueltmann, Andrew Hinson and Graeme Morton. Guelph, ON: Centre for Scottish Studies, 2009.

———. "Migrant Scots in a British City: Toronto's Scottish Community, 1881–1911." PhD thesis, University of Guelph, 2010.

Historical Methods: A Journal of Quantitative and Interdisciplinary History. Special issue on the Canadian Families Project. 33, no. 4 (2000).

Historical Methods: A Journal of Quantitative and Interdisciplinary History. Special issue on the Canadian Century Research Infrastructure. 40, no. 2 (2007).

Holdsworth, Deryck. "Historical Geography: New Ways of Imaging and Seeing the Past." *Progress in Human Geography* 27, no. 4 (2003): 486–93.

Hooke, Katharine. *From Campsite to Cottage: Early Stoney Lake.* Peterborough: Peterborough Historical Society, 1992.

Howe, C. D., and J. H. White. *Trent Watershed Survey.* Toronto: Commission of Conservation, 1913.

Inwood, Kris, and Jim Irwin. "Land, Income and Regional Inequality: New Estimates of Provincial Incomes and Growth in Canada, 1871–1891." *Acadiensis* 31, no. 2 (2002): 157–84.

Jacobs, Jane. *Edge of Empire: Postcolonialism and the City.* New York: Routledge, 1996.

Jacobsen, J. A. *Alaskan Voyage, 1881–83: An Expedition to the Northwest Coast of America.* Translated from the German text of Adrian Woldt by Erna Gunther. Chicago: University of Chicago Press, 1977.

Jasen, Patricia. *Wild Things: Nature, Culture, and Tourism in Ontario, 1790–1914.* Toronto: University of Toronto Press, 1995.

Jenkins, William. "In Search of the Lace Curtain: Residential Mobility, Class Transformation, and Everyday Practise among Buffalo's Irish, 1880–1910." *Journal of Urban History* 35, no. 7 (2009): 970–97.

Jones, Christopher F. "A Landscape of Energy Abundance: Anthracite Coal Canals and the Roots of American Fossil Fuel Dependence, 1820–1860." *Environmental History* 15 (July 2010): 449–84.

Katz, Michael. *The People of Hamilton, Canada West: Family and Class in a Mid-Nineteenth-Century City.* Cambridge, MA: Harvard University Press, 1975.

Katz, Michael B., Michael J Doucet, and Mark J. Stern. "Population Persistence and Early Industrialization in a Canadian City: Hamilton, Ontario, 1851–1871." *Social Science History* 2, no. 2 (1978): 220.

Keddie, Grant. *Songhees Pictorial: A History of the Songhees People as seen by Outsiders, 1790–1912.* Victoria, BC: Royal British Columbia Museum, 2003.

Keeling, Arn. "'Born in an Atomic Test Tube:' Landscapes of Cyclonic Development at Uranium City, Saskatchewan." *The Canadian Geographer* 54, no. 2 (2010): 228–52.

Kelly, Kenneth. "The Impact of Nineteenth Century Agricultural Settlement on the Land." In *Perspectives on Landscape and Settlement in Nineteenth Century Ontario*, edited by J. David Wood, 64–77. Toronto: McClelland & Stewart, 1975.

Kirkconnell, Watson. *County of Victoria: Centennial History.* Lindsay, ON: Victoria County Council, 1967.

Kline, Ronald R. *Consumers in the Countryside: Technology and Social Change in Rural America.* Baltimore, MD: Johns Hopkins University Press, 2000.

Knowles, Anne Kelly. "GIS and History." In *Placing History: How Maps, Spatial Data and GIS are Changing Historical Scholarship*, edited by Anne Kelly Knowles, 7–8. Redlands, CA: ESRI Press, 2008.

Knowles, Anne Kelly, ed. *Placing History: How Maps, Spatial Data, and GIS Are Changing Historical Scholarship*. Redlands, CA: ESRI Press, 2008.

Kuletz, Valerie L. *The Tainted Desert: Environmental and Social Ruin in the American West*. New York: Routledge, 1998.

Kunstler, James Howard. *The Geography of Nowhere: The Rise and Decline of America's Man-Made Landscape*. New York: Touchstone, 1993.

Lafitau, Joseph François. *Moeurs des sauvages ameriquains comparées aux moeurs des premiers temps*. 4 vols. Paris: Chez Saugrain l'aîné et al., 1724.

Lafreniere, Donald, and Jason Gilliland. "Beyond the Narrative: Using H-GIS to Reveal Hidden Patterns and Processes of Daily Life in Nineteenth Century Cities." Paper presented to the Association of American Geographers, Seattle, April 2011.

———. "A Socio-Spatial Analysis of the Nineteenth-Century Journey to Work in London, Ontario." Paper presented to the Social Science History Association, Chicago, Illinois, November 2010.

Lai, David Chuenyan. *Chinatowns: Towns within Cities in Canada*. Vancouver: UBC Press, 1988.

———. *Forbidden City within Victoria: Myth, Symbol and Streetscape of Canada's Earliest Chinatown*. Victoria: Orca Books, 1991.

Lane, Hannah M. "Tribalism, Proselytism, and Pluralism: Protestant, Family, and Denominational Identity in Mid-Nineteenth-Century St Stephen, New Brunswick." In *Households of Faith: Family, Gender, and Community in Canada, 1760–1969*, edited by Nancy Christie. Montreal and Kingston: McGill-Queen's University Press, 2002.

Lapping, Mark. "Stone Walls, Woodlands, and Farm Buildings: Artifacts of New England's Agrarian Past." In *A Landscape History of New England*, edited by Blake A. Harrison and Richard William Judd. Cambridge, MA: MIT Press, 2011.

Lattimer, J. E. *Taxation in Prince Edward Island, A Report*. Charlottetown: Department of Reconstruction, 1945.

Lavery, Doug, and Mary Lavery. *Up the Burleigh Road … beyond the boulders*. Peterborough: Trent Valley Archives, 2006, 2007.

Lee, A.C.K., and R. Maheswaran. "The Health Benefits of Urban Green Spaces: A Review of the Evidence." *Journal of Public Health* 33, no. 2 (2011): 212–22.

Lefebvre, Henri. *Espace et politique, Le droit à la ville II*. Paris: Anthropos, 1976.

———. *The Production of Space*. Translated by Donald Nicholson-Smith. Oxford: Blackwell, 1991.

———. "Space, Social Product and Use Value." In *Critical Sociology: European International Perspectives,* edited by J. W. Freiberg, 286. New York: Irvington/Wiley, 1979.

Lemon, James. *Toronto since 1918: An Illustrated History*. Toronto: Lorimer, 1985.

Liu, M. L., and H. Q. Tian. "China's Land Cover and Land Use Change from 1700 to 2005: Estimations from High-resolution Satellite Data and Historical Archives." *Global Biogeochemical Cycles* 24, no. 3 (2010), dx.doi.org/10.1029/2009GB003687.

Loucks, O. L. "A Forest Classification for the Maritime Provinces." *Proceedings of the Nova Scotia Institute of Science* 25 (1962): 85–167.

Lower, A.R.M. *The North American Assault on the Canadian Forest: A History of the Lumber Trade between Canada and the United States.* Toronto: Ryerson Press, 1938.

Luria, Sarah. "Geotexts." In *Geohumanities: Art, History, Text at the Edge of Place*, edited by Michael Dear, Jim Ketchum, Sarah Luria, and Doug Richardson, 67. London: Routledge, 2011.

Lutz, John Sutton. *Makuk: A New History of Aboriginal–White Relations.* Vancouver: UBC Press, 2007.

Maas, J., R. A. Verheij, S. de Vries, P. Spreeuwenberg, F. G. Schellevis, and P. P. Groenewegen. "Morbidity Is Related to a Green Living Environment." *Journal of Epidemiology and Community Health* 63, no. 12 (2009): 967–73.

Macfarlane, Daniel. "To the Heart of the Continent: Canada and the Negotiation of the St. Lawrence Seaway and Power Project, 1921–1954." PhD dissertation, University of Ottawa, 2010.

———. "Rapid Changes: Canada and the St. Lawrence and Seaway Project." Research Paper, Program on Water Issues (POWI), Munk School of Global Affairs (University of Toronto), 2010, http://www.powi.ca/pdfs/other/Macfarlane-POWI%20paper.pdf.

MacIntyre, Colin. "The Environmental Pre-History of Prince Edward Island, 1769–1970: A Reconnaissance in Force." MA thesis, University of Prince Edward Island, 2010.

MacLeod, Roderick, and Mary Anne Poutanen. "Proper Objects of This Institution: Working Families, Children and the British and Canadian School in Nineteenth-Century Montreal." *Historical Studies in Education* 20, no. 2 (2008): 22–54.

Manitoba. *Report of the Minister of Agriculture of the Province of Manitoba*, 1880–83. Table XIII.

Manore, Jean L. *Cross-Currents: Hydroelectricity and the Engineering of Northern Ontario.* Waterloo: Wilfrid Laurier University Press, 1999.

Maritime Resource Management Service Inc. *Proposal for a Prince Edward Island Inventory Pilot Study: Prepared for P.E.I. Dept. of Energy and Forestry.* Amherst: Maritime Resource Management Service Inc., 1987.

Massey, Doreen. "Space-Time and the Politics of Location." *Architectural Digest* 68: 3&4 (March April 1998): 34.

Mawani, Renisa. *Colonial Proximities: Crossracial Encounters and Juridical Truths in British Columbia, 1871–1921.* Vancouver: UBC Press, 2009.

———. "Legal Geographies of Aboriginal Segregation in British Columbia: The Making and Unmaking of the Songhees Reserve, 1850–1911." In *Isolation, Places and Practices of Exclusion*, edited by Carolyn Strange and Alison Basher, 173–90. London: Routledge, 2003.

McCalla, Douglas. *Planting the Province: The Economic History of Upper Canada, 1784–1870* Toronto: University of Toronto Press, 1993.

McCann, Larry, Ian Buck, and Ole Heggen. "Family Geographies: A National Perspective." In *Household Counts: Canadian households and families in 1901*, edited by Eric Sager and Peter Baskerville, 180–97. Toronto: University of Toronto Press, 2007.

McDonald, S. M., and W. M. Glen. *1958 Forest Inventory of Prince Edward Island*, P.E.I. Department of Environment, Energy and Forestry, 2006.

McGowan, Mark. "Coming Out of the Cloister: Some Reflections on the Developments in the Study of Religion in Canada, 1980–1990." *International Journal of Canadian Studies* 1–2 (1990): 175–202.

McInnis, Marvin. "Ontario Agriculture, 1851–1901: A Cartographic Overview." In *Canadian Papers in Rural History*, vol. 5, edited by Donald H. Akenson, 290–301. Gananoque, ON: Langdale Press, 1986.

———. "Output and Productivity in Canadian Agriculture, 1870–71 to 1926–27." In *Long-Term Factors in American Economic Growth*, edited by Stanley L. Engerman and Robert E. Gallman, 751. Chicago: University of Chicago Press, 1992 [reprint].

———. "Perspectives on Ontario Agriculture, 1815–1930." *Canadian Papers in Rural History* 8 (1992): 17–127.

McLaren, Allan A. *Religion and Social Class: The Disruption Years in Aberdeen*. London: Routledge and Kegan Paul, 1974.

McKee, K. L., and W. H. Patrick, Jr. "The Relationship of Smooth Cordgrass (*Spartina alterniflora*) to Tidal Datums: A Review." *Estuaries* 11, no. 3 (1988): 143–44.

McRoberts, Omar M. *Streets of Glory: Church and Community in a Black Urban Neighbourhood*. Chicago: University of Chicago Press, 2003.

Merchant, Carolyn. *Ecological Revolutions: Nature, Gender, and Science in New England*. Chapel Hill: University of North Carolina Press, 1989.

Mitchell, Peta. "'The Stratified record on which we set our feet': The Spatial Turn and the Multilayering of History, Geography and Geology." In *Geohumanities: Art, History, Text at the Edge of Place*, edited by Michael Dear, Jim Ketchum, Sarah Luria, and Doug Richardson, 81. London: Routledge, 2011.

Mitman, Gregg. *Breathing Spaces: How Allergies Shape Our Lives and Landscapes*. New Haven, CT: Yale University Press, 2007.

Moir, John. *Enduring Witness: A History of the Presbyterian Church in Canada*. Toronto: Presbyterian Publications, 1974.

Moldofsky, Byron. "The CCRI Geographic Files: Introduction and Examples." In *The Dawn of "Canada's Century:" The Hidden Histories*, edited by Gordon Darroch.. Montreal and Kingston: McGill-Queen's University Press (forthcoming).

Molotch, Harvey. "The City as a Growth Machine: Toward a Political Economy of Place." *American Journal of Sociology* 82, no. 2 (1976): 309–32.

Montagnes, Ian. *Port Hope: A History*. Port Hope, ON: Ganaraska Press, 2007.

Montgomery, L. M. *Jane of Lantern Hill*. Toronto: McClelland & Stewart, 1937.

Morgan, Lewis H. *League of the Ho-de'-no-sau-nee or Iroquois*. North Dighton, MA: JG Press, 1995 [1851].

Morny, D. L. *Farewell My Bluebell: A Vignette of Lowertown*. Ottawa: 1998.

Mulvany, Charles Pelham. *Toronto: Past and Present*. Toronto: W. E. Caiger, 1884.

"A Nation-Wide Inventory of our Forest Resources." *Natural Resources Canada* 8 (March 1929): 2.

Nelles, H. V. "Introduction." In *Philosophy of Railroads*, by T. C. Keefer, ix–lxiii. Toronto: University of Toronto Press, 1972.

———. *The Politics of Development: Forests, Mines, and Hydro-Electric Power in Ontario,*

1849–1941. Montreal and Kingston: McGill-Queen's University Press, 2005 [MacMillan, 1974].

Nicholson, N. L., and L. M. Sebert. *The Maps of Canada: A Guide to Official Canadian Maps, Charts, Atlases and Gazetteers.* Folkestone, England: Wm. Dawson; Hamden, CT: Archon Books, 1981.

Novak, Mathew, and Jason Gilliland. "'Buried beneath the waves': Using GIS to examine the social and physical impacts of a historical flood." *Digital Studies* 1, no. 2 (2010), http://www.digitalstudies.org/ojs/index.php/digital_studies/issue/view/20.

Nowak, D. J., D. E. Crane, and J. C. Stevens. "Air Pollution Removal by Urban Trees and Shrubs in the United States." *Urban Forestry and Urban Greening* 4 (2006): 115–23.

Nowak, D. J. "Historical Vegetation Change in Oakland and its Implications for Urban Forest Management." *Journal of Arboriculture* 19 (1993): 313–19.

Nye, David. *Electrifying America: Social Meanings of a New Technology, 1880–1949.* Cambridge, MA: MIT Press, 1990.

Olson, Sherry. "Ethnic Partition of the Workforce in 1840s Montreal." *Labour / Le Travail* 53 (Spring 2004): 1–66.

———. "Occupations and Residential Spaces in Nineteenth Century Montreal." *Historical Methods* 22 (Summer 1989): 81–99.

Olson, Sherry, and Patricia A. Thornton. *Peopling the North American City: Montreal, 1840–1900.* Montreal and Kingston: McGill-Queen's University Press, 2011.

Ontario. Department of Energy and Resources Management. *Otonabee Region Conservation Report: Summary.* Toronto, 1965.

Oswald, Diane L. *Fire Insurance Maps: Their History and Application.* College Station, TX: Lacewing Press, 1997.

Parr, Joy. *Domestic Goods: The Material, the Moral and the Economic in the Postwar Years.* Toronto: University of Toronto Press, 1999.

———. "'Lostscapes': Found Sources in Search of a Fitting Representation." *Journal of the Association for History and Computing* 7, no. 2 (2004), http://hdl.handle.net/2027/spo.3310410.0007.101.

———. *Sensing Changes: Technologies, Environments, and the Everyday, 1953–2003.* Vancouver: UBC Press, 2010.

Parsons, H. M. *Biographical Sketches and Review, First Presbyterian Church in Toronto and Knox Church, 1820–1890.* Toronto: Oxford Press, 1890.

Peluso, Nancy Lee. "Whose Woods Are These? Counter-Mapping Forest Territories in Kalimantan, Indonesia." *Antipode* 27, no. 4 (1995): 383–406.

Peponis, J., E. Hadjinkolaou, C. Livieratos, and D. A. Fatouros. "The Spatial Core of Urban Culture." *Ekistics* 334/335 (1989): 43–55.

Perry, Adele. *On the Edge of Empire: Gender, Race, and the Making of British Columbia, 1849–1871.* Toronto: University of Toronto Press, 2001.

Pham, Thi-Thanh-Hiên, Philippe Apparicio, Anne-Marie Séguin, and Martin Gagnon. "Mapping the Greenscape and Environmental Equity in Montreal: An Application of Remote Sensing and GIS." In *Mapping Environmental Issues in the City: Arts and Cartography Cross Perspectives* (Lecture Notes in Geoinformation and Cartography), edited by S. Caquard, L. Vaughan, and

W. Cartwright, 30–48. Berlin: Springer, 2011.

Pham, Thi-Thanh-Hiên, Philippe Apparicio, Anne-Marie Séguin, Shawn Landry, and Martin Gagnon. "Spatial Distribution of Vegetation in Montreal: An Uneven Distribution or Environmental Inequity?" *Landscape and Urban Planning* 107, no. 3 (2012): 214–24.

Pham, Thi-Thanh-Hiên, Philippe Apparicio, Shawn Landry, Anne-Marie Séguin, and Martin Gagnon. "Predictors of the Distribution of Street and Backyard Vegetation in Montreal, Canada." *Urban Forestry & Urban Greening* 12, no. 1 (2013):18–27; http://dx.doi.org/10.1016/j.ufug.2012.09.002.

Phillips, Stephanie. "The Kahnawake Mohawks and the St. Lawrence Seaway." Master's thesis, McGill University, 2000.

Philpotts, L. E. *Aerial Photo Interpretation of Land Use Changes in Fourteen Lots in Prince Edward Island.* Economics Division, Canada Department of Agriculture, 1958.

Piper, Liza. *The Industrial Transformation of Subarctic Canada.* Vancouver: UBC Press, 2009.

Platt, Harold L. *The Electric City: Energy and the Growth of the Chicago Area, 1880–1930.* Chicago: University of Chicago Press, 1991.

Pocock, D.C.D. "Sight and Knowledge." *Transactions of the Institute of British Geographers* 6 (1981): 385–93.

Pooley, Colin G. "Space, Society and History." *Journal of Urban History* 31 (2005): 753–61.

Pred, Alan. "Place as Historically Contingent Process: Structuration and the Time-Geography of Becoming Places." *Annals of the Association of American Geographers* 74, no. 2 (1984): 279–97.

Prince Edward Island. *Journal of the Legislative Assembly of the province of Prince Edward Island.* 1915.

———. Public Archives and Records Office. "Report of the Superintendant of Census Returns, Charlottetown, August 23, 1871." In Manitoba, *Report of the Minister of Agriculture of the Province of Manitoba,* 1880–83. Montreal: Gazette Printing, 1881.

Purdy, Al. "The country north of Belleville." In *The Cariboo Horses.* Toronto: McClelland & Stewart, 1965.

Pyne, Stephen J. *Awful Splendour: A Fire History of Canada.* Vancouver: UBC Press, 2007.

Québec. "Base de données topographiques du Québec (BDTQ) à l'échelle de 1/20 000." Normes de production, version 1.0. Québec: Direction de la cartographie topographique, Ministère des Ressources naturelles et de la Faune, 1999.

Rajala, Richard A. *Feds, Forests, and Fire: A Century of Canadian Forestry Innovation.* Transformation Series 13, Canadian Science and Technology Museum, 2005.

Ramankutty, N., and J. A. Foley. "Characterizing Patterns of Global Land Use: An Analysis of Global Croplands Data." *Global Biogeochemical Cycles* 12 (1998): 667–85.

———. "Estimating historical changes in land cover: North American croplands from 1850 to 1992." *Global Ecology and Biogeography* 8, no. 5 (1999): 381–96.

Ramankutty, N., J. A. Foley, and N. J. Olejniczak. "Land-Use Change and Global Food Production." In *Land Use and Soil Resources,* edited by Ademola K. Braimoh and Paul L. G. Vlek, 23–40. Dordrecht:

Springer Science+Business Media B. V., 2008.

Ramankutty, N., E. Heller, and J. Rhemtulla. "Prevailing Myths about Agricultural Abandonment and Forest Regrowth in the United States." *Annals of the American Association of Geography* 100, no. 3 (2010): 502–12.

Richardson, Douglas. "Spatial Histories: Geohistories." In *Geohumanities: Art, History, Text at the Edge of Place*, edited by Michael Dear, Jim Ketchum, Sarah Luria, and Doug Richardson, 210. London: Routledge, 2011.

Roy, Patricia E. *The Oriental Question: Consolidating a White Man's Province, 1914–41.* Vancouver: UBC Press, 2003.

———. *The Triumph of Citizenship: The Japanese and Chinese in Canada, 1941–67.* Vancouver: UBC Press, 2007.

———. *A White Man's Province: British Columbia Politicians and Chinese and Japanese Immigrants, 1858–1914.* Vancouver: UBC Press, 1990.

Roy, Sudipto, Jason Byrne, and Catherine Pickering. "A Systematic Quantitative Review of Urban Tree Benefits, Costs and Assessment Methods across Cities in Different Climatic Zones." *Urban Forestry and Urban Greening* 11, no. 4 (2012): 351–63.

Russell, Peter A. "Forest into Farmland: Upper Canadian Clearing Rates, 1822–1839." *Agricultural History* 57, no. 3 (1983): 338.

Raymond, C. W., and J. A. Rayburn. "Land Abandonment in Prince Edward Island." *Geographical Bulletin* 19 (1963): 78–86.

Redfield, A. C. "Development of a New England Salt Marsh." *Ecological Monographs* 42, no. 2 (1972): 201–37.

Reeves, Wayne. "From Acquisition to Restoration: A History of Protecting Toronto's Natural Places." In *Special Places: The Changing Ecosystems of the Toronto Region*, edited by Betty I. Roots, Donald A. Chant, and Conrad E. Heidenreich, 229–41. Vancouver: UBC Press, 1999.

Reid, Gerald. *Kahnawà:ke: Factionalism, Traditionalism, and Nationalism in a Mohawk Community.* Lincoln: University of Nebraska Press, 2004.

Richardson, A. H. *A Report on the Ganaraska Watershed.* Toronto: Dominion and Ontario Governments, 1944.

Robertson, J. Ross. *Robertson's Landmarks of Toronto: A Collection of Historical Sketches of York from 1793 to 1837 and of Toronto from 1834 to 1904.* Toronto: J. R. Robertson, 1904.

Robinson, Donald C. E., Werner A. Kurz, and Christine Pinkham. *Estimating the Carbon Losses from Deforestation in Canada.* Prepared for the National Climate Change Secretariat, Canadian Forest Service. Vancouver: ESSA Technologies, 1999.

Robnik, Diane. *The Mills of Peterborough County.* Peterborough: Trent Valley Archives, 2006.

Rueck, Daniel. "When Bridges Become Barriers: Montreal and Kahnawake Mohawk Territory." In *Metropolitan Natures: Urban Environmental Histories of Montreal*, edited by Stéphane Castonguay and Michèle Dagenais, 228–44. Pittsburgh: University of Pittsburgh Press, 2011.

Rundstrom, Robert A. "GIS, Indigenous Peoples, and Epistimological Diversity." *Cartography and Geographic Information Systems* 22, no. 1 (1995): 45–57.

———. "Mapping, Postmodernism, Indigenous People and the Changing Direction of North American Cartography." *Cartographica* 28, no. 2 (1991): 1–12.

Said, Edward. *Orientalism*. New York: Vintage, 1978.

Sager, Eric, and Peter Baskerville, eds. *Household Counts: Canadian Households and Families in 1901*. Toronto: University of Toronto Press, 2007.

Sandlos, John. *Hunters at the Margin: Native People and Wildlife Conservation in the Northwest Territories*. Vancouver: UBC Press, 2007.

Sandwell, R. W. "History as Experiment: Microhistory and Environmental History." In *Method and Meaning in Canadian Environmental History*, edited by Alan MacEachern and William Turkel, 122–36. Toronto: Thomas Nelson, 2008.

———. "Missing Canadians: Reclaiming the A-Liberal Past." In *Liberalism and Hegemony: Debating the Canadian Liberal Revolution*, edited by Jean-François Constant and Michel Ducharme, 246–73. Toronto: University of Toronto Press, 2009.

———. "Notes towards a History of Rural Canada, 1870–1940." In John R. Parkins, and Maureen G. Reed, eds. *Social Transformation in Rural Canada: Community, Cultures, and Collective Action* (Vancouver: UBC Press, 2012), 21-42.

———. "Rural Households, Subsistence and Environment on the Canadian Shield, 1901–1940." Paper delivered at *Bringing Subsistence out of the Shadows: An Environmental History Workshop on Subsistence Relationships*. Nipissing University, North Bay, Ontario, 3 October 2009.

———. "Rural Reconstruction: Towards a New Synthesis in Canadian History." *Histoire Sociale / Social History* 27, no. 53 (1994): 1–32.

Schroeder, H., J. Flannigan, and R. Coles. "Residents Attitudes toward Street Trees in the UK and US Communities." *Arboriculture and Urban Forestry* 32 (2006): 236–46.

Schuurman, Nadine. "Formalization Matters: Critical GIS and Ontology Research." *Annals of the Association of American Geographers* 96, no. 4 (2006): 726–39.

Scott, James C. *Seeing Like a State: How Certain Schemes to Improve the Human Condition Have Failed*. New Haven, CT: Yale University Press, 1998.

Sellers, P., and M. Wellisch. *Greenhouse Gas Contribution of Canada's Land-Use Change and Forestry Activities: 1990–2010*. Prepared by MWA Consultants for Environment Canada, July 1998.

Sérodes, J.-B., and M. Dubé. "Dynamique sédimentaire d'un estran à spartines (Kamouraska, Québec)." *Le naturaliste canadien* 110 (1983): 11–26.

Seeley, H. E. "The Use of Air Photographs for Forestry Purposes." *Forestry Chronicle* 12 (December 1935): 287–93.

Sherwood, Jay. *Furrows in the Sky: The Adventures of Gerry Andrews*. Victoria: Royal BC Museum, 2012.

Siebert, Stefan, Felix T. Portmann, and Petra Döll. "Global Patterns of Cropland Use Intensity." *Remote Sensing* 2 (2010): 1625–43.

Sitwell, O.F.G. "Difficulties in the Interpretation of the Agricultural Statistics in the Canadian Censuses of the Nineteenth Century." *Canadian Geographer* 13, no. 1 (1969): 72–76.

Smedley, Agnes, and Brian Smedley. *Race in North America: Origin and Evolution of a Worldview*. Boulder, CO: Westview, 2012.

Smil, V. "China's Agricultural Land." *The China Quarterly* 158 (June 1999): 414–29.

Sobey, Douglas G. *Early Descriptions of the Forests of Prince Edward Island: A Source-Book, Vol. II. The British and Post-Confederation Periods – 1758–c.1900.* Charlottetown: 2006.

Sobey, D. G., and W. M. Glen. *The Forest Vegetation of the Prince Edward Island National Park and the Adjacent Watershed Area Prior to European Settlement as Revealed from a Search and Analysis of the Historical and Archival Literature.* Phase 2; Analysis and Discussion. Contracted Report for Prince Edward Island National Park, 2002.

———. "A Mapping of the Present and Past Forest-types of Prince Edward Island." *Canadian Field-Naturalist* 118, no. 4 (2004): 504–20.

Stanley, Timothy J. *Contesting White Supremacy: School Segregation, Anti-Racism, and the Making of Chinese Canadians.* Vancouver: UBC Press, 2011.

Stilgenbauer, F. A. *The Geography of Prince Edward Island.* PhD dissertation. University of Michigan, Ann Arbor, 1929.

Stanger-Ross, Jordan. "An Inviting Parish: Community without Locality in Postwar Italian Toronto." *Canadian Historical Review* 87, no. 3 (2006): 381–407.

Steinberg, Theodore. *Nature Incorporated: Industrialization and the Waters of New England.* Amherst: University of Massachusetts Press, 1991.

St-Hilaire, Marc, Byron Moldofsky, Laurent Richard, and Mariange Beaudry. "Geocoding and Mapping Historical Census Data: The Geographical Component of the Canadian Century Research Infrastructure." *Historical Methods* 40, no. 2 (2007): 76–91.

Strong-Boag, Veronica. *The New Day Recalled: Lives of Girls and Women in English Canada, 1919–1939.* Toronto: Copp-Clark, 1988.

Suire, Y. *Le Marais poitevin: une écohistoire du XVIᵉ à l'aube du XXᵉ siècle.* La Roche-sur-Yon: Centre vendéen de recherches historiques, 2006.

Sweeny, Robert C. H. "Property and Gender: Lessons from a 19th-Century Town." *Journal of Canadian Studies* (London) 22 (2006/7): 9–34.

———. "Risky Spaces: the Montreal Fire Insurance Company, 1817–20." In *The Territories of Business*, edited by C. Bellavance and P. Lanthier, 9–23. Sainte-Foy: Les Presses de l'Université Laval, 2004.

Sweeny, Robert C. H., and Sherry Olson. "MAP: Montréal l'avenir du passé, Sharing Geodatabases Yesterday, Today and Tomorrow." *Geomatica* 57, no. 2 (2003): 145–54.

Taylor, H. W., J. Clarke, and W. R. Wightman. "Contrasting Land Development Rates in Southern Ontario to 1891." In *Canadian Papers in Rural History*, vol. 5, edited by Donald H. Akenson, 50–72. Gananoque, ON: Langdale Press, 1986.

Theberge, Clifford, and Elaine Theberge. *At the Edge of the Shield: A History of Smith Township, 1818–1980.* Peterborough: Smith Township Historical Committee, 1982.

Thomson, Don W. *Men and Meridians: The History of Surveying and Mapping in Canada, Vol. 1, Prior to 1867.* Ottawa: Queen's Printer, 1966.

———. *Men and Meridians: The History of Surveying and Mapping in Canada, Vol. 2, 1867 to 1917.* Ottawa: Queen's Printer, 1967.

———. *Skyview Canada: A Story of Aerial Photography in Canada*. Ottawa: R.B.W. Ltd., 1975.

Tobey, Ronald C. *Technology as Freedom: The New Deal and the Electrical Modernization of the American Home*. Berkeley: University of California Press, 1996.

Tobias, Terry N. *Chief Kerry's Moose: A Guidebook to Land Use and Occupancy Mapping, Research Design and Data Collection*. Vancouver: Union of B. C. Indian Chiefs and Ecotrust Canada, 2000.

Tooke, T. R., B. Klinkenberg, and N. C. Coops. "A Geographical Approach to Identifying Vegetation-Related Environmental Equity in Canadian Cities." *Environment and Planning B: Planning and Design* 37 (2010): 1040–1056.

Trigger, Rosalyn. "God's Mobile Mansions: Protestant Church Relocation and Extension in Montreal, 1850–1914." PhD thesis, McGill University, 2004.

Turkel, William. *The Archive of Place: Unearthing the Pasts of the Chilcotin Plateau*. Vancouver: UBC Press, 2007.

Ulrich, S. "View through a Window May Influence Recovery from Surgery." *Science* 224 (1984): 420–21.

Urquhart, M. C., ed. *Historical Statistics of Canada*. Toronto: Macmillan, 1965.

Van de Putte, Bart and Andrew Miles. "A Social Classification Scheme for Historical Occupational Data." *Historical Methods* 38, no. 2 (2005): 61–92.

van den Bogaert, Harmen Meyndertsz. *A Journey into Mohawk and Oneida Country, 1634–1635: The Journal of Marmen Meyndertsz van den Bogaert*. Translated by Charles T. Gehring and William A. Starna. Edited by Charles T. Gehring and William A. Starna. Syracuse, NY: Syracuse University Press, 1988.

van Dijk, Teun Adrianus. *News as Discourse*. Hillsdale, NJ: L. Erlbaum, 1988.

Van Kooten, Kornelis G. "Bioeconomic Evaluation of Government Agricultural Programs on Wetlands Conversion." *Land Economics* 69, no. 1 (1993): 27–38.

Vaughan, Laura. "The Spatial Syntax of Urban Segregation." *Progress in Planning* 67, no. 3 (2007): 205–94.

Verger, F. *Marais et wadden du littoral français: etude de géomorphologie*. Bordeaux: Biscaye Frères, 1968.

———. *Zones humides du littoral français: estuaries, deltas, marais et lagunes*. Paris: Belin, 2009.

Walton, J. T., D. J. Nowak, and E. J. Greenfield. "Assessing Urban Forest Canopy Cover." *Arboriculture and Urban Forestry* 34, no. 6 (2008): 334–40.

Ward, W. Peter. "Stature, Migration and Human Welfare in South China, 1850–1930." *Economics & Human Biology* (2012), DOI: 10.1016/j.ehb.2012.10.003.

———. *White Canada Forever. Popular Attitudes and Public Policy Towards Orientals in British Columbia*. 3d. ed. Montreal and Kingston: McGill-Queen's University Press, 2002.

Wellen, Christopher. "Ontologies of Cree Hydrography: Formalization and Realization." Master's thesis, McGill University, 2008.

Wetherhell, Margaret, and Jonathan Potter. *Mapping the Language of Racism: Discourse and the Legitimation of Exploitation*. New York: Columbia University Press.

White, Richard. "Foreword." In *Placing History: How Maps, Spatial Data, and GIS are Changing Historical Scholarship*, edited by Anne Kelly Knowles, ix–xi. Redlands, CA: ESRI Press, 2008.

White, Richard W. *Urban Infrastructure and Urban Growth in the Toronto Region: 1950s to the 1990s.* Toronto: Neptis Foundation, 2003.

Williams, Michael. *Americans and Their Forests: A Historical Geography.* Cambridge: Cambridge University Press, 1992.

Wilson, Catharine A. *A New Lease on Life: Landlords, Tenants and Immigrants in Ireland and Canada.* Montreal and Kingston: McGill-Queen's University Press, 1994.

Withers, Charles. "Place and the 'Spatial Turn' in Geography and History." *Journal of the History of Ideas* 70, no. 4 (2009): 637–58.

Wolfe, Patrick. "Land, Labour, and Difference: Elementary Structures of Race." *American Historical Review* 106, no. 3 (June 2001): 866–905.

Wood, J. David. *Making Ontario: Agricultural Colonization and Landscape Re-creation before the Railway.* Montreal and Kingston: McGill-Queen's University Press, 2000.

Wynn, Graeme. *Canada and Arctic North America: An Environmental History.* Santa Barbara, CA: ABC-CLIO, 2007.

———. "Notes on Society and Environment in Old Ontario." *Journal of Social History* 13, no. 1 (1979): 49–65.

———. "Timber Production and Trade to 1850." In *Historical Atlas of Canada*, edited by R. Louis Gentilcore, Plate 11. Toronto: University of Toronto Press, 1987.

York, Geoffrey. *People of the Pines: The Warriors and the Legacy of Oka.* Toronto: Little, Brown, 1991.

Young, R. F. "Managing Municipal Greenspace for Ecosystem Services." *Urban Forestry and Urban Greening* 9 (2010): 313–21.

Zavitz, E. J. *Report on the Reforestation of Waste Lands in Southern Ontario.* Toronto: L. K. Cameron, 1908.

Zeller, Suzanne. *Inventing Canada: Early Victorian Science and the Idea of a Transcontinental Nation.* Toronto: University of Toronto Press, 1987.

Zucchi, John. *A History of Ethnic Enclaves in Canada.* Ottawa: Canadian Historical Association, 2007.

Notes on Contributors

Colleen Beard is Head, Map Library at Brock University, James A. Gibson Library.

Stephen Bocking is Professor and Chair of the Environmental and Resource Science/Studies Program, Trent University.

Jennifer Bonnell is an L.R. Wilson Assistant Professor of Canadian History at McMaster University.

Jim Clifford is an Assistant Professor in the History Department at the University of Saskatchewan.

Joanna Dean is an Associate Professor in the Department of History at Carleton University.

François Dufaux is an architect and Associate Professor in the School of Architecture at the Université Laval.

Patrick A. Dunae is a Research Associate at Vancouver Island University and Adjunct Associate Professor in the History Department at the University of Victoria.

Marcel Fortin is Geographic Information Systems and Map Librarian at the Map and Data Library, University of Toronto.

Jason Gilliland is Director of the Urban Development Program and Associate Professor in the Department of Geography, the School of Health Studies, and the Department of Paediatrics at the University of Western Ontario.

William (Bill) M. Glen is a forest and woodland consultant in Bonshaw, Prince Edward Island.

Megan Harvey is a PhD Candidate in the Department of History at the University of Victoria.

Matthew G. Hatvany is Full Professor in the Department of Geography, Université Laval, and member of the Centre interuniversitaire d'études québécoises (CIÉQ-Laval).

Sally Hermansen is Associate Dean of the Faculty of Arts and Senior Instructor in the Department of Geography at the University of British Columbia.

Andrew Hinson is a SSHRC Postdoctoral Fellow at St Michael's College at the University of Toronto.

Don Lafreniere is a Vanier Canada Scholar and PhD Candidate at the Department of Geography, University of Western Ontario.

John S. Lutz is an Associate Professor in the Department of History at the University of Victoria.

Daniel Macfarlane is a Visiting Scholar in Canadian Studies at Carleton University.

Joshua D. MacFadyen is a SSHRC postdoctoral fellow in History at the University of Western Ontario.

Jennifer Marvin is GIS Librarian and Coordinator of the Data Resource Centre at the University of Guelph.

Cameron Metcalf is a Systems Librarian at the University of Ottawa Library.

Byron Moldofsky is the Manager of the GIS and Cartography Office in the Department of Geography and Program in Planning, University of Toronto.

Jon Pasher is a Physical Scientist in Wildlife and Landscape Science at Environment Canada.

Sherry Olson is Professor of Geography at McGill University and a member of the Centre interuniversitaire d'études québécoises (CIÉQ).

Daniel Rueck is a Postdoctoral Research Fellow at the Robarts Centre for Canadian Studies and the Department of History at York University.

Ruth Sandwell is an Associate Professor in the Department of Curriculum Teaching and Learning at the Ontario Institute for Studies in Education at the University of Toronto.

Henry Yu is an Associate Professor in the Department of History and Principal of St. John's Graduate College at the University of British Columbia.

Barbara Znamirowski is Head of the Maps, Data, and Government Information Centre (MaDGIC) at Trent University Library.

Index

Free Church of Scotland, 66
fuel
    cooking, 137
    storage, 59
    use, xviii, 214–15, 239–64, 265n12. *See also* energy consumption
Fujian, 227

G

Gaffield, Chad, xii, 285n14
gas, including natural gas, 53, 59, 167, 174, 222, 242, 243, 245, 250, 265n12
gasoline, 244, 263, 268n53
General Register of Chinese Immigration, 225–28
geocoding, 64, 284
geographers, ix, xi, xii, xvi, xvii, xviii, 11, 18, 19, 28, 105, 139, 146, 176, 184, 193, 198, 206, 219
    historical, xi, xii, xvii, 85, 182, 184
    urban, 1
Geographic Information Systems (GIS) software, x, xi, xiii, 24n40, 27–28, 33, 56, 64, 91–92, 139, 145, 147, 177n4, 186, 229, 240, 279
geohumanities, 2, 19
Geological Survey of Canada (GSC), 202, 205
georectified, 35, 39. *See also* georeferencing
georeferencing, xi, xii, 5, 17, 23n39, 24n40, 28, 33, 41, 45, 48, 50, 90–91, 93, 94, 96, 107n31, 114, 139–40, 185, 203. *See also* georectified; historical photographs, geotagging of
Gephi (software), 235, 236
Gilliland, Jason, xii, xiii, xvi, xixn12, xixn13, 21n3, 21n9, 22n26, 23n31, 24n41 176n1, 177n9, 178n16, 179n23
Glen, William, viii, xvii, 221n20, 221n25, 221n26, 222n34,222n35, 223n58
Global Positioning System (GPS), 37, 38, 145, 183, 185, 186, 187, 190, 193, 231, 232
Goad, Charles E. , xv, 175. *See also* fire insurance plans
Gold Mountain, 228, 234
Google Earth (software), vii, xiii, xvi, 27–42, 57–58, 94
Google Maps, 27, 28, 29, 30, 42n3, 42n7, 92, 94. *See also* mashups
Google Maps API, 92, 95. *See also* mashups
Gould, Stephen J., 193, 195n31
Grand Trunk Railway, 37–38, 167. *See also* railways
Great Britain, 273
Great Britain Historical GIS Project, xi, 284, 285n9
Gréber, Jacques, 117, 118

Gregory, Ian, xi, xviiin4, xviiin10, 107n28, 108n39, 108n69, 268n54, 284, 286n29
groundtruthing, 33
Guangdong province, China, 227, 230, 232

H

Haida Gwaii (formerly Queen Charlotte Islands), British Columbia, 5
Hale, Jordan, vii, 240
Haliburton County, Ontario, 84–88, 102–6
Harris, Cole, 21n10, 106n6
Harvey, Megan, xvi, 21n3, 23n34
Hatvany, Matthew, viii, xvii, 194n2, 194n6, 194n7, 194n8, 195n12
Hawaiian, 4
Hayes, Derek, vii
heritage conservation, 105
Hermansen, Sally, viii, xviii, 229, 234
Heungsan County, China, 232
Hinson, Andrew, xvii, 79n14, 79n15, 79n7
Historians
    amateur, 99
    American, xi
    architectural, xvii, 176
    Canadian, xvi, 44, 241
    cultural, 1, 2,
    economic, 219
    environmental, 28, 44, 85, 90, 91, 199, 202, 218, 219. *See also* environmental history
    health, 226
    local, 38, 40, 85, 96,
    public, 41
    social, 62, 176, 241, 279,
    spatial, 234
    urban, 73, 78,
    urban forest, 112
Historians and Geographic Information Systems use, ix–xiii, xiv, 18, 19, 27–29, 41, 44–45, 73, 84–85, 90, 99, 105, 144–47, 176, 193–94, 198, 202, 218, 219, 228–29, 240, 242
historic site designation, 34
historical geography, xvi, 90, 99, 271
Historical GIS (HGIS)
    challenges in using, ix, x, xiii, xv, 17, 18, 19, 44, 47, 49, 58, 93, 94, 96, 105, 106
    in history of architecture, 175
    limitations, 124, 147
historical photographs, 2, 32–35, 38–40, 90–91, 93, 96, 118–21, 125, 126n18, 143, 155, 233. *See also* historical photographs, geotagging of
historical photographs, geotagging of, 35, 39. *See also* georeferencing
Historical Statistics of the United States, 221
Hoisan county, China, 232

households, 5, 17–20, 70–75, 77, 173, 178, 228, 241–43, 246–51, 256, 259–63, 279–80, 284
Howe, C. D., 107, 108, 206
Hudson's Bay Company, 5
hydroelectric power, 29, 130, 133, 241–56, 260, 265n12, 267n35. *See also* electrification

I

Iceland, 273
imagined geographies, xviii, 227–28
immigrants, 3, 4, 16, 20, 200, 229–33, 245
immigration,
    Chinese, 11, 13, 225–37. *See also* Chinese Canadians
    inland, 86, 89
improved land, 197–200, 206–10, 217–19, 220n4, 219n8, 222n42, 267n37
income tax, 210
incubator, 261
Indian Act, Canadian, 135, 140
Indian Quarter, 6, 20
Indian Reserve, 5, 9
Indigenous land practices, xvii, 129–32, 143–49
Indigenous peoples, xvii, 2–13, 20, 129–49, 149n3, 245. *See also* Aboriginal peoples; Amerindians; North American Indians
industrial classification, 59
industrial development, 43–44, 47–48, 55–56, 64, 83–85, 88–90, 103, 130, 132, 167, 174, 192, 232, 239, 244–46, 258
Integrated Public Use Microdata Series Project (IPUMS), 273
intermarriage, 6, 20. *See also* mixed-race families
interviews,
    audio, 33, 35
    video, 233, 237
Iroquois or Haudenosaunee, 132, 143

J

Jacobs, Jane, 4, 22n11
Jacobsen, Johannes, 6, 22n20

K

Kahnawá:ke (Kahnawake), Québec, viii, xvii, 129–49
Kamouraska County, Québec, xvii, 182–94. *See also* Bay of Kamouraska, Québec
Kataratiron (Joseph Jacob), 138
Katz, Michael, 72, 74, 80n26, 80n25
kernel density technique, GIS, 77
Khakwani, Oliver, viii, 234

Trigger, Rosalyn, 69, 79n4, 80, 176
Trust and Loan Company of Upper
    Canada,, 168
Turkel, William J., vii, xiii, 107–108n38,
    221n30, 264n2

U

Un-Forbidden City, 11, 13, 14, 15, 16, 20.
    See also Forbidden City
Underwriters' Survey Bureau, xv. *See also*
    fire insurance plans
unidirectional, 193
United States, x, xi, xii, xiii, 29, 71, 88,
    102, 103, 121, 123, 184, 190, 199,
    200, 206, 213, 225, 228, 242, 263,
    273, 284
University of British Columbia (UBC), viii,
    xiii, 226, 229, 232, 233, 234, 237
University of British Columbia Asian
    Library, 232
University of Toronto Map and Data
    Library, vii, 52, 57
urban forest, 111–25
urban form, xii, xv, xvi, xvii, 1– 20, 44–47,
    49, 111–25, 153–76, 228
urban renewal, 112, 118, 121, 124, 125
urbanization, 239, 242, 246, 258

V

vacuum cleaners (electric), 252, 254, 255

Vankoughnet, Lawrence, 136, 150n16,
    150n18, 150n20, 150n24, 150n25,
    150n26, 150n28, 150n29, 150n30,
    150n34, 150n35, 150n36, 151n39,
    151n41, 151n42, 151n44
Victoria, British Columbia, xvi, 1–20, 228,
    229, 266
ViHistory Project (Vancouver Island
    History Project), 17

W

Walbank map, 139, 140, 145
Walbank Survey, 130–32, 139, 145
Walbank, William Mclea, 130–40, 144,
    149, 150n10, 150n16, 150n18,
    150n20, 150n25, 150n26, 150n29,
    150n30, 150n34, 150n35, 150n36,
    151n39, 151n41, 151n42, 151n43,
    151n44, 151n45, 152n74
Ward, Peter W., viii, 14, 21, 23, 226, 237
water, running, 242, 250, 251, 252, 253
watersheds, ix, 33, 45, 46, 48, 52, 57, 84,
    104, 198, 202, 207, 211-214, 219
Welland Canal, 31, 33, 34, 36
Welland Canals, historic, xvi, 29, 33, 35
West River Watershed, Prince Edward
    Island, 212, 213
Western University Map Library, vii
wetlands, 45, 47, 50, 94, 182, 184, 202, 203,
    210, 212, 215
White, Richard, xi, xii, xviiin5, 1, 21n2,
    59n4, 107n27
Whiteread, Rachel, 1, 2, 20, 21
Wilmot River Watershed, Prince Edward
    Island, 212, 213

Winnipeg, Manitoba, 114, 260, 266n29
women, 5, 13, 135, 199, 228, 241, 242,
    251, 273
women's work, 242
wood, 5, 96, 100, 132, 134, 136, 138, 147,
    212, 214, 242, 243. *See also* timber;
    lumber
    firewood, 134, 138, 242, 243
    fuelwood, 136, 137, 214, 215, 242, 243
    hardwood, 102, 203, 212, 214, 215, 216
    mixedwood, 212, 213, 214
    pulpwood, 214
    softwood, 203, 212, 214, 215, 216
    studwood, 214
woodland, 46, 112, 113, 199, 221n38
woodlots, 88, 103, 121, 122, 163, 199, 206,
    214, 216, 218
Wright, Miriam, 230

Y

Yu, Henry, viii, xviii, 226, 228, 229, 233,
    234, 235
Yuen, Eleanor, viii, 232

Z

Zeller, Suzanne, 200, 220n15
Zhang, Feng, 226
Znamirowski, Barbara, vii, viii, xvii
zoom, as a form of analysis, 37, 56, 107n31,
    154–55, 171, 175, 187